Lecture Notes in Artificial Intelligence   8929

Subseries of Lecture Notes in Computer Science

LNAI Series Editors

Randy Goebel
University of Alberta, Edmonton, Canada
Yuzuru Tanaka
Hokkaido University, Sapporo, Japan
Wolfgang Wahlster
DFKI and Saarland University, Saarbrücken, Germany

LNAI Founding Series Editor

Joerg Siekmann
DFKI and Saarland University, Saarbrücken, Germany

T0224069

Pompeu Casanovas   Ugo Pagallo
Monica Palmirani   Giovanni Sartor (Eds.)

AI Approaches to the Complexity of Legal Systems

AICOL 2013 International Workshops, AICOL-IV@IVR
Belo Horizonte, Brazil, July 21-27, 2013
and AICOL-V@SINTELNET-JURIX
Bologna, Italy, December 11, 2013
Revised Selected Papers

 Springer

Volume Editors

Pompeu Casanovas
Autonomous University of Barcelona, UAB Institute of Law and Technology
Bellaterra, Spain
and
Royal Melbourne Institute of Technology
Melbourne, VIC, Australia
E-mail: pompeu.casanovas@uab.cat

Ugo Pagallo
University of Turin, Turin Law School, Turin, Italy
E-mail: ugo.pagallo@unito.it

Monica Palmirani
University of Bologna, CIRSFID, Bologna, Italy
E-mail: monica.palmirani@unibo.it

Giovanni Sartor
European University Institute, Florence, Italy
and
University of Bologna, CIRSFID, Bologna, Italy
E-mail: giovanni.sartor@eui.eu

ISSN 0302-9743   e-ISSN 1611-3349
ISBN 978-3-662-45959-1   e-ISBN 978-3-662-45960-7
DOI 10.1007/978-3-662-45960-7
Springer Heidelberg New York Dordrecht London

Library of Congress Control Number: 2014958006

LNCS Sublibrary: SL 7 – Artificial Intelligence

Typesetting: Camera-ready by author, data conversion by Scientific Publishing Services, Chennai, India

Printed on acid-free paper

Springer is part of Springer Science+Business Media (www.springer.com)

Preface

AICOL stands for Artificial Intelligence Approaches to the Complexity of Legal Systems. This volume presents the revised selected papers of the two last AICOL Workshops. The first took place as part of the 26th IVR Congress in Belo Horizonte, Brazil, during July 21–27, 2013. The latter was held in Bologna as a Joint Special Workshop of JURIX2013 (December 11, 2013) on Social Intelligence with the cooperation of the European Network for Social Intelligence (http://www.sintelnet.eu/, EU Project).

The present volume follows the previous AICOL volumes: AICOL I-II, published in 2010, including papers from the first AICOL conference in Beijing (24th IVR Congress, September 15-20, 2009, China), and the follow-up in Rotterdam (JURIX-09, Rotterdam, November 16-18, The Netherlands); and AICOL III, published in 2012, resulting from the third AICOL conference, held in Frankfurt am Main (25th IVR, August 15-20, 2011, Germany).

Like its predecessors, this volume embodies the philosophy of the AICOL conferences, that is, providing a meeting point for different researchers, such as legal theorists, political scientists, linguists, logicians, and computational and cognitive scientists, eager to discuss and share their findings and proposals. In this sense, the keywords "complexity" and "complex systems" sum up the perspective chosen to describe recent developments in AI and law, legal theory, argumentation, the Semantic Web, and multi-agent systems.

As the reader can easily check, AICOL incorporates in its fourth edition the perspective of social intelligence, the intertwined human–machine perspective on cognition, agency, and institutions. This promising approach brings together the analytical and empirical perspective of social sciences. Stemming from this starting point, the volume is divided into four main sections: (i) Social Intelligence and Legal Conceptual Models, (ii) Legal Theory, Normative Systems and Software Agents, (iii) Semantic Web Technologies, Legal Ontologies and Argumentation, (iv) Crowdsourcing and Online Dispute Resolution (ODR).

Finally, a special thanks is due to the excellent Program Committee for their hard work in reviewing the submitted papers. Their criticism and very useful comments and suggestions were instrumental in achieving a high-quality publication. We also thank the authors for submitting good papers, responding to the reviewers' comments, and abiding by our production schedule.

November 2014

Pompeu Casanovas
Monica Palmirani
Ugo Pagallo
Giovanni Sartor
Program Chairs

Organization

Organizer Committee

Danièle Bourcier	Centre d'Études et de Recherches de Science Administrative et Politique, Université de Paris II, France
Pompeu Casanovas	UAB Institute of Law and Technology, Barcelona, Spain and Centre for Applied Social Research-RMIT, Melbourne, Australia
Monica Palmirani	CIRSFID - University of Bologna, Italy
Ugo Pagallo	University of Turin, Italy
Giovanni Sartor	European University Institute and University of Bologna, Italy

Program Committee

Michal Araszkiewicz	Jagiellonian University, Poland
Guido Boella	University of Turin, Italy
Daniele Bourcier	Centre d'Études et de Recherches de Science Administrative et Politique, Université de Paris II, France
Pompeu Casanovas	UAB Institute of Law and Technology, Barcelona, Spain and Centre for Applied Social Research-RMIT, Melbourne, Australia
Enrico Francesconi	ITTIG-CNR Florence, Italy
Guido Governatori	NICTA, Australia
Renato Iannella	Semantic Identity, Australia
Arno Lodder	VU University Amsterdam, The Netherlands
Pablo Noriega	Artificial Intelligence Institute of the Spanish Council for Scientific Research, Spain
Ugo Pagallo	University of Turin, Italy
Monica Palmirani	CIRSFID - University of Bologna, Italy
Enric Plaza	Artificial Intelligence Institute of the Spanish Council for Scientific Research, Spain
Marta Poblet	RMIT University, Australia
Antoni Roig	UAB Institute of Law and Technology, Barcelona, Spain

Giovanni Sartor   European University Institute and University
                               of Bologna, Italy
Burkhard Schafer   University of Edinburgh, UK
Daniela Tiscornia   ITTIG-CNR Florence, Italy
Anton Vedder   TILT, University of Tilburg, The Netherlands
Fabio Vitali   DISI, University of Bologna, Italy
Radboud Winkels   Leibniz Center for Law, University
                               of Amsterdam, The Netherlands
Adam Wyner   University of Aberdeen, UK
John Zeleznikow   School of Information Systems, Victoria
                               University, Australia

Table of Contents

Introduction

Social Intelligence and Legal Conceptual Models

Legal Theory, Normative Systems and Software Agents

Semantic Web Technologies, Legal Ontologies and Argumentation

Crowdsourcing and Online Dispute Resolution (ODR)

Law, Social Intelligence, nMAS and the Semantic Web: An Overview

Pompeu Casanovas[1,2], Ugo Pagallo[3], Monica Palmirani[4], and Giovanni Sartor[5,6]

[1] Centre for Applied Social Research-RMIT, Melbourne, Australia
pompeu.casanovas@uab.cat
[2] Royal Melbourne Institute of Technology, Melbourne, La Trobe St. 124, VIC 3000
pompeu.casanovas@rmit.edu.au
[3] Torino Law School, University of Torino, Lungo Dora Siena 100, 10153 Torino, Italy
ugo.pagallo@unito.it
[4] CIRSFID, University of Bologna, via Zamboni 33, 40126 Bologna, Italy
monica.palmirani@unibo.it
[5] European University Institute, Florence, Italy
giovanni.sartor@eui.eu
[6] CIRSFID, University of Bologna, via Zamboni 33, 40126 Bologna, Italy

Abstract. This introduction presents the principles and fundamentals of the AICOL scientific initiative and in particular the main contributions of the current volume, underlining the interdisciplinary approach and the variety of adopted methodologies.

Keywords: AI & Law, social intelligence, legal theory, complex systems, Semantic Web, legal ontologies, nMulti-Agent Systems.

1 Presentation

The outcomes from AICOL IV/V are compliant with different kind of objectives. Firstly, the aim is to introduce and develop models of legal knowledge, concerning its organization, structure and content, especially in order to promote mutual understanding and communication between different legal systems and cultures. By achieving more precise models of legal concepts —from multilingual dictionaries to taxonomies and legal ontologies, namely formal models of legal conceptualization— we intend to enhance our comprehension of legal cultures, identifying their commonalities and differences. Moreover, by increasingly profiting from computer support in managing legal knowledge, we aim at both drawing on convergences and bridging differences for deeper understanding of today's legal challenges.

Secondly, focus is on the comparison of multiple formal approaches to the law, supporting both internal and the external viewpoints on legal phenomena: logical models, cognitive theories, argumentation frameworks, graph theory, complexity theory, cybernetics, game theory, etc. The purpose is to stress possible convergences in the realm of, say, conceptual structures, argumentation schemes, emergent

P. Casanovas et al. (Eds.): AICOL IV/V 2013, LNAI 8929, pp. 1–10, 2014.

behaviors, learning evolution, adaptation, simulation, and more. By promoting a fruitful interaction between some of the most relevant contributions to AI research on contemporary legal systems, attention is drawn to the most recent research in the field, e.g., the use of sentiment analysis in crowd-sourcing for anticipating geopolitical crises, e-discovery in legal firms and tribunals, gamification in legal environment, and so forth.

Thirdly, AICOL addresses the ways in which the current information revolution impacts on basic pillars of today's legal and political systems, in such fields as e-democracy, e-government, transnational governance, etc. What is at stake concerns changes and developments that occur at a rapid pace, as the law transforms itself, in order to respond and progress alongside the advances of technology. Consider some canonical representations, such as Hans Kelsen's idea of the law as a set of rules enforced through the menace of physical sanctions: "if A, then B" [1]. Whilst the ubiquity of the internet has magnified the troubles with the enforcement of the law, the legitimacy of the state's action is contested, as states claim to unilaterally regulate extraterritorial conduct by imposing norms on individuals who have no say in the decisions affecting them. In addition to the traditional hard and soft law-tools of governance, such as national rules, international treaties, codes of conduct, guidelines, or the standardization of best practices, it is no surprise how the new scenarios of the information revolution increasingly suggest that the dynamics of current societies can be governed through codes, architectures, and AI systems, so as to embed legal rules and safeguards into technology.

In this new socio-technological context, issues of legal reasoning, concepts, sources of the law, different meanings of complexity have to be taken into account. As to the models of legal knowledge and formal approaches to the law, special attention should be paid to that which Seth Lloyd, drawing on research by Charles Bennett in the 1980s and, furthermore, Ray Solomonoff and Gregory Chaitin in the 1960s, dubs as "logic depth" [2]. Here, the subject matter appears increasingly complex as the quantity of information grows and its theoretical compression decreases, in order to represent such object via a computer program. Then, the notion of complexity which refers to some formal approaches to the law that aim to address the emergence of spontaneous orders, e.g., work in social intelligence and crowd-sourcing, should be traced back to seminal research by Friedrich Hayek and the very difference between deliberate human arrangements and unintentional orders [3]. What makes this side of the law specifically complex has to do with the ways in which only the dynamics of social interaction, rather than the master plan of legislators and policy makers, can achieve satisfactory results in several fields of today's legal systems. Remarkably, this is also the opinion of several experts in information and communication technology (ICT)-law, that conceive the internet as a "self-governing realm of individual liberty, beyond the reach of government control" [4].

Finally, some facets of this latter research in spontaneous orders, much as work in legal theory and on how the information revolution affects current legal and political systems, suggest a further notion of complexity. Think about some crucial concepts, as responsibility, enforcement, validity, representation, deliberation, and more, and how they are changing. As a consequence of complexity, finding the right balance between, say, "representation and resolution, while implementing the agreement to agree on the basis of ethical principles that are informed by universal human rights, is a current major challenge for liberal democracies in which ICTs will increasingly strengthen the representational side" [5]. From this latter point of view, we may say that the more an issue is complex, the more it affects, or impacts on, the whole infrastructure and environment of the system with which we are dealing.

Clearly, such challenges can be properly tackled at the previous levels of complexity. Whereas the profound transformation of such concepts as, say, democracy and representation, challenges the system as a whole, it also affects models of legal knowledge as well as formal approaches to the law. Therefore, the level of complexity does not hinge on whether focus is on the different ways in which legal reasoning, or legal concepts, or the sources of the system, work. Rather, what is crucial is how we address such issues, according to a given problem. Thus the multiple topics addressed in the AICOL meetings and their results are here presented in connection with four main parts, stemming from the broader conceptual ones and ending up with the specific field of crowdsourcing and ODR: (i) Social Intelligence and Conceptual Legal Models, (ii) Legal Theory, Normative Systems and Software Agents, (iii) Semantic Web Technologies, Legal Ontologies and Argumentation, (iv) Crowdsourcing and Online Dispute Resolution (ODR).

2   The Quest for Social Intelligence

As it is classically defined in social and cognitive psychology, *social intelligence* can be conceived as the mental ability to understand the motives, emotions, intentions and actions of other people and to motivate and influence the behavior of (groups of) people. Still, this definition does not focus specifically on the artificial, technical, cultural, economic and political interfaces that the emergence of Web 2.0 and 3.0 fosters and anchors.

Collective intelligence is one of the most intriguing dimensions of the so-called "social web" *emotional intelligence* —the ability to produce and use empathy— is another one. And we can figure out that economic and institutional organizations are also related to this formula: "Social Intelligence is all about understanding and combining Social Media (Networking) and Business Intelligence".[1] These different aspects, which are present as information processing, can be modeled for institutional design combining the result of empirical findings, technical languages, and formal representations.

This volume aims to discuss how social intelligence approaches can shed light on AI and law, legal theory, argumentation, conflict resolution, the semantic web, and normative multi-agent systems. This can be done in all steps of the legal process — drafting, contracting, judging— and all uses of social and legal norms —applying, arguing, implementing, and enforcing the law. Besides, there is an ongoing discussion about modeling the evolving concept of law within the new environment of the

---

[1] http://www.scoop.it/t/social-intelligence

Internet of Things and the new governance and ethical challenges faced by such institutions as the EU (data protection, security, identity, etc.).

Three myths have to be faced. The first is the belief that individual and collective knowledge are different in nature. The second is that artificial agents never will reach the level of complexity of human beings. The third is that only humans can be legally ruled, for law is a special way of existence of regulatory systems.

Admittedly, individual and collective behavior show different features, although the comprehension of social intelligence means understanding individual intelligence. However, a multi-agent notion of social intelligence suggests that we should go beyond the individual level of analysis. Therefore, social intelligence includes both the objective effects of social action and the cognitive properties of individual and social action, much as the relationships between the two [6].

Modeling from this theoretical perspective, several consequences follow for the legal design and shaping of both artificial and natural societies. Perhaps the most important consequence is that legal and institutional designs are not only a way to figure out an autonomous realm of norms, but a theoretical way of understanding how normative, institutional and legal systems emerge and work interactively in social and artificial contexts.

Moreover, along with the developments of the Cloud, the Internet of Things and the new stages of the Semantic Web, we are all living in a *hybrid* and *intertwined* world, in which it makes no sense making a divide between a virtual and non-virtual reality. Social contexts are interactively shaped. In the words of Castelfranchi [7]: "No collective action would be possible without shared and/or ascribed mental contents. (...). Our social minds for social interactions are coordination artifacts and social institutions". It is clear that legal models and legal theory cannot be set apart.

3   Normative Systems, Software Agents

Quite recently, Pablo Noriega, Julian Padget, Harko Verhagen, and Mark d'Inverno [9] have proposed a general tripartite view that highlights the interplay between the institutional models that prescribe the behavior of participants, the corresponding implementation of these prescriptions and the actual performance of the system. Among the main challenges for the development of Artificial Socio-Cognitive Systems they expressly mention the *synergy* with philosophy of law —and, we might add, legal theory:

> A systematic study of ASCS will most likely require the convergence of several disciplines. The topic of social coordination is currently being inspected (within the Sintelnet project) from different standpoints: games, social simulation, analytical sociology, cognitive and social psychology, formalisms for informal phenomena, crowd-based applications, institutional theory and philosophy of law. These activities are already fostering collaborations with a strong synergistic component. This experience points in the direction of new academic communities that are likely to spawn conferences and periodic publications and eventually develop curricula and training.

This is a shared vision. Some time ago, Boella, van der Torre and Verhagen [9] set ten challenges for normative Multi-Agent Systems (nMAS) being developed towards this interactive direction. There are further proposals. Gordon, Governatori and Rotolo [10] have focused on requirements for rule interchange languages following the normative structure of core legal theory. Others, mainly authors committed to agreement technologies, are opening up the field to contracting, negotiating and decision-making theories [11].

With the Web of Data, attention to legal details and regulatory constraints are increasingly a broad research topic. Some, e.g. Espinosa and Fornés [12], have surveyed the state of the art on the intersection between privacy and MAS. They have classified the risks regarding the information-related activities that these studies aim to prevent in terms of information collection, information processing, and information dissemination.

We can assume these legal components as external constraints coming from the outer environment: on this basis, privacy, data protection and security constitute an inescapable challenge for the design of institutions and regulatory models. Yet, from the inner point of view, non-standard deontic logic and legal argumentative reasoning appear crucial to integrate all these different aspects into a coherent and consistent stance. This twofold side of the problem is also at stake with recent developments of the Semantic Web.

4   Semantic Web Developments

The Semantic Web has entered into a new stage due to the need for semantic linked data developments. The so-called 5 Star Linked Open Data settled by Tim Berners-Lee [13] refers, according to W3C, to an incremental framework for deploying data. The 5 Star Linked Data system is cumulative, and each additional star presumes the data meets the criteria of previous steps. We reproduce here for the sake of clarity this already well-known scheme [14]:

☆ Publish data on the Web in any format (e.g., PDF, JPEG) accompanied by an explicit Open License (expression of rights).
☆☆ Publish structured data on the Web in a machine-readable format (e.g., Excel instead of images).
☆☆☆ Publish structured data on the Web in a documented, non-proprietary data format (e.g., CSV, KML instead of Excel).
☆☆☆☆ Publish structured data on the Web as RDF (e.g. Turtle, RDFa, JSON-LD, SPARQL) using URIs to identify things.
☆☆☆☆☆Link your data to other people's data to provide context.

According to the ongoing research carried out by the W3C, Star Linked Open Data includes an Open License (expression of rights) and assumes works as publications on the public Web [14]. But we should notice that this opening to the public space of published data and metadata immediately raises legal problems in private and commercial law —licensing, patents, intellectual and industrial property.— much as concerning the relationship with global markets and global governance. This means that the notion of *public* space is at stake too.

Very likely, the opportunity to choose the specific way of publishing will contribute to the redefinition of this notion. Open source cannot be confused with public space, and the regulation of data and the protection of citizens are deeply intertwined. Rights, institutions and governance are the different dimensions for a new legal framework in which different jurisdictions collide. Again, the connection between law and the Semantic Web constitute an inescapable new challenge for the community that can be grasped either from an external or from an internal point of view.

Law has been usually taken into account by Semantic Web developers as a requirement or preliminary condition for web services and regulatory ontologies. Accordingly, languages for expressing rights (Rights Expression Languages, i.e. REL, plus ODRL, ODRL-S, MPEG-21...), privacy, identity, authentication, integrity, security, and trust, legally or institutionally oriented, are increasingly a hot topic in the Web of Data [15]. However, the technicalities of such languages as REL, ODRL, etc., are not simply neutral. Rather, they contribute to transform and reshape the meaning of the rights and interests assumed as preliminary conditions or requirements for the development of the Semantic Web.

5   On the Content of this Volume

This new volume of the AICOL-Workshops addresses the issues put forward in the former sections. As already stated, and for the sake of clarity, we have divided the papers into four main sections: (i) Social Intelligence and Legal Conceptual Models, (ii) Legal Theory, Normative Systems and Software Agents, (iii) Semantic Web Technologies, Legal Ontologies and Argumentation, (iv) Crowdsourcing and Online Dispute Resolution (ODR). It should be noticed that these categories are not discrete: several papers can fit into the nMAS section and into the Semantic Web part as well, for they build up ontologies or delve into semantic languages. This only shows the close relations between them.

5.1   Social Intelligence and Legal Conceptual Models

Ugo Pagallo addresses the sources of law, and connects some features of the information revolution to social intelligence and to some legal mechanisms to avoid lack of protection (burdens of proof, duty of knowledge, and limits to the use of self-enforcing technologies). Stemming from a broad legal perspective, Fernando Galindo advocates for interdisciplinary approaches in the making of ICT regulations. He raises the specific problem of the consequences of the introduction of Smart Cities and design of services that will constitute the infrastructure of those Cities. These two contributions stress the need for a flexible understanding of the way legal norms and rules should be conceived, applied and eventually enforced in these new environments. The paper by Eleonora Bassi, David Leoni, Stefano Leucci, Juan Pane, and Lorenzino Vaccari, in the context of the Trentino Open Data Project, proposes a semantic open source stack to preserve data protection and privacy rights for publishing anonymised deliberations edited with the NormeinRete software for government open data.

Following the same line of arguments, Pompeu Casanovas and John Zeleznikow stress the importance of ethical principles —mainly fairness— for Online Dispute Resolution. They raise the comparative question of the synergy and structural coincidence between general information principles in several fields (privacy, data protection, linked open data...), stemming from the related notion of Semantic Web Regulatory Models. Then, the paper by Andrea Ciambra and Pompeu Casanovas suggests a way of building composite indicators to test the institutional strengthening of such models.

5.2   Legal Theory, Normative Systems and Software Agents

The second section points at the connection between legal theory, normative systems and software agents. As already shown, one of the urgent issues to be solved is how to technically connect legal conceptual models, deontic logic and normative Multi-agent Systems (nMAS).

The first two papers raise the issue of dynamicity and time in legal theory. They both focus on legal normative knowledge. Monica Palmirani and Luca Cervone state that modifications in legal norms create a very intricate network of citations, not always easy to be tracked and properly accessed. They are providing a theoretical model based on indexes for measuring the complexity of each modificatory action, and they set as well a diagram system to visualize indexes of the resultant legal order per year and document. The authors have created an active impact indicator per document, to reveal the dynamic complexity introduced by modificatory actions in the legal order. Similarly, Michał Araszkiewicz asserts in his paper that the meta-information concerning admissibility of certain changes to legal systems and, specially, to constitutional principles, should become a standard element of databases of statutory legal knowledge. This proposal is presented as a contribution to the theory of hybrid legal knowledge systems, encompassing both rule-based and case-based elements, and tracing its roots back to some previous works already carried out in the tradition of AI & Law modeling.

The remaining three papers are centered on software agents. Taking inspiration from some existing models coming from socio-legal and social object theories, Alessio Antonini, Cecilia Blengino, Guido Boella and Leendert van der Torre tackle the inner relation between legal norms, principles and roles. They set a social ontology to represent entities related to normative systems to be encased into Eunomos, a norm management system to facilitate the spotting and management of legal content using legal statutes or cases. The next paper, by Guido Boella, Silvano C. Tosatto, Sepideh Ghanavati, Joris Hulstijn, Llio Humphreys, Robert Muthuri, André Rifaut, and Leendert van der Torre, introduces Eunomos, along with LEGAL-URN. The former processes the normative content of texts and legal documents.

The latter, factors in legal requirements as part of strategic business planning. The combination of both systems is able to technically reconstruct, reason and cope with the problem of business regulatory compliance, which is one of the classic problems in legal theory.

The last paper of this section is a legal one. Attention is drawn to the analysis of criminal liability of software agents. Pedro Freitas, Francisco Andrade and Paulo Novais consider several solutions (i.e. Perpetration-via-Another Liability Model, Natural-Probable-Consequence Liability Model, Direct Liability Model), to conclude that the inner conceptual structure of criminal law gains benefit from the challenge raised by software agents.

5.3   Semantic Web Technologies, Legal Ontologies and Argumentation

The third section of the volume deals with the development of the Semantic Web, the construction of legal ontologies, and their use in legal argumentation and in the regulation, interoperability, management and monitoring of web services and linked data.

Knowledge acquisition, first. Natural Language Processing (NLP) provides an array of techniques and tools to be applied to legal corpuses and databases. The paper by Makoto Nakamura, Yasuhiro Ogawa, and Katsuhiko Toyama, aims at the production of a Japanese legal terminology for translators, with proper explanations and accessible citations. Surface pattern recognition, extraction of legal terms and definitions, XML tagging, and annotation are used. The paper shows some experimental results on the proposed methodology.

In the second contribution to the Semantic Web framework, Marcelo Ceci presents a formalization of legal concepts and argumentation patterns occurring in judicial decision making. In praise of this objective, he uses a set of metadata associated with judicial concepts and an ontology library. He is currently combining the features of WBL2 with description logics and defeasible rules in the framework of Carneades argumentation graphs. The paper depicts the reasoning path and legal interpretations carried out by the judge in a specific case.

The third paper of this section, by Elie Abi-Lahoud, Leona O'Brien, and Tom Butler, addresses the problem of regulatory compliance, not from the normative system point of view —as faced by Boella and van der Torre in this same volume— but from the ontological perspective. Authors show the existing need of representing the legal knowledge of the complex field of financial documents and regulations, leaning on ontologies. They identify a list of challenges to be faced that require human subject matter expertise in their understanding. It is suggested the use of Semantics of Business Vocabulary and business Rule (SBVR), supported by a series of examples from a completed experiment on a piece of regulation from the US Bank Secrecy Act.

The last contributions to this section consist of ontological applications to solve interoperability problems in two main EU Projects. Enrico Francesconi, Ginevra Peruginelli, Ernst Steigenga, and Daniela Tiscornia introduce the CODEX Project. This project concerns file and exchange cross-border legal procedures between all the

European states. The authors offer an overview of the e-Delivery platform architecture. The latter is a Large Scale Pilot project in the domain of e-Justice, to help citizens, professionals and administrations with an easier access to transnational justice. The second EU Project, CAPER, has the aim to provide interoperability to European Law Enforcement Agencies (police) so as to foster fast and secure exchange of information to fight organized crime. Jorge González-Conejero, Rebeca Varela-Figueroa, Juan Muñoz-Gómez, and Emma Teodoro present the European LEAs Interoperability Ontology (ELIO), which models the structure of legal crimes according to the Europol taxonomy, and the knowledge directly gathered from LEAs.

5.4 Crowdsourcing and Online Dispute Resolution (ODR)

Among the most thrilling areas in social technology during the past five years are those concerning citizen participation and democratization mechanisms. The volume closes with two contributions from the well-settled field of ODR, and two further papers from the emergent field of crowdsourcing.

Context, environment, ambiance, offer the first key. Paulo Novais, Davide Carneiro, Francisco Andrade, and José Neves, look at the function of sensitive-context technology and its importance for conflict resolution and ODR. They address the issue of improving the communication layer of the framework, by including contextual information that is meaningful for the conflict management and the resolution process. Josep Suquet, Pompeu Casanovas, Xavier Binefa, Oriol Martínez, Adrià Ruiz, and Jordi Ceballos present the prototype of CONSUMEDIA, an ODR platform with some functionalities such as the recognition of emotions in the mediation room that might enhance the professional work of mediators.

The third paper of Marta Poblet, Esteban García-Cuesta and Pompeu Casanovas addresses the different definitions of crowdsourcing and offers a review of the state of the art platforms applied in the different phases of disaster management. A model based on a taxonomy of crowdsourcing roles and tasks is suggested.

Last, but not least, Nuno Luz, Nuno Silva and Paulo Novais propose a method to define a set of ground rules for the assisted construction of workflow definition ontologies from domain ontologies. That is, a method for the construction of micro-task workflows from legal domain ontologies.

Acknowledgments. We warmly thank Cristiana Teixeira Santos for her invaluable help in the editing process. This volume has been partially supported by the following projects. EU Projects: SINTELNET FP7-ICT-2009-C-286380; CAPER FP7-EU Agreement 261712; EU Project Erasmus Mundus Doctorate on Law, Science and Technology, 520250-1-2011-1-IT-ERA MUNDUS-EMJD; National Projects: DER2012-39492-C02-01- CROWDSOURCING; CONSUMEDIA IPT-2011-1015-430000; CROWDCRISIS CONTROL IPT-2012-0968-390000.

References

1. Kelsen, H.: General Theory of the Law and the State. Harvard University Press, Cambridge (1949), Trans. A. Wedberg
2. Lloyd, S.: Measures of Complexity: A Nonexhaustive List. IEEE Control Systems 21(4), 7–8 (2001)
3. Hayek, F.A.: Law, Legislation and Liberty: A New Statement of the Liberal Principles of Justice and Political Economy. Chicago University Press, Chicago (1982)
4. Solum, L.B.: Models of Internet Governance. In: Bygrave, L.A., Bing, J. (eds.) Internet Governance: Infrastructure and Institutions, pp. 48–91. Oxford University Press, New York (2009)
5. Floridi, L.: The Fourth Revolution – The Impact of Information and Communication Technologies on Our Lives. Oxford University Press (2014)
6. Conte, R.M.: Social Intelligence Among Autonomous Agents. Computational & Mathematical Organization Theory 5(3), 203–228 (1999)
7. Castelfranchi, C.: Minds as Social Institutions. Phenomenology and Cognitive Sciences 13(1), 121–143 (2014)
8. Noriega, P., Padget, J., Verhagen, H., d'Inverno, M.: The Challenge of Artificial Socio-Cognitive Systems. In: The 17th International Workshop on Coordination, Organisations, Institutions and Norms, AAMAS-2014 (2014),
   http://aamas2014.lip6.fr/proceedings/workshops/
   AAMAS2014-W22/AAMAS2014-W22-index.html
9. Boella, G., van der Torre, L., Verhagen, H.: Ten challenges for normative multi-agent systems, Dagstuhl Seminar Proceedings 08361, Programming Multi-Agent Systems (2008), http://drops.dagstuhl.de/opus/volltexte/2008/1636
10. Gordon, T.F., Governatori, G., Rotolo, A.: Rules and Norms: Requirements for Rule Interchange Languages in the Legal Domain. In: Governatori, G., Hall, J., Paschke, A. (eds.) RuleML 2009. LNCS, vol. 5858, pp. 282–296. Springer, Heidelberg (2009)
11. Ossowski, S. (ed.): Agreement Technologies, LGTS. Springer, Dordrecht (2013)
12. Such, J.M., Espinosa, A., Fornés, A.: A survey of privacy in multi-agent systems. The Knowledge Engineering Review 29(3), 314–344 (2013)
13. Berners-Lee, T.: Linked Data,
    http://www.w3.org/DesignIssues/LinkedData.html
14. W3C Working Group Note 27 June 2013, Linked Data Glossary,
    file:///C:/Users/TOSHIBA/Desktop/W3C-Linked%20Data%20and%20the%20SW/
    Linked%20Data%20Glossary.htm
15. Presutti, V., d'Amato, C., Gandon, F., d'Aquin, M., Staab, S., Tordai, A. (eds.): ESWC 2014. LNCS, vol. 8465. Springer, Heidelberg (2014)

The Legal Roots of Social Intelligence and the Challenges of the Information Revolution

Ugo Pagallo

University of Turin, Turin, Italy
ugo.pagallo@unito.it

Abstract. The paper traces current research on social intelligence back to the everlasting debate on the sources of law and the formalization of social, as opposed to individual, intelligence as the binding force of social customs. After the crisis of the Westphalian model, the legal role of social intelligence can be appreciated nowadays in accordance with new forms of customary and transnational law, much as social norms that a myriad of communities have developed online. Since rearrangements of the legal sources are intertwined with distributions of power, however, what is especially at stake today concerns the sovereign claim to regulate extraterritorial conduct, much as imposing norms on individuals that have no say in the decisions affecting them, through the mechanisms of design, code, and architecture. Current tussles on the future of the internet and its governance show that it would be deadly wrong to take today's legal role of social intelligence for granted.

Keywords: Governance, ICT-driven societies, IT law, Legal customs, Social intelligence, Sources of law, Spontaneous orders, Westphalian model.

1 Introduction

Over the past years "social intelligence" has become a buzzword of contemporary scientific research by fostering a large set of empirical and theoretical studies on information technologies (ITs)-enabled social situations, self-organizing evidence-based policies, agent-based computing, self-organizing normed-governed systems, contract based systems, computational justice, and more. The overall idea is to explore the interplay of ITs, philosophy, humanities, and the social sciences, as the European network for social intelligence (Sintelnet)'s webpage is keen to inform us. In light of current work on "social intelligence" and the aim to explore the new horizons opened up by the information revolution, in such fields as social, collective and emotional intelligence, smart data and the semantic web, intentional and collective action, natural language processing, and the like, it seems fruitful to dwell on the legal features of this work. Thanks to this stance, we can appreciate both sides of what scholars used to sum up as dialectics in Middle Ages, namely endurances (genus proximum) and breakthroughs (diffentia specifica) in the legal field vis-à-vis the information revolution and IT-enabled social intelligence.

On the one hand, what seems to be firm in the legal domain has been stressed time and again in the fields of IT law, AI and the law, robotics, etc. Consider the remarks

P. Casanovas et al. (Eds.): AICOL IV/V 2013, LNAI 8929, pp. 11–25, 2014.
© Springer-Verlag Berlin Heidelberg 2014

of the "unexceptionalists" in the field of IT law, so that principles and provisions of the legal tradition would be capable of tackling all of the new legal issues emerging with this technology. In the phrasing of Jack Goldsmith in Against Cyberanarchy (1998), "a nation's right to control events within its territory and to protect the citizens permits it to regulate the local effects of extraterritorial acts" and, moreover, whilst the flow of the information on the internet transcends, most of the time, conventional borders of national legal systems, the transnational legal impact of the internet should be conceived as "identical to transnational activity mediated by other means, such as mail or telephone or smoke signal" [1]. The claim that the internet, or robotics, or AI, etc., neither create nor modify legal concepts, such as the principle of territoriality, the effects doctrine, and the like, is still popular among scholars [2].

On the other hand, the traditional representation of the legal order as grounded on the principle of national sovereignty – so that "in the absence of consensual international solutions, prevailing concepts of territorial sovereignty permit a nation to regulate the local effects of extraterritorial conduct" [1] – is questioned because there are no clear national boundaries in cyberspace. This leads to the illegitimate situation where a state pretends to regulate extraterritorial conduct by imposing norms on individuals who have no say in the decisions affecting them or conversely, the flow of information on the internet can determine the ineffectiveness of state action because citizens would be affected by conducts that the states are simply unable to regulate. In the wording of an unrepentant "exceptionalist" as David Post, "border-crossing events and transactions, previously at the margins of the legal system and of sufficient rarity to be cabined off into a small corner of the legal universe... have migrated, in cyberspace, to the core of that system" [3]. Like in other fields of scientific research, such as physics, biology, or engineering, scale matters.

Going back to work in social intelligence, what is then today's state-of-the-art? Does IT-enabled social intelligence affect basic pillars of the law or, vice versa, according to traditional outlooks on IT law, AI and the law, or robotics, IT-enabled social intelligence neither creates nor modifies legal concepts? Moreover, is there a middle ground in between such extremes?

In order to offer a hopefully comprehensive view of these issues, the paper is presented in five parts. Next, in Section 2, focus is on the genus proximum, namely the traditional representation of what is conceived today as social intelligence in terms of legal customs, social norms, and spontaneous orders, as a source of the law. In Section 3, attention is drawn to the reasons why this traditional representation eclipsed with the so-called Westphalian paradigm, and why this latter model broke down in the mid 1900s. On this basis, the paper introduces the analysis of the differentia specifica, that is how the information revolution and IT-enabled social intelligence may impact on the legal field. In Section 4, this viewpoint is deepened with the reasons why national law-making activism is increasingly short of breath, and why constitutional powers of national governments have been joined – and even replaced – by the network of competences and institutions summarized by the idea of governance. The legal tools of governance are then examined in Section 5, so as to appreciate the role that social norms, much as spontaneous orders, play in current legal systems. By assessing how the information revolution reshapes the sources of the law, Section 6 takes into account models of political legitimacy and democratic processes, much as republican institutions that shall respect equal worth of all individuals. Whilst it is admittedly an open question how such institutions have to be built, or even conceived in cyberspace [4, 5, 6], the conclusion insists on "the goal that could successfully orient our political strategy in terms of transparency and tolerance" [7]: what is at stake concerns the right balance between legal representation and political resolution.

2   Legal Customs and Spontaneous Orders

The genus proximum of the analysis between legal science and social intelligence is given by the concept of custom, or customary law. The legal formalization of social, as opposed to individual, intelligence can properly be traced back to ancient Roman law and its notion of custom as a source of the system (fons iuris). Since Roman law existed for some twelve hundred years, that is from the foundation of Rome to the rule of Justinian, it is somehow natural that the meaning and definition of custom had evolved throughout the centuries. For the sake of conciseness, it suffices to sum up this evolution with the Latin saying "opinio iuris ac necessitatis." The reason why individuals act in a certain way, that is in accordance with the customs of a given society, is the belief (opinio) that such action had to be carried out because that is the social practice of the community and, therefore, it represents a legal obligation (necessitas). As such, lest we revert to the realm of myths, no specific individual had ever invented, or imposed, such social patterns: just on the contrary, these social patterns should be interpreted in the phrasing of Friedrich Hayek [8], as an unintentional phenomenon, or spontaneous order.

Against the tenets of social constructivism, e.g. Thomas Hobbes's philosophy of law and the Cartesian tradition, Hayek reckons that human intelligence has emerged and developed by following such unintentional rules of conduct, rather than the other way around, that is as an intelligent species that determines and establishes, as such, its own social norms. In the phrasing of Rules and Order (1973), i.e. the first volume of Law, Legislation and Liberty (ed. 1982), "these rules of conduct have thus not developed as the recognized conditions for the achievement of a known purpose, but have evolved because the groups who practiced them were more successful and displaced others... The problem of conducting himself successfully in a world only partially known to man was thus solved by adhering to rules which had served him well but which he did not and could not know to be true in the Cartesian sense" [8].

Among the advocates of this tradition that stress the key role of ignorance in human evolution and link the latter to the function that social intelligence has in legal and political affairs, Hayek lists a number of scholars: John Milton, John Locke, John Stuart Mill and Walter Bagehot on human ignorance, much as Adam Smith, David Hume and Adam Ferguson on human evolution, presented as a "process of cumulative development" [9]. Still, according to Hayek, this tradition should be properly understood in light of ancient Roman law and, more particularly, in accordance with the preliminary remarks of Cicero in the second book of De republica:

> Cato... used to say that the government of Rome was
> superior to that of other states; because in them the
> great men were mere isolated individuals, who
> regulated their constitutions according to their own
> ipse dixits, their own laws, and their own ordinances.
> ... Our Roman constitution, on the contrary, did not
> spring from the genius of an individual, but of many;
> and it was established, not in the lifetime of a man, but
> in the course of ages and centuries (trans. by Francis
> Barham, available at "The Online Library of
> Liberty").

On this basis, Hayek suggests that we should distinguish between two different kinds of legal sources, namely between kosmos and taxis, that is between spontaneous orders and human political planning. Although this differentiation is not new – for example, Italian legal scholars use to distinguish between material sources of law, such as customs, and formal sources, such as statutes and codes – Hayek's distinction has a normative aim. As he affirms in chapter 2 of Rules and Order, "one of our main contentions will be that very complex orders, comprising more particular facts than any brain could ascertain or manipulate, can be brought about only through forces inducing the formation of spontaneous orders" [8]. In other words, there are a number of fields concerning human interaction in which only the unintentional dynamics of social intelligence, rather than the master plan of legislators and policy makers, can achieve satisfactory results. Remarkably, this is also the opinion of several experts in IT law today [e.g. 4], who conceive the internet as a "self-governing realm of individual liberty, beyond the reach of government control" [5].

For the moment, however, let us dwell on the descriptive side of this story, in order to understand why the traditional representation of what is conceivable as social intelligence in terms of legal customs and social norms eclipsed with the so-called Westphalian paradigm, and why this latter model broke down some seventy years ago. The normative analysis of today's sources of the law is postponed until Section 5.

3   The Paradigm of Westphalia and Its Crisis

The Westphalian paradigm, so called after the 1648 series of peace treaties signed in Germany to conclude the Thirty Years War, pivots around the principle of sovereignty and, in Hayek's jargon, taxis as the main, or even unique, source of the law. From a theoretical viewpoint, the reference model is given by Hobbes's work and his critiques of the natural law tradition, the then popular dualism between gubernaculum and iurisdictio, that is between the seat of power and the sources of law, much as the ancient idea of customary law as the main source of the entire system. From the Hobbesian perspective, there is no legal room for social intelligence and unintentional orders, because this sort of natural spontaneity leads to the conflicts and warfare of the state-of-nature, where the man is a wolf to his fellow man (homo homini lupus). Correspondingly, the way in which individuals can overcome this chaotic condition – which is either provoked by the lack of rules, or triggered by the multiple, or even opposite, versions of uncertain customs – is represented by the social covenant. Pace Cicero's ideal of the commonwealth (res publica), the only basis for a peaceful human interaction is given by a contract, that is constructivism. In the words of chapter 18 of Hobbes's Leviathan:

> "A Commonwealth is said to be instituted when a multitude of men do agree, and covenant, every one with every one, that to whatsoever man, or assembly of men, shall be given by the major part the right to present the person of them all, that is to say, to be their representative" [10].

Over the past century, Hobbes has been considered as the father of the modern legal and political thought; and, all in all, there are good reasons to follow this historiographical tradition [11]. Suffice it to recall three of such reasons. First, what the law is hinges on the will of the sovereign. Second, in the field of international law, no one is set to judge the decisions of sovereign states, since the law is made up by the rules effectively established by national sovereigns. Third, customary law should not be conceived as a legal source any longer, because both their international and national bases ought to be grounded on the will of the sovereign. Going back to chapter 18 of Hobbes's Leviathan, "it is annexed to the sovereignty the right of making war and peace with other nations and Commonwealths; that is to say, of judging when it is for the public good, and how great forces are to be assembled, armed, and paid for that end" [10].

From Hobbes's work and the Westphalian paradigm, of course, it does not follow a plain correspondence between theory and practice, between model and history. Moreover, some tenets of this political representation are still controversial: for instance, scholars still discuss whether Hobbes should be conceived as a "liberal" thinker [12]. According to some interpretations of the Leviathan, citizens have indeed the faculty to decide whether they should obey certain of the sovereign's commands in the "foresight of their own preservation." After all, this was the interpretation of some contemporaries of Hobbes, such as Filmer, Clarendon, and Bishop Bramhall in The Catching of the Leviathan (1658), where the latter dubs Hobbes's book as a "Rebel's catechism." Contemplate what the famous and problematic sentence of chapter 21 of Leviathan states: "When therefore our refusal to obey frustrates the end for which the sovereignty was ordained, then there is no liberty to refuse; otherwise, there is." The same ambiguity applies to how the sources of the legal system should be grasped. On one hand, by tracing them back to the will of the sovereign, the model paves the way for future positivistic, and even totalitarian outcomes: as remarked in chapter 26 of Leviathan, "the law is a command, and a command consisteth in declaration or manifestation of the will of him that commandeth." On the other hand, once we assume that that command must be expressed "by voice, writing, or some other sufficient argument of the same," the principle corresponds to the clause of irresponsibility in the criminal law field, which is summed up, in continental Europe, with the formula of the "principle of legality," i.e., "no crime, no punishment without a criminal law" (nullum crimen nulla poena sine lege).1

Yet, despite this ambivalence, what the Westphalian paradigm stands for in this context is pretty clear, namely a monistic doctrine of the legal sources that triumphed throughout the 1800s, just to decline around the mid 1900s. This decline can be expressed with the words of Philip Jessup and the seminal 1956 lectures at Yale Law School that shed light on a law neither national, nor international, but transnational, that is, in order "to include all law which regulates actions or events that transcend national frontiers. Both public and private international law are included, as are other rules which do not wholly fit into such standard categories" [13]. Whether or not this process has to be traced back to the belle époque [14], it seems uncontroversial that the more a set of issues becomes systemic, the less such problems can be tackled at a national level. Although this inverse relationship was noted over and over the last century, the information revolution has dramatically accelerated this very process. As a result, from a legal and political viewpoint, a new Locke would have to change the title of his masterpiece, and dub it nowadays "Two Treatises of Governance." Next section explores why.

4   From Government to Governance

The information revolution is affecting our understanding about the world and about ourselves: we are interconnected informational beings that share with biological organisms and engineered artefacts "a global environment ultimately made of information," i.e., what Luciano Floridi calls "the infosphere" [15]. A crucial feature of this new environment has to do with the complex ways in which multi agent (human/artificial) systems interact. This complexity challenges concepts and ways of reasoning through which, so far, we have grasped basic tenets of the law and politics. A key point of the analysis concerns the use of ICTs: whereas, over the past centuries, human societies have been ICT-related but mainly dependent on technologies that revolve around energy and basic resources, today's societies are progressively dependent on ICTs and moreover, on information as a vital resource. In a nutshell, we are dealing with ICT-driven societies [7].

What this huge transformation means, from a legal and political viewpoint, can be illustrated with the ubiquitous nature of the information on the internet. The flow of this information transcends conventional boundaries of national legal systems, as

---

1 In the wording of Article 7 of the 1950 European Convention on Human Rights, "[n]o one shall be held guilty of any criminal offence on account of any act or omission which did not constitute a criminal offence under national or international law at the time when it was committed." However, as lawyers know, there is a savings provision pursuant to art. 7(2) of the Convention, which states: "This article shall not prejudice the trial and punishment of any person for any act or omission which, at the time when it was committed, was criminal according the general principles of law recognized by civilized nations." The aim of this provision is to cover such exceptional cases as the Nuremberg trial against the Nazis.

shown by cases that scholars address as a part of their everyday work in the fields of data protection, computer crimes, digital copyright, e-commerce, and so forth. This flow of information jeopardizes traditional assumptions of legal and political thought, since the idea of the law as a set of rules enforced through the menace of physical sanctions [e.g. 16] often falls short in coping with the new challenges of the information revolution: identity thefts, spamming, phishing, viruses, and cyber attacks have increased over the past decade, regardless of harsh national laws like the US anti-spam act from 2003. Furthermore, a number of issues, such as national security, cyber-terrorism, availability of resources and connectivity, concern the whole infrastructure and environment of today's ICT-driven societies and thus, these issues have to be tackled at international and transnational levels. Whereas constitutional powers of national governments have been joined – and even replaced – by the network of competences and institutions summarized by the idea of governance, sovereign states, although still relevant, should be conceived as one of the agents in the public arena.

In [17], eight meanings of governance are discussed: in this section, it suffices to quote two of them. On the one hand, according to the World Bank, the idea of governance concerns "the process and institutions through which decisions are made and authority in a country is exercised" [17]. On the other hand, Hyden, Court and Mease refer to "the formation and stewardship of the formal and informal rules that regulate the public realm, the arena in which state as well as economic and societal actors interact to make decisions" [17]. On this basis, the notion of governance can be furthered as a matter of "good" governance. In the case of the World Bank, focus should be on inclusiveness and accountability established in three key areas, namely, i) "selection, accountability and replacement of authorities"; ii) "efficiency of institutions, regulations, resource management"; and, iii) "respect for institutions, laws and interactions among players in civil society, business, and politics." In the case of Hyden, Court and Mease, the concept of good governance can be measured along six dimensions, i.e., "participation, fairness, decency, efficiency, accountability, and transparency," in each of the following arenas: "civil society, political society, government, bureaucracy, economic society, judiciary."

Drawing on such definitions, we can appreciate how the system of the legal sources appears far more complex than it used to be under the traditional Westphalian model and the dichotomy between national and international law. By including Jessup's "other rules which do not wholly fit into such standard categories" [13], the current sources comprise such fields of transnational law as the internal legal regimes of multinational organizations and today's lex mercatoria, enterprises and labour unions as private actors in international labour law, much as human rights law, sports law and, of course, IT law [18, 19, etc.]. Whilst some propose a parallel between the old medieval system of European common law (ius commune) and the new system of plural legal sources [20], others refer to Jessup's "other rules" as a sort of global law without the state [21]. Yet, in both cases, there is room for the return of customary law as a fundamental component of the whole system and, hence, a new legal role for social intelligence and spontaneous orders. Next section dwells on this scenario in light of the new dichotomies between hard law and soft law, and between game players and game designers. The overall idea is to lay down, so to speak, the statics of the system, that is its new legal sources and tools. On this basis, Section 6 aims to deepen the dynamics of the system, namely the processes that characterize and challenge today's ICT-driven societies from a normative viewpoint.

5   The Legal Tools of Governance

There are four major differences between the system of legal sources of the Westphalian model and today's governance of ICT-driven societies. First, this latter system of legal sources is tripartite, rather than bipartite: in addition to the traditional sources of national law and international law, in which the only relevant actors used to be the sovereign states, the system includes the sources of transnational law and the agency of non-state, or private (as opposed to public), actors.

Second, the new system of legal sources incorporates customary law as a key part of the system. To be sure, traditional international law has always hinged on customary rules, such as the principle pact sunt servanda, that is "agreements must be kept." Yet, this customary basis of international law has suggested time and again, that international law is a rudimental sort of legal system or, at least, it should be deemed as mere positive international morality. On the contrary, customs of transnational law provide the solid basis for such fields as current lex mercatoria, or transnational corporate and business law, in accordance with the thesis of Hayek on kosmos, unintentional orders and the role of social intelligence.

Third, pace Kelsen's definition of law mentioned above in section 4, we should further distinguish between binding and non-binding rules, that is between hard law and soft law-tools of governance. In other words, in addition to the traditional hard law-rules of the legal system, such as national statutes, codes, or international agreements, we have to add recommendations, codes of conduct, guidelines, and the standardization of best practices. Although scholars often equate the hard rules of the law with the effectiveness of national legal systems, so that the norms of both international and transnational law would be less and less binding, this is not necessary so. On the one hand, among the sources of national law, there is room for forms of soft law such as, say, the recommendations and opinions of data protection authorities. On the other hand, once we consider such a field as the current network of internet governance, it is noteworthy that several of the effective binding rules have their source in the field of transnational law, spontaneous orders, and the decision of non-state actors, rather than the traditional activism of national lawmakers.

Fourth, the new scenarios of the information revolution have suggested national and international lawmakers, and private companies alike, more sophisticated forms of legal enforcement, complementing the traditional hard rules of the law and softer forms of legalized governance via the mechanisms of design, codes, and IT architectures. Admittedly, such a shaping is not necessarily digital: consider the installation of speed bumps in roads as a means to reduce the velocity of cars, lest drivers opt to destroy their own vehicles. Yet, scale again matters, in that many impasses of today's legal and political systems are increasingly tackled by embedding normative constraints into ICTs through the design of interfaces, self-enforcing technologies, default settings, and so forth. Whereas, in their work on The Design with Intent Method, Lockton, Harrison and Stanton describe 101 ways in which products can influence the behaviour of their users [22], it suffices to focus on three different ways in which we may evaluate this new role of governance actors as game designers, rather than game players, of current social interaction.

The first aim which design may have is to encourage the change of social behaviour. Think about the free-riding phenomenon on peer-to-peer (P2P)-networks, where most peers tend to use these systems to find information and download their favourite files without contributing to the performance of the system. Whilst this selfish behaviour is triggered by many properties of P2P applications, like anonymity and hard traceability of the nodes, designers have proposed ways to tackle the issue through incentives based on trust (e.g., reputation mechanisms), trade (e.g., services in return), or alternatively slowing down the connectivity of the user who does not help the process of file-sharing [23]. In addition, design mechanisms can induce the change of people's behaviour via friendly interfaces, location-based services, and so forth. These examples are particularly relevant because encouraging individuals to change their behaviour prevents risks of paternalism, when the purpose of design is to encourage such a change of behaviour by widening the range of choices and options. At its best, this latter design policy is illustrated by the open architecture of a web "out of control" [24].

The second aim concerns how to decrease the impact of harm-generating behaviour, rather than changing individual conduct via design mechanisms. This further goal is well represented by efforts in security measures that can be conceived as a sort of digital airbag: as it occurs with friendly interfaces, this kind of design mechanism prevents claims of paternalism, because it does not impinge on individual autonomy, no more than traditional airbags affect how people drive. Contrary to design mechanisms that intend to broaden individual choices, however, the design of digital airbags may raise issues of strong moral and legal responsibility, much as conflicts of interests. A typical instance is given by the processing of patient names in hospitals via information systems, where patient names should be kept separated from data on medical treatments or health status. How about users, including doctors, who may find such mechanism too onerous? Furthermore, responsibility for this type of mechanisms is intertwined with the technical meticulousness of the project and its reliability, e.g., security measures for the informative systems of hospitals or, say, an atomic plant.

Then, there is the most critical aim of design, namely to prevent harm generating-behaviour from occurring through the use of self-enforcing technologies, such as DRMs in the field of intellectual property protection, or some versions of automatic privacy by design [e.g. 25]. Serious issues of national security, connectivity and availability of resources, much as child pornography or cyber-terrorism, may suggest endorsing such type of design mechanism, though the latter should be conceived as the exception, or last resort option, for the governance of ICT-driven societies. Contemplate some of the ethical, legal, and technical reasons that make problematic the aim of design to automatically prevent harmful conduct from occurring. As to the ethical reasons, specific design choices may result in conflicts between values and, vice versa, conflicts between values may impact on the features of design: we have evidence that "some technical artefacts bear directly and systematically on the realization, or suppression, of particular configurations of social, ethical, and political values" [26]. As to the legal reasons against this type of design policy, the development and use of self-enforcing technologies risk to curtail both collective and individual autonomy severely. Basic tenets of the rule of law would be at risk, since people's behaviour would unilaterally be determined on the basis of technology, rather than by choices of the relevant political institutions: what is imperilled is "the public understanding of law with its application eliminating a useful interface between the law's terms and its application" [27].

Finally, attention should be drawn to the technical difficulties of achieving such total control through design: doubts are cast by "a rich body of scholarship concerning the theory and practice of 'traditional' rule-based regulation [that] bears witness to the impossibility of designing regulatory standards in the form of legal rules that will hit their target with perfect accuracy" [28]. Indeed, there is the technical difficulty of applying to a machine concepts traditionally employed by lawyers, through the formalization of norms, rights, or duties: after all, legal safeguards often present highly context-dependent notions as, say, security measures, personal data, or data controllers, that raise a number of relevant problems when reducing the informational complexity of a legal system where concepts and relations are subject to evolution [29]. To the best of my knowledge, it is impossible to program software so as to prevent forms of harm generating-behaviour even in such simple cases as defamations: these constraints emphasize critical facets of design that suggest to reverse the burden of proof when the use of allegedly perfect self-enforcing technologies is at stake. In the wording of the US Supreme Court's decision on the Communications Decency Act ("CDA") from 26 June 1997, "as a matter of constitutional tradition, in the absence of evidence to the contrary, we presume that governmental regulation... is more likely to interfere with the free exchange of ideas than to encourage it."

6   Between Representation and Resolution

The previous section has focused on the statics of the systems, namely the hard law and soft law-tools of governance, much as the variety of design mechanisms, through which governance actors may attempt to rule the dynamics of today's ICT-driven societies. However, in order to grasp the specificity of societies that progressively are dependent on information as a vital resource, let us prevent a twofold misunderstanding. At times, scholars address the challenges of the information revolution to the traditional models of political legitimacy and democratic processes as if the aim were to find the magic bullet. Vice versa, others have devoted themselves to debunk these myths, such as a new direct online democracy, a digital communism, and so forth, by simply reversing the paradise of such techno-enthusiasts [30]. All in all, we should conceive today's information revolution in a sober way, that is as a set of constraints and possibilities that transform or reshape the environment of people's interaction.

On the one hand, this profound transformation affects norms, competences, and institutions of today's governance, much as people's autonomy and the right of the individuals to have a say in the decisions affecting them: consider the debate on the role that national sovereign states should have in today's internet governance, vis-à-vis such transnational and technical organizations as, for example, the internet corporation for the assignment of names and numbers (ICANN). Moreover, contemplate how a myriad of communities have emerged and developed their own legal systems online [6, 31, 32, etc.]. Theoretically, five models of internet governance may be conceived of [5]: the model of cyberspace and spontaneous ordering, the model of transnational institutions and international organizations; the model of code and internet architecture; the model of national governments; and, finally, the model of market regulation. Whereas, in the phrasing of Solum, "no single model provides the solution to all the problems that Internet regulation can address," it follows that "the best models of Internet governance are hybrids that incorporate some elements from all five models" [5].

Yet, on the other hand, a normative approach is vital, so as to order thinking about making governance policies for current ICT-driven societies. As Luciano Floridi suggests in his contribution to The Onlife Manifesto, focus should be on the foundations of an "efficient" and "intelligent" multi-agent system, the model of which may represent a goal that could successfully orient our political strategy in terms of transparency and tolerance: "Finding the right balance between representation and resolution, while implementing the agreement to agree on the basis of ethical principles that are informed by universal human rights, is a current major challenge for liberal democracies in which ICTs will increasingly strengthen the representational side" [7]. Time and again throughout this paper, attention has been drawn to the rearrangement of the national law sources vis-à-vis the strengthening of the representational side via the crisis of the Westphalian model (Section 3), much as the return of customary law and a new role for social intelligence and spontaneous orders as a fundamental component of the system (Sections 4 and 5). At the end of the day, this rearrangement should be conceived as that which actually is, namely a huge redistribution of power. Therefore, how should we strike the right balance between representation and resolution?

First, the self-organizing properties of current social interaction, on which I have insisted in this paper, should be prioritized. In accordance with the Supreme Court's CDA ruling, which concluded the previous section, this means that the burden of proof falls on national and international lawmakers, much as governance actors, whose aim is to rule the processes of ICT-driven societies. After all, this is what occurred at the World Conference on International Telecommunications (WCIT-12), held in Dubai, United Arab Emirates, in which several national governments had to illustrate the (preposterous) reasons why they should have the right to manage the internet, by divesting "ICANN of its authority and bring domain-name administration within the scope of a government-only agency like the International Telecommunications Union (ITU)" [33]. Luckily, this new attempt to impose the bankrupt theory of the Westphalian system finally failed, much as the US Stop Online Piracy Act (SOPA) and the Protect IP Act (PIPA) bills did in winter 2011-2012.

Second, once the need for some sort of regulation is proven, governance actors should really know the subject matter which they intend to govern. Although this latter proviso may appear as a truism, this is the bread and butter of scholars dealing with the regulation of cyberspace [31]; on making laws for cyberspace [6]; etc. Think again of WCIT-12 and debate prior to the Dubai conference, on the economic modelling of the internet and the proposal of the European Telecommunications Network Operators' Association (ETNO), a group of European telecommunications providers led by Telecom Italia, Telefónica España, France Telecom, and Deutsche Telekom. Leaving the technical details of the proposal for a new economic model of the internet aside, it is noteworthy that the decision of the WCIT-12 conference Chair was to move the debate into the ITU and more particularly, into the ITU division (ITU-T) that designs telecommunications standards. "By analogy, it would be the equivalent of taking one's tax questions to an architect rather than a certified public accountant or other tax expert. To be sure, an architect is educated, licensed, and may even have a personal opinion about taxes and money – and even how certain construction techniques might be cheaper or result in tax rebates. However, to state the obvious: architects build and design things, while accountants deal with taxes and money" [33].

Third, once the subject matter of the governance regulation is properly known, it is likely that both binding and non binding rules will increasingly concern the architecture, code, or design of the system, rather than traditional legal rules that have to be enforced through the menace of physical sanctions. Here, the three design mechanisms discussed above in Section 5 are critical. When the aim is to broaden the range of people's choices, so as to encourage the change of their behaviour, such design policy looks legally and politically sound: this approach to design prevents threats of paternalism that hinge on the regulatory tools of technology, since it fosters collective and individual autonomy. Likewise, the aim of design to decrease the impact of harm-generating behaviour through the use of digital airbags, such as security measures or user friendly interfaces, respects collective and individual autonomy, because this approach to design does not impinge on people's choices, no more than traditional airbags affect how individuals behave on the highways. Yet, to complement the hard and soft-law tools of governance by design entails its own risks, when the aim is to prevent harm-generating behaviour from occurring. Although many impasses of today's legal and political systems can be properly addressed by embedding legal safeguards into ICT and other kinds of technology, we already mentioned some of the several legal, ethical and technical reasons why the use of allegedly perfect self-enforcing technologies raises serious threats of paternalism and, even, of authoritarianism. Whether DRMs in the field of digital copyright, automatic versions of the principle of privacy by design, or Western systems of filters in order to control the flow of information on the internet, the result is the modelling of individual conduct [34, 35]. Recent statutes, such as HADOPI in France, or DEA in UK, show how new ways of protecting citizens even against themselves do materialize.

7   Conclusions

This paper has traced current work on social intelligence back to the everlasting debate on the sources of law, so as to examine some crucial challenges of the information revolution, namely if, and to what extent, there is legal room for processes of social intelligence and in Hayek's jargon, whether unintentional and spontaneous orders can be deemed as sources of today's legal systems. In section 2, attention was drawn to ancient Roman law and the formalization of social, as opposed to individual, intelligence as the binding force of legal customs. Then, in section 3, focus was on the eclipse of this representation in light of some tenets of Hobbes's legal philosophy, and the paradigm of Westphalia that triumphed throughout the 1800s, just to decline around the mid 1900s. This latter process was summarized in section 4, in accordance with the evolution from the role of government and national sovereign states, that is the core of the Westphalian model, to the complex network of processes, sources, and institutions summed up by today's governance of ICT-driven societies. Whilst section 5 examined the statics of the system, namely the legal tools of governance, section 6 contextualized them in light of current debate on how to govern ICT-driven societies and, more particularly, matters of internet governance.

As to the statics of the system and differences between the Westphalian model and the current system of legal sources, the paper insisted on the legal role of social intelligence through new forms of customary and transnational law, much as social norms that a myriad of communities have developed online. However, current tussles on the future of the internet and its governance showed that it would be deadly wrong to take such a new legal role for granted. Although national law-making activism is increasingly short of breath, the backlash of sovereign states on today's kosmos is understandable, once we recall that rearrangements of legal sources are intertwined with a redistribution of power. The challenges of the information revolution do not only concern whether traditional state action over ICT-driven societies is more or less effective. In addition, such challenges regard how national states aim to regulate and control both territorial and extraterritorial conduct by imposing norms on individuals that have no say in the decisions affecting them, through the mechanisms of design, codes, and architectures. A procedural approach has been suggested, so that: i) the burden of proof should fall on national and international lawmakers that aim to intervene in the self-organizing properties of current social interaction; ii) governance actors should really know the field in which they intend to intervene, once the need for regulations is proven; and, iii) self-enforcing technologies should represent the exception, or last resort option, for coping with the impact of the information revolution. From a normative viewpoint, these are the conditions for a right balance between representation and resolution.

References

1. Goldsmith, J.: Against Cyberanarchy. University of Chicago Law Review 65, 1199–1250 (1998)
2. Wu, T.: Who Controls the Internet. Oxford University Press, New York (2006)

3. Post, D.G.: Against "Against Cyberspace". Berkeley Technological Law Journal 17, 1365–1383 (2002)
4. Post, D.: In Search of Jefferson's Moose: Notes on the State of Cyberspace. Oxford University Press, New York (2009)
5. Solum, L.B.: Models of Internet Governance. In: Bygrave, L.A., Bing, J. (eds.) Internet Governance: Infrastructure and Institutions, pp. 48–91. Oxford University Press, New York (2009)
6. Reed, C.: Making Laws for Cyberspace. Oxford University Press, Oxford (2012)
7. Floridi, L.: The Fourth Revolution – The Impact of Information and Communication Technologies on Our Lives. Oxford University Press (2014)
8. Hayek, F.A.: Law, Legislation and Liberty. A New Statement of the Liberal Principles of Justice and Political Economy. University of Chicago Press, Chicago (1982)
9. Hayek, F.A.: The Constitution of Liberty. University of Chicago Press, Chicago (1960)
10. Hobbes, T.: Leviathan, R.T. (ed.). Cambridge University Press, Cambridge (1999)
11. Bobbio, N.: Thomas Hobbes and the Natural Law Tradition. University of Chicago Press, Chicago (1993) transl. by D. Gobetti
12. Strauss, L.: The Political Philosophy of Hobbes. Its Basis and its Genesis. Oxford University Press, Oxford (1936)
13. Jessup, P.C.: Transnational Law. Yale University Press, New Haven (1950)
14. Amin, S.: The Political Economy of the Twentieth Century. Monthly Review 52(2), 1–17 (2000)
15. Floridi, L.: The Ethics of Information. Volume II of Principia Philosophiae Informationis. Oxford University Press (2013)
16. Kelsen, H.: General Theory of the Law and the State. Harvard University Press, Cambridge (1949)
17. Grindle, M. Good Enough Governance Revisited (2005), http://www.odi.org.uk/events/docs/1281.pdf (retrieved on 21 May 2012)
18. Murray, A.D.: Information Technology Law: The Law and Society. Oxford University Press, Oxford (2010)
19. Lloyd, I.J.: Information Technology Law. Oxford University Press, Oxford (2011)
20. Coing, H.: Von Bologna bis Brüssels: Europäische Gemeinsamkeit, Gegenwart und Zukunft, Kölner Juristische Gesellschaft, vol. IX, Bergish Gladbach, Köln (1989)
21. Teubner, G.: Breaking Frames: The Global Interplay of Legal and Social Systems. American Journal of Comparative Law 45, 145–169 (1997)
22. Lockton, D., Harrison, D.J.: The Design with Intent Method: A Design Tool for Influencing User Behaviour. Applied Ergonomics 41(3), 382–392 (2010)
23. Glorioso, A., Pagallo, U.: The Social Impact of P2P Systems. In: Shen, X., Yu, H., Buford, J., Akon, M. (eds.) Handbook of Peer-to-Peer Networking, pp. 47–70. Springer, Dordrecht (2010)
24. Berners-Lee, T.: Weaving the Web. Harper, San Francisco (1999)
25. Cavoukian, A.: Privacy by Design: The Definitive Workshop. Identity in the Information Society 3(2), 247-251 (2010)
26. Flanagan, M., Howe, D., Nissenbaum, H.: Embodying Values in Technology: Theory and Practice. In: van der Hoven, J., Weckert, J. (eds.) Information Technology and Moral Philosophy. Cambridge University Press, New York (2008)
27. Zittrain, J.: Perfect Enforcement on Tomorrow's Internet. In: Brownsword, R., Yeung, K. (eds.) Regulating Technologies: Legal Futures, Regulatory Frames and Technological Fixes, pp. 125–156. Hart, London (2007)

28. Yeung, K.: Towards an Understanding of Regulation by Design. In: Brownsword, R., Yeung, K. (eds.) Regulating Technologies: Legal Futures, Regulatory Frames and Technological Fixes, pp. 79–108. Hart, London (2007)

29. Pagallo, U.: As Law Goes By: Topology, Ontology, Evolution. In: Casanovas, P., Pagallo, U., Sartor, G., Ajani, G. (eds.) AICOL-II/JURIX 2009. LNCS, vol. 6237, pp. 12–26. Springer, Heidelberg (2010)

30. Morozov, E.: The Net Delusion. Perseus, New York (2011)

31. Murray, A.D.: The Regulation of Cyberspace: Control in the Online Environment. Routledge-Cavendish, New York (2007)

32. Schultz, T.: Private Legal Systems: What Cyberspace Might Teach Legal Theorists. Yale Journal of Law and Technology 10, 151–193 (2007)

33. Cerf, V., Ryan, P., Senges, M.: Internet Governance is Our Shared Responsibility. Publication forthcoming in A Journal of Law and Policy for the Information Society 10, ISJLP (2014), http://www.ssrn (August 13, 2013)

34. Pagallo, U.: Cracking down on Autonomy: Three Challenges to Design in IT Law. Ethics and Information Technology 14(4), 319–328 (2012)

35. Pagallo, U.: Good Onlife Governance: On Law, Spontaneous Orders, and Design. In: Floridi, L. (ed.) The Onlife Manifesto: Being Human in a Hyperconnected Era. Springer, Dordrecht (2015)

Methods for Law and ICT:
An Approach for the Development of Smart Cities

Fernando Galindo

University Zaragoza, Zaragoza, Spain
cfa@unizar.es

Abstract. The paper summarizes the methods followed over thirty years (1984-2014) in the study of the subject "Law and information and communication technology (ICT)". From the beginning the emphasis was placed on that the most appropriate approach is to put into action an interdisciplinary activity aimed at solving specific-real problems. The same methodology was used to develop juridical proposals able to integrate the use of the innovation brought by ICTs, as "electronic commerce or government", in daily life. The paper concludes showing that the methodology is appropriate to participate in the creation and development of technological innovations such as the construction of services for "Smart Cities".

Keywords: Law and information and communication technologies, Methods, Interdisciplinaritie, Smart Cities.

1   Introduction

Occasionally it is worth remembering. It is not enough to cater to the context, the requirements and the significance of the daily problems to solve them. The memory also tends to give sense, perspective, to any problem that is immediately stranger. That is why we will express here methods, styles of work and researches that have allowed providing solutions to problems that are involved in relation to the topics "Law and information and communication technology (ICT)". This is made in order to find in these procedures, insights that can serve other people to solve these problems in the coming years.

The specific problem we want to solve for now and the near future is the legal regulation of the consequences of the introduction of so-called Smart Cities from the time of the design of the services that will constitute the infrastructure of those Cities. A project, in which the author of this paper is involved, intends to help to solve these kinds of problems with an adequate approach. The project is entitled City 2.020,[1] and

---

[1] The Project Ciudad2020 is the INNPRONTA Project IPT-20111006, funded by the Spanish Centre for Industrial Technological Development, the University of Zaragoza participates in the project as advisor of the Atos Research @ Innovation (ATI) Division of the firm Atos. The link of the Project is located in: http://www.innprontaciudad2020.es/ (consulted on March 5, 2014).

P. Casanovas et al. (Eds.): AICOL IV/V 2013, LNAI 8929, pp. 26–40, 2014.

it is the basis for building technical solutions, programs and services for citizens who will live in an hypothetical intelligent city of the year 2.020. The well-intentioned forecasts say, for example, that these kinds of services or programs will ensure that the environment is preserved, that driving on roads and cities will be easier and less dangerous as today, and that administrative services provided to citizens by means of ICT will make their daily lives more comfortable [1, 2].

For these purposes it is interesting to present here several solutions. These may serve also in order to improve the lives of citizens and promote their participation in political institutions in response to democratic principles, problems resulting from the relationship established between ICT and Law in the last thirty years (1984-2014), attending to the experiences of the research groups in which we have participated, and the basic characteristics of the social context in which these have been developed. This is that we propose to do in this paper, highlighting features and notes of significant research carried out at that time.

That is why this paper is focused, first, on presenting research that has been busy building information systems that allow the storage and retrieval of legal texts, making specific systems accessible to the largest possible part of citizens.

The second part of the paper is on the legal approach developed with the emergence and expansion of the Internet, the communications technologies, with their advantages and disadvantages. The objective of the research was the providing of policies and technical proposals to ensure a smooth implementation of technological resources. This is presented through the exhibition of the implementation of electronic trade and government services developed in compliance with the principles and norms of the rule of law.

From the past experiences it has proposed a legal approach to build Smart Cities, being this increasingly target horizon accepted as research, development and innovation activity in services or systems. The objective is to promote a widespread use of ICT for the majority of citizens through the use of technological devices such as mobile phones or similar (tablets, for example). This is the third part of the paper.

Finally, the conclusion comes.

2   Thesaurus

In the second half of the eighties of the twentieth century, when it came to exploring the possibilities offered by ICT in the juridical area, initiated its application to the management of the administration of justice in courts and tribunals [3], it seemed appropriate to address the topic Law - ICT as research's object. The choice of the research was twofold: firstly to allow to study some consequences of the introduction of ICT in legal activities, in this case of a judicial nature, on the other to generate interest in the subject to future lawyers. The idea was to consider the relevance of theories advocated conducting legal studies from the characteristics of the language of the rules, as a formal expression of Laws passed in Parliament by the elected representatives of the citizens.

Existed at that time also a doctrine that believed that mathematic theories and ICT techniques could assist the automatic generation of rules to particular cases [4, 5].

The establishing of collections, conveniently studied and refined, of expressive words on legal problems, as compiled dictionaries or thesauri, was the first phase of the research, for the subsequent generation of standards.

This was even more feasible if it happened, as we wanted to do with the investigations undertaken, which specific aspects of legal-philosophical discussion of the moment were considered, as it was the case with the possibilities offered by several variants of offered theories of law. Especially the studies coming from the theory of legal argument, that considered the different characteristics of the activities on interpretation and enforcement of the rule of law, made by juridical professionals [6, 7].

To this end, together with several professors from Mathematics (Algebra and Statistics), and another of the dogmatic subjects that are explained in the Faculties of Law since the nineteenth century (specifically Civil Law, Criminal Law and Administrative Law), we began (from the Philosophy of Law) to conduct interdisciplinary research related to the most profitable results of ICT at that time: its ability to store and retrieve legal documents [8], and to represent this knowledge using logic programming languages. It was the birth of the research group at the University of Zaragoza named after (from 2003) "Data protection and electronic signature".[2]

With this, it became clear that the aim of the research was beyond recovery documentation. The Group wanted to use also the possibilities offered by programs called "expert systems" to access and retrieve legal documents of local interest.

This research was called the building of a "smart legal thesaurus" [9].

The consequences of these activities were excellent in regards to the construction of legal databases, comprising legislation, jurisprudence and doctrine onto a subject of special interest in the Autonomous Community, as the historical and current Law of Aragon. It was a new "channel" of knowledge.[3]

Different was the result of investigations regarding the construction of expert systems, which facilitate citizens and specialists access to specific legal texts relating to the exercise of their own rights authorized by the Law. These systems were built, but they could not move from the prototype stage, being expressed as programming languages in the artificial intelligence style [10]. These programs responded through dialogues to possible questions regarding what are some of the specific rules, other than the Spanish general, at the age of majority in the autonomous community of Aragon.

Later, another research was developed whose object was to represent not so much the content as the rules of procedural character in a court or tribunal for a certain issue [11]. The use of this model increased the efficiency of the research, but it still could prove not useful in the general absence of resources and interest to go to the "industrial" test of the same, as then we needed.

The conclusion reached in these investigations was referring to that his approach was preliminary, technologically speaking, as it barely existed, at least in Spain, interest in the application thereof. At that time the storing of legal documentation on digi-

---

[2] http://www.lefis.org/pdfe/ (consulted on March 5, 2014).

[3] See the chanel in: http://www.unizar.es/derecho/standum_est_chartae/ (consulted on March 5, 2014).

tal media for companies and public institutions began. It was noted that at the time was not possible to obtain practical results, or programs coming from such research applications.

It is important to say that, however, it was indeed obtained as a result of these investigations, detailed studies and approaches to the expression of the characteristics of the activity named access to legal texts, and other professional activities of jurists [12]. It was obtained also the redefinition of the concept of law, coming to define this as a communicative activity [13].

The relevance of this line of research, reflected in several projects, seen from today, found especially in demonstrating the need of interdisciplinary research. It was because the object of study has so many facets that it was impossible to consider all of them from either another area in isolation. Another conclusion referred to the legal field only was related also to interdisciplinarity. Ie while it was necessary in these investigations lawyers who knew the "dogmatic" field considered from an academic perspective, it also became necessary the participation of lawyers that know how to put the law in action, this is practical jurists expert in the resolution of legal conflicts. Consistent with the above was also required the participation of a philosophy of law that was aware of the general characteristics of scientific/technical thinking, legal philosophy (that deals with the three-dimensional aspect of the law: integration of values, rules and facts) [14] and the basic principles and applications of the "dogmatic" arguments carried out by the Science of Law.

3   Digital Signature

The development of telecommunications and its integration with information systems enabled the Internet and the developing of their applications, especially systems that allowed the email and the large-scale carrying of electronic commerce. This was from the second half of the nineties.

Indeed, as the Internet became operational, beginning to put into action the real-time communications through the use of ICT, to initially be able to send and receive messages via email, or buy products from suppliers of goods and services via the Internet (which soon was called e-commerce), the research group considered as objects of their activities these:

1. The further development of proposals to assist juridical activities using information retrieval systems, supported by the most sophisticated telecommunications in relation to specific domains [15], and

2. The development of proposals to ensure the rights of specific individuals to communicate their thoughts and decisions freely at the time of transmitting and receiving information using resources that made possible more and better recovery of legal information, as adapted to the specific needs of identified users. These were the work aimed at ensuring the management of the identification of senders and receivers of messages, and the preservation of the integrity of the message content through the use of electronic signatures [16].

The investigations were carried out since the second half of the nineties, while taking place the emergence of the innovations.

The innovation in research on the development of retrieval systems of legal documentation, oriented in the outlined direction: organization of the management of public institutions (this will be nominated more late "electronic Government"), was influenced by the fact that in Spain the free access to this documentation for all citizens occurred by initiative of the public institutions responsible for creating them. This took place gradually since the second half of the nineties. In other countries, there were no such initiatives. This did not bring to Spain the need of the establishment of research centers with the function to make public, unofficially, legal texts by using ICTs.

The emergence of these research centers is easily understood in other countries as expressed below.

According to the principles of the State of Law the texts of the Law are publicly available once representatives of the citizens develop them. If they were only accessible by the use of systems of legal documentation retrieval and their owners were companies whose services were to be paid by users, this activity could not satisfy the principles of free access to legal texts. Therefore, the research centers in other countries catered to break this dynamic marketing dissemination, promoting their advertising through the establishment of "unofficial" documentary collections open to access them by all citizens "on line".[4]

In contrast, in Spain, since 1995, legislative and administrative regulations have been made accessible through official channels to anyone, free-form, by Internet. Since then, the Official Gazette ("Boletin Oficial del Estado"), the official organ of publication of state regulations, is available to the public. The same applies practically with respect to, regulations promulgated by the Autonomous Communities.

These initiatives are formally generalized from January 1, 2004.

In regard to the court documents, the General Council of the Judiciary established in 1996 the Judicial Documentation Centre responsible entity to make public the decisions of the Supreme Court, the High Courts of Justice of the Autonomous Communities and other judicial entities.[5]

The results were, as it has been said, that the research group focused their work rather than on the construction of generic recovery systems as legal documentation, in the building of support systems to legal decisions made by concrete institutions, or by the administration, the courts and legal practitioners using new ICT systems such as email and the Internet, whose use began to spread throughout the period considered (second half of the nineties) [17].

The strategy allowed the study to look at organizational and functional changes that these uses claimed of different organizations and legal agents. That is why from the early years of this decade, the group focused on what was started at that time earning him the expression e-government. It had as consequence the study of the expansion of the use of the techniques of governance, or effectiveness, with respect to the juridical activities of public administrations [18].

---

[4] This is the movement "law via the Internet". The last conference is located in: http://www.jerseylvi2013.org/

[5] See: http://www.poderjudicial.es/cgpj/es/Temas/Documentacion-Judicial (consulted on March 5, 2014).

It was the time of the enactment of European Directives and state Laws, focusing on the regulation of "electronic signature", the construction of public key infrastructures and regulation of services of the information society, which allowed to make a reality that legislation.[6]

A few later (more precisely: from 1997 to 2000) began the work of the group jointly with notaries, registries of property, clerks of the courts and lawyers on the features that should give confidence to the use of ICT and Internet. The group proposed the construction of institutions or systems of certification authorities (CAs) of public key, dedicated to promote the trust in the use of ICT attending to the rule of law.[7]

At this time the group has formed CAs and PKIs jointly with notaries,[8] which formed the basis for the establishment of unofficial institutions of certification administered by the research group.

In the judiciary, along with attorneys and judges, the group gave the first steps in the construction of a system that allows the secure telematics transmission of documents from the offices of the attorneys to the courts.[9]

These activities had the character of Research + Development + innovation. In most of the activities companies participated. The content dealt with standards, assessments, knowledge of technologies and their social and economic implications, and deep knowledge of the legal system and its application, once all of them collaborated in the formation of provisions that would later be issued by the relevant entities.

4   Smart Cities

4.1   Introduction: Smart Cities and Services.

At the begin of the second decade of the two thousand: from two thousand eleven, the

---

[6] Directive 1999/93/EC of the European Parliament and of the Council of 13 December 1999 on a Community framework for electronic signatures, Directive 2000/31/EC of the European Parliament and of the Council of 8 June 2000 on certain legal aspects of information society services, in particular electronic commerce, in the Internal Market ('Directive on electronic commerce') and the Directive 2002/58/EC of the European Parliament and of the Council of 12 July 2002 concerning the processing of personal data and the protection of privacy in the electronic communications sector ('Directive on privacy and electronic communications').

[7] The most significant project was called AEQUITAS: "The admission as evidence in trials of criminal character of digitally signed electronic products". It was supported by the European Union. The final report (1998) is located in: http://cordis.europa.eu/infosec/src/study11.htm (consulted on March 5, 2014).

[8] The activities implemented jointly with the Spanish notary was at the origin of the Notarial Certification Agency: http://www.ancert.com (consulted on March 5, 2014).

[9] These activities were the precedent of the implantation of the LexNet system, which is a platform for secure exchange of information between the courts and a wide variety of legal practitioners in their daily work, that need to exchange legal documents (notices, letters and demands). See: https://www.administraciondejusticia.gob.es (consulted on March 5, 2014).

research group "Data protection and electronic signature" works in providing legal advice to the design of a platform that aims to provide services to the citizens that will live in the so-called "Smart Cities".

The platform is the content of the industrial project of I + D + i , entitled City 2.020. It is an Innpronta project. The project aims to achieve by the building of services to citizens, progress in the areas of energy efficiency, Future Internet, Internet of things, human behavior, environmental sustainability and mobility and transport. The project estimates that the design and implementation of these services will build the city of the future, a city that will satisfy the characteristics of sustainability, intelligence and efficiency. The project itself says that conceives, designs and implements a new paradigm of sustainable and efficient city, supported on three key areas: Energy, Transport and Environmental Control.[10]

What kind of services are being developed in the City 2020 project? There are several examples that are either in operation or initially. They will be in use in a short time. We speak later on some of them. We indicate before the existence and characteristics of the existent "infrastructure" that enable their development.

An infrastructure is constituted by the increasingly large information that is accessible in "standard Internet format", or because the users publish them: this is the case of communications made through social networks like Twitter, or because, with regard to public information of all kinds, governments are standing in its openness in a respectful way (more or less) with current legislation, making it accessible to all who want to use it.[11] The latter has been increased by the expansion of the acceptance of the political principle of transparency in the activities of the government, therefore prescribed, for example, by Spanish Law 19/2013 of 9 December, and the obligation that the Governments have to give general access of the information stored and treated in their daily lives, in an adequate way to the content of the advertised information.

The services. It is for this that it should develop services that detect, depending on what the temperature inside and outside of a home is, for example, the time when the heating or cooling must be it off automatically, and from knowledge habits of the owners or tenants of the house have to respect and forecasts made by meteorologists on changing temperatures and can even make autonomous decisions. The same goes for lighting homes and public roads or streets. In both cases the use of services, thereby saving energy, seek the goal of the City 2020 project.[12]

It likewise development services provided through the use of traffic lights and other traffic signs or appropriate sensors, designed to control the traffic density, open to the movement of emergency services or transportation route vehicles engaged in distribution of goods. These systems have developed also simulations based on historical events of what happened in a specific period of time, allowing the forecasting by "smart" statistics on what might happen at one time or another.

---

[10] See: http://www.innprontaciudad2020.es/ (consulted on March 5, 2014).

[11] See the heterogeneity of the information that have been published in:
http://datos.gob.es/datos/ (consulted on March 5, 2014).

[12] The services created by the City 2020 project can be found in:
http://www.innprontaciudad2020.es/index.php/es/
documentacion-ficheros-relativos-al-proyecto/4-entregables
(consulted on March 5, 2014).

There are computer programs capable of indicating whether or not seats exist in a parking while making forecasts of occupancy for the future, allowing citizens to book by a certain date and time.

There are designs of services providing information on the existence or not of bicycles in a station for the storage and collection of these vehicles, as well as existing services near a particular station restaurant, food shops, shopping centers, museums...

Other services / programs offered cultural agendas tailored to the tastes of users. Other programs / services built travel routes or sightseeing in cities as a concrete response to the tastes, interests and age of the user...

These examples are proof on the building of services and programs that assist citizens. The organizations that may use these programs are called Smart Cities.

4.2   Legal Solutions: Generalities.

What indications should be done from a legal perspective to the design of these services?

Of course the first one is that they ought be built respecting the rights of citizens enshrined in the Constitution and all the rules that make up the ordered treatment, which are highly developed at present given the progressive implementation of the use of ICTs in daily life. This implies to require that juridical elements must be present in the design of services or, briefly, that the content of democratic order must be respected. This is a requirement to fulfill at the time of design, if we consider that the foreseen facilities advice of future relevant social changes that will affect to the acquired rights of those who already realize the activities that the new services will make in the future.

This means that if we look, for example, that the services or programs in construction must have information captured by sensors as, for example, changes occurring in nature, their use must be allowed by those who are holders of the sensors and make the required analysis and interpretation of the sensed data to build information. The owners of the sensors and performing interpretations of the data must take responsibility for the quality of information and the consequences of unforeseeable effects of the programs that process this information.

If the information relates to activities of daily living, ie to personal identification and information of life of citizens, their use must be voluntarily accepted by the citizens. It is preferable that the use of personal data from the smart services becomes transformed them into anonymous format in order to preserve as far as possible the right of every citizen to the privacy of their personal life recognized by the Constitution. The implementation of the prescribed safety measures as the anonymisation are recognized as part of the law in this area. Thus, the best is that all information is treated in the form of patterns of individual behavior of the service users. These may be made available to whoever, and be generated by other users / citizens who acquire the services. Responsibility likewise be bound to companies that create or maintain services.

Another thing. No one can escape the social significance of services / applications are being designed and put into practice in order to achieve the ultimate goal of the implementation of the Smart City. As it has been expressed the objective of the development of the "smart services" is the transforming of the service delivery done so far by companies, organizations and individual persons having professional expertise, in the provision of other services developed and delivered automatically from the data sensed by automata or previously experienced by the same users or others [19].

The question to be resolved is how it should be designed programs / services / automata / "artifacts" of smart cities in order to be able to meet with them the needs for which are made while preserving the rights and duties of everyone involved in the process of design, supply, acquisition and use of these services / programs that are guaranteed by the regulation for democratic legal systems?

To answer this we must consider that there are three elements necessary to elaborate programs:

1. Building databases or designing programs.

2. Communication between users and databases.

3. The requirements of the regulations for the construction and use of programs /services.

There are in the next sections some of the characteristics of these elements, considering in particular those relating to legal regulation.

4.3   The Design of the Programs and the Value of Communications

Sensors and open data provide information needed for building applications or services. Citizens with their behavior may also be considered sensors: the "citizen sensor" by providing exemplary personal information and the use of exemplary model of service behavior, which is reflected in the databases. The citizens are, of course, service users also.

A model means that the information generated by users using the services stored in anonymised form as to be able to use it for a future request for services of a similar nature made by the same user or with other features, will or interest similar to those of who generated the service / program.

The use of media or communication channels, under appropriate security, it is essential in the case of the supply of services in parts intelligent cities. This is because applications and services are designed considering that access to systems occurs through the use of Smartphones, tablets or personal computers. Ie it is expected that users of these services are not legal entities: public organizations or companies, but citizens.

Companies are usually service providers.

Public administrations are suppliers of open data.

The latter can also be recipients of the services themselves or, especially, the information generated by the use of smart services providers whose data are open. With both administrations can manage activities more effective and offer democratic public services for which they are responsible, as they are required to meet the demands and needs of users citizens thereof, balanced spending and public funds, following guidelines of good governance that are required from the Spanish reform of art. 135 of the Constitution in 2011. With this information, in addition, public authorities can make forecasts for the organization of the provision of open data for specific periods and agreements.

4.4   The Regulation

When we talk about the legal regulation of the systems / programs for the Smart Cities the paradox is that we talk about something unknown, because there is no legal regulation on Intelligent Cities outside the legislation that enhances the performance of research and development on the same, or interest agreements on the same question agreed between administrative organizations, municipalities fundamentally, who propose the design of programs and services for citizens on account of their general obligation to support the public R & D and its ability to create jobs.[13]

Therefore the question arises: how to go forward with the legal aspects?

The truth is that it is not easy to do something in this respect by the positivist theories of law, those who merely do an exegesis of the law using these theories or the dogmatic science of law. According to their rules and principles, if there are no laws the jurist can not make legal considerations. The general policy for these theories is that the commentaries on the laws can be made only by the legislative power with new laws. He makes the laws. This is not the problem with other legal theories as the communicative study of the Law. This theory studies juridical activities and their accommodation or not to the norms, principles and values of a democratic society [20, 21].

From this latter perspective, which is what we have in mind here, it is necessary to make statements about the rules which must be addressed in the design and implementation of programs / smart services, given the relevant character and important social function of this kind or foreseen services and the legal / evaluative requirements that they need to meet even when we are talking now of a phenomenon generally limited to R + D + i . We must also say that this legal perspective must be made at this time of design because otherwise the services / systems could not be used in the future due to the manifest illegality in which would incur those who will design and utilize in the event that systems generate defective services.[14]

The proposal, also, is not new: as it was mentioned in the third section of this paper, the implementation of electronic commerce and government also required the prior establishment of legal rules allowed the operation of the programs / systems in accordance with the rules and principles of the legal systems in order to overcome the limitations encountered in the use of technology. There are more arguments of legal character also. There are new rules of direct application for the construction of new services for Smart Cities. This is the case with the regulations on reuse of public information. It is the same with the requirements of transparency to the functioning of public institutions. Another case is the regulation on the access to public information. The same has relation to the requirements for compliance with the rules of good government. All these rules pay attention to the problems identified by the setting in the activities of the technological innovations that involves the construction of intelligent services.

---

[13] This is the case of "smartcity" the Spanish network of Smart Cities. See the website at: http://www.redciudadesinteligentes.es/ (consulted on March 5, 2014).

[14] The basic juridical requirements in the City 2020 project are located in: http://www.innprontaciudad2020.es/index.php/es/documentacion-ficheros-relativos-al-proyecto/white-papers/28-proteccion-de-datos-personales (consulted March 5, 2014).

To add to this, it is the obvious consideration that the regulations to which we must also address the design of services for smart cities is the existing norms to regulate the functioning of society and ICT from some time (the seventies in the twentieth century): data protection , security measures, electronic signature, electronic access of citizens to public services, preservation of intellectual and industrial property and general measures of law provided for the preservation and attribution of responsibility.

4.5   The Preparation of Juridical Proposals According to Democratic Principles

The legal / juridical principles and rules of law are not sufficiently satisfied when political institutions as Cities form associations in order to boost the "intelligent cities movement" or the European Union itself, when established and updated Directives on the reuse of information from public Administrations, confined to expand the possibilities of economic growth which the construction of "services / programs for the Smart Cities" let. This kind of initiatives   considers that this development is able to create wealth or, especially, jobs. No doubt: this is "an" obligation of their function, but they are not "all" their obligations to citizens. We refer us here to public institutions in the State of Law. They are responsible for promoting all kinds of democratic activities, ie, in our case, the performance of new designs or programs due to thereby ensure compliance with several rules of law, as we detail below, but it is not only due to satisfaction on job creation .

This is because it must be remembered that a democratic political system and the institutions that comprise it, is not only justified by job creation but it is because it addresses [22]:

(1). The guarantee and promotion of three mechanisms, today early fundamental legal, recognized in the constitutions and made reality in the daily life of the countries where the same work,

(2). The fulfillment of a prerequisite for the exercise of mechanisms and principles: the access to information, and

(3). The adoption as policy action of the governance.

The last (3) has real relevance because, as we will see, governance is one of the main philosophies or policies to be followed in the implementation of democratic principles and the design of R + D + i projects about services / Smart programs. The reason is relevant, specially, when there are no rules governing directly the phenomenon. The governance principles / caution must be observed in the construction and operation of automated services because they cannot solve themselves all the complexity that happens in reality in the field of application of smart services. This lack of regulation is another justification for focusing more legal proposals in relation to the existing regulations. We speak on this in the next section.

4.6   Governance

The foregoing consideration does not prevent the recognition of a recent common political practice: the exercise of the democratic powers of the authorities through what is called governance. Governance is defined by the dictionary as "Art or the manner of governing that has as objective the achievement of sustainable economic, social and institutional development, promoting a healthy balance between the state, civil society and the market economy."

This means recognizing the expansion in the public sphere, as own uses or practices of the rulers (including the expression to all public officials into action the three political powers), the principles, techniques, or uses of government 's own field of the business rules. This is the same as saying: the setting in motion of the efficiency and market rules as a criterion of preference or concomitant with the activities and juridical criteria of the public administration in the State of Law.

This style of action or policy does not preclude the statements established in the preceding paragraphs, it is the fact that the respective power must be exercised in consistent form with the implementation of the principles inherent to the put in practice of the rule of Law, summarizing the action of democracy, by legal mandate, governs the actions of public authorities also in the knowledge society [23, 24]. This is also predicated on the field considered in this paper: prevention with respect to the making of usual activities with aid of automatic / intelligent services programs. Here, as we see below, it is run in a manner similar to what happens with the complex application of Law by lawyers / judges as it is generally recognized. This is the same as saying that in the design of smart programs, thinking about putting them into reality, it is necessary the realization of the weighting mechanism, the self-governance, rather than the "automatic" or logical application of subsumption.

It should be recalled, in summary form, for these purposes, the basic message sent by some scholars, philosophers of law, with respect to judicial decisions, the application of law, or, in general, the law practice from the early twentieth century [25].

Since that time, just since the beginning of the obligation of the German judges to implement the German Civil Code under its responsibility in all cases that citizens pose them, it emerged as critical considerations on the idea that the application of law by judges was reduced to the immediate execution of the subsumption of the particular case in the Law, as presumed the liberal principles and Codes. Ehrlich, along with writers and judges that joined the Movement of Free Law, showed that the process of implementation of the law could not be reduced to the subsumption, once the irremediable loopholes make that the most of judgments are "free" creations of the same judges, in order not to incur the corresponding responsibility not to take decisions in cases submitted to them by legal imperative, whose assumptions and solutions not coincide with the prescriptions of the Law [26].

From these considerations emerged throughout the twentieth century to the present multiple reflections directed to complete the process of judicial application of law with other explanations.

Some of the proposed solutions were: knowledge of conceptions and social convictions (proposal by Ehrlich through sociology: the living law), consideration that the judicial process and legal reasoning are integrated by topics or common places that help to [27] the application of the Law, the establishment of regulatory systems to the application made by the use of logic counting on the construction of the normative pyramid rationally that extends the legal field [28], the proposal that the study of the Laws is made from the "pre-understanding" of their content [29,30,31]), the study of the judicial application of the Law in response to the broad scope and content of the arguments that occurs in the juridical process [32,33], consideration of the agreement of the jurical activities with the social legitimacy: the consensus, which are aimed by the Laws and the state organization as a whole (all the three branches or powers) in the democratic societies [34], the consideration that all human activities are carried out in response to a knowledge of reality produced in the maintained contact with reality by "autopoietic procedures" [35], and not only by intellectual development of scientific proposals...

These and other proposals were occupied, in short, to put emphasis on the circumstances of the judicial application, in order to explain them and give more complex understanding than those provided by the subsumption or formal discourse, that are centered on legal texts solutions. The governance principles help also to review the trust in the automation of services, and produce and suggest that these programs / services / systems / devices are constructed in a way that respects the rules and legal principles existing at the moment that we can say here that are synthetically expressed in these: the protection of personal data, the regulation on transparency and open data use, the warranty liability for the proposed services to users and the compliance with administrative requirements in the case of the provision of "Smart" juridical services.

5   Conclusion

From the above summary presentation of several experiences / projects of research occurred over thirty years, we can say to anyone interested in conducting research on Law and ICT field, that no such activities can be reduced to be an exegetical or analytical study of specific rules, or the collection and processing of information on the effectiveness of the rules of the Law, for example. Nor can they be to present the characteristics of technological development and innovations or damages resulting from the use or misuse of ICT devices. The occupation in this area requires, above all, to have enough to make joint efforts with professionals who have been trained in different skills to which they are subject to the law school training preparation. As has been indicated is the participation in the R + D + i, larger or smaller, in any case intended for solving real problems, which lets to take perspective and substantiate arguments about problems implied by the relationship between Law and ICTs.

Interdisciplinaritie is therefore another requirement. Well understood interdisciplinaritie, Ie conscious use of a language to handle cultural tools such as the resources that social science standards let, and not from its own particular language of experts in a particular area of knowledge. The necessary degree of interdisciplinaritie is necessary to talk with different training specialists, experts in the use of the usual tools of the social sciences.

These practices and usages must not lose, perhaps this is the most important conclusion, the perspective that the researchers are working out social and legal problems in a broad sense: Law and consensus, Values, Justice, Efficiency, Governance... and not on normative issues only.

Acknowledgement. The paper is based, partially, on the activities developed in the Project Ciudad 2020, INNPRONTA Project IPT-20111006, funded by the Spanish Centre for Industrial Technological Development, the University of Zaragoza participates in the project as advisor of the Atos Research @ Innovation (ATI) Division of the firm Atos.

References

1. Gibson, D.V., et al. (eds.): The Technopolis Phenomenon: Smart Cities, Fast Systems, Global Networks. Rowman & Littlefield Publishers Inc., Lanham (1992)
2. Paskaleva, K.: Enabling the smart city: the progress of city e-governance in Europe. International Journal of Innovation and Regional Development 1(4), 405–422 (2009)
3. de Catalunya, G.: Gestión automatizada en el ámbito de la Justicia. Generalitat de Catalunya, Barcelona (1983)
4. Sánchez-Mazas, M.: Modelli aritmetici per l'informatica giuridica. Rivista Informatica i Diritto. 1 (1978)
5. Debessonet, C.: An automated intelligent system based on a model of a legal system. Rutgers Computer and Technology Law Journal 10, 31–58 (1984)
6. Aarnio, A., Alexy, R., Peczenick, A.: The foundation of legal reasoning. Rechtstheorie 279, 133-158, 257-279, 423–448 (1981)
7. Habermas, J.: Theorie des kommunikativen Handelns. Suhrkamp, Frankfurt (1981)
8. Bing, J.: Handbook of legal information Retrieval. Norwegian Research Center for Computers and Law, Oslo (1984)
9. Galindo, F.: El derecho, habla o lenguaje Una aproximación a su carácter a través de la informatización de una disposición jurídica. In: Vide, M. (ed.) Lenguajes Naturales y Lenguajes Formales, pp. 284–292. Facultad de Filología, Barcelona (1986)
10. Capper, P., Susskind, R.: Latent damage law - The expert system: a study of computers in legal problem solving. Butterworths, London (1988)
11. Galindo, F.: Expert Systems as Tools for the Explanation of the Legal Domain: The ARPO Experiences. Expert systems with applications. An International Journal 4, 363–367 (1992)
12. Galindo, F.: El acceso a textos jurídicos. Introducción práctica a la filosofía del derecho. Mira Editores, Zaragoza (1993)
13. Galindo, F.: The communicative concept of law. Journal of Legal Pluralism and Unofficial Law 41, 111–129 (1998)
14. Reale, M.: Teoría tridimensional del derecho: una visión integral del derecho. Tecnos, Madrid (1997)
15. Baaz, M., Galindo, F., Quirchmayr, G., Vazquez, M.: The appplication of Kripke-Type structures to regional development programs. In: Wagner, R.R., Lazanský, J., Mařík, V. (eds.) DEXA 1993. LNCS, vol. 720, pp. 523–528. Springer, Heidelberg (1993)
16. Pastor, J., Delgado, J., Galindo, F.: Criptografía, privacidad y autodeterminación informática. Universidad de Zaragoza, Zaragoza (1995)
17. Galindo, F., Muñoz, J.F., Rivas, A.: Elaboración de un plan estratégico para la implantación del correo electrónico en la Diputación General de Aragón. In: TECNIMAP 1995. IV Jornadas sobre las tecnologías de la información para la modernización de las administraciones públicas. Ministerio para las Administraciones Públicas, Madrid (1995)
18. Galindo, F., Quirchmayr, G. (eds.): Advances in Electronic Government. Pre-Proceedings of the Working Conference of the International Federation of Information Processing. Seminario de Informática y Derecho. Universidad de Zaragoza, Zaragoza (2000)
19. Pagallo, U.: The Laws of Robots. Crimes, Contracts, and Torts. Law, Governance and Technology Series, vol. 10, pp. 183–192. Springer, Heidelberg (2013)
20. Galindo, F.: The communicative concept of law. Journal of Legal Pluralism and Unofficial Law 41, 111–129 (1998)
21. Douglas-Scott, S.: Law after Modernity. Hart Publishing Ltd, Oxford (2013)
22. Honneth, A.: El Derecho de la libertad. Esbozo de una eticidad democrática. Katz Editores, Madrid (2014).

23. Katz, E.D.: Engineering the Endgame. Michigan Law Review 109, 349–386 (2010)
24. Burca, G., Kilpatrick, C., Scott, J. (eds.): Critical Legal Perspectives on Global Governance. Liber Amicorum David M. Trubek. Hart Publishing, Oxford (2014)
25. Robles, G.: Teoría del Derecho. Thomson Civitas, Madrid (2010)
26. Ehrlich, E.: Freie Rechtsfindung und freie Rechtswissenschaft. Scientia Verlag. Leipzig (1903)
27. Viehweg, T.: Topik und Jurisprudenz. Beck, München (1974)
28. Kelsen, H.: Teoría general del Estado. UNAM, México (1979)
29. Esser, J.: Principio y norma en la elaboración jurisprudencial del derecho privado. Bosch, Barcelona (1961)
30. Engisch, K.: La idea de concreción en el derecho y en la ciencia jurídica actuales. EUNSA, Pamplona (1968)
31. Gadamer, H.G.: Verdad y método: fundamentos de una hermeneútica filosófica. Sígueme. Salamanca (1977)
32. Perelman, C.: La lógica jurídica y la nueva retórica. Civitas, Madrid (1979)
33. Alexy, R.: Begriff und Geltung des Rechts. Alber, Freiburg (1992)
34. Habermas, J.: Faktizität und Geltung. Beiträge zur Diskurstheorie des Rechts und der demokratischen Rechtsstaats. Suhrkamp, Frankfurt (1993)
35. Maturana, H.: El árbol del conocimiento: las bases biológicas del conocimiento humano. Editorial Universitaria, Santiago de Chile (1988)

Opening Public Deliberations: Transparency, Privacy, Anonymisation

Eleonora Bassi[1], David Leoni[1], Stefano Leucci[1],
Juan Pane[1], and Lorenzino Vaccari[2]

[1] University of Trento, Italy
{bassi,pane}@disi.unitn.it,
{david.leoni,stefano.leucci}@unitn.it
[2] Autonomous Province of Trento, Italy
lorenzino.vaccari@provincia.tn.it

Abstract. The open data movement is demanding publication of data withheld by public institutions. Wide access to government data improves transparency and also fosters economic growth. Still, careless publication of personal data can easily lead to privacy violations. Due to these concerns, the Italian law states that even public deliberations must be anonymised for long term publication. In the context of the Trentino Open Data Project (Italy), we first analyse privacy legislation and anonymisation techniques. Then, we propose a semantic open source stack based on entity and word sense disambiguation techniques for publishing anonymised deliberations edited with Norme in Rete software.

Keywords: open data, public sector information, digital administration, public deliberations, privacy, anonymisation, semantics, legal texts.

1 Introduction

Governments around the world are starting to recognize the value of the data kept in public institutions. The open data movement pushes for such data disclosure, as it allows broader public scrutiny and also boosts economies often choked by excessive bureaucracy. In this paper we analyze the problem of disclosing public deliberations as open data while preserving individual privacy. Which are the European and Italian legal frameworks in transparency and open data? Is it possible to use existing XML standards for legal documents? How can we assist the identification of personal data inside deliberations with semantic technologies? In the following, we try to answer these questions. We move from an overview of the European and Italian legal framework on open data (Section 2), in order to introduce our topic and its prominence after the new Italian rules on transparency (Section 3). In Sections 4 and 5 we analyse some technical and legal issues. In Section 6 we expose some anonymisation techniques and discuss their utility in our context. Finally, in Section 7 we propose a semantic open source stack to handle publication of anonymised deliberations in the Trentino Open Data Project.

P. Casanovas et al. (Eds.): AICOL IV/V 2013, LNAI 8929, pp. 41–53, 2014.
© Springer-Verlag Berlin Heidelberg 2014

2   The European and Italian Legal Framework on Transparency and Open Data

In the past ten years, open data initiatives became every day more important for the digital information market: the main ambition is to enforce the innovation of public sector in order to enhance the transparency of public administrations activities and the participation of citizens. This goal is reached by publishing data previously withheld from public scrutiny, thus greatly improving governments accountability.

2.1   The European Legal Framework

Open data policies became a legislative program in Europe since the last 90: the European legislator adopted D-2003/98/EC [10], introducing rules that allow and encourage the reuse of public sector information (PSI), that is the information gathered and owned by public sector bodies (PSBs), in order to remove barriers such as discriminatory practices, monopoly markets and a lack of transparency [2],[15],[17],[19]. PSI is a very wide notion that often includes personal data. According to the PSI Directive, also personal data could be reused, but in a way that shall not affect the level of protection of the individuals according to D-95/46/EC (Privacy Directive). The difficulty to solve the problem of compatibility of these two directives (PSI Directive and Privacy Directive) striking a fair balance between all the fundamental rights and interests involved (transparency, freedom of information, right to privacy, access to public documents and reuse) made the case for the reuse of personal data a crucial point for the European legislator and Data Protection Authorities [7],[9],[16],[18],[24]. This matter affects the issues we are analyzing in this paper: how to assure data protection for the case of reuse of provisions and deliberations from PSBs (as pointed out by the Commission Decision of 12 December 2011 on the reuse of Commission documents(2011/833/EU)). The European legislator - both in the PSI Directive of 2003 and in its revision of 2013 (D-2013/37/EU) - preferred not to decide how to balance those different disciplines, and the consequent practical and technical measures for assuring a legitimate reuse of personal data - with the only exception of personal data from intelligent transportation system databases for which D-2010/40/UE (ITS Directive) prescribes that full anonymisation should be adopted.

2.2   The Italian Legal Framework

In 2006 the Italian legislator adopted the D. Lgs. n. 36/2006 that transposed the PSI directive: some local administrations implemented the European and national rules on PSI reuse, but updated them following the main core of European best practices on open data and the hints offered by the revision process of the PSI Directive. Finally, in the last year, Italy adopted a new framework of rules on transparency, accountability and the disclosure of data from public administrations (D. Lgs. 33/2013). Although this new Decree is not directed primarily to the implementation of European rules on public sector information and open data, but to improve the functioning of public administrations, accountability and transparency, it requires the publication as open data of a large number of datasets and official documents, including deliberations. Thus, in the new Italian legal framework, the problem we analyze in this paper assumes an important role for enabling transparency and accountability through open data measures preserving privacy rights.

3   Transparency, Public Availability and Disclosure of Public Deliberations

Deliberations are concrete and particular acts of public administrations necessary to the exercise of their activities. Publicity is a prerequisite for the validity of the act that allows the ability to know. According to the Italian law on local government (Art. 124, D. Lgs. 267/2000), all the deliberations of the municipalities and the provinces are published by publication on the city register, at the headquarters institution, for fifteen consecutive days, except for special provisions of law.

3.1   From Paper to Bits

The Italian Digital Administration Code (D. Lgs. 82/2005) provides that the electronic version produces legal effects of publicity in the cases and in the manners expressly provided by the law. The L. 69/2009 on simplification and competitiveness in public administration establishes the rule that from January 1st, 2012, the publication of acts and administrative measures which have the effect of legal publicity are read as acquitted with the publication of information on their web sites by government and public bodies.

3.2   Problems of Interpretation

The recent introduction of the Decree 33/2013 creates problems of interpretation. Art. 7 provides that data subject to mandatory disclosure are published in open format pursuant to the Italian Digital Administration Code. The problem arises with regard to the time criterion of publication: Art. 8, D. Lgs. 33/2013, provides that data subject to mandatory disclosure under the current regulations are published for a period of five years and in any case until the published acts produce their effects. We have here a conflict of interpretation between the D. Lgs. 267/2000 and the recent D. Lgs. 33/2013: fifteen days (Art. 124, D. Lgs. 267/2000) or five years (Art. 8, D. Lgs. 33/2013)? The question could be solved by the principle of succession of laws in time which prefers the idea of the D. Lgs. 33/2013 (five years). But the principle of specialty could be used to solve the problem: according to the special rule, that is an exception to the general one, the D. Lgs. 267/2000 would keep its effects. As a possible solution, the act must be published for fifteen days (inclusive of all personal data contained within), and thereafter for the next five years it will be published in anonymous form, in accordance with art. 4, D. Lgs. 33/2013. Deliberations will remain available in their entirety to persons who advance an instance of access according to the requirements of L. 241/1990.

4   Opening Public Deliberations: Some Technical Remarks

The main problem in opening public deliberations concerns the structure of texts, which is not uniform across different Italian administrations. Several projects aim to solve this issue: the most complete and useful specifications for structuring legal texts are Norme in Rete and AkomaNtoso. The first is supported by the drafting environment xmLeges and the second by the application AT4AM, which is currently in use at the European Parliament. Since the Italian policy made by the Agenzia per l'Italia Digitale [11] recommends usage of NormeInRete mark-up schema, we decided to adopt xmLeges editor which best supports it.

4.1   NIR Project

NIR, developed by CNIPA (Italian National Center for Information Technology in the Public Administration) in conjunction with the Italian Ministry of Justice, ITTIG-CNR (Institute of Legal Theory and Techniques of the Italian National Research Council), University of Bologna and Italian Parliament, proposed the adoption of XML as a standard for representing legal documents using also additional meta information and a uniform cross referencing system (URN), providing documents with characteristics of interoperability and effectiveness of use. Another goal was to foster the building of legal texts access facilities for both citizens and legal experts. The standard for legal document description was created to increase degree of depth in text hierarchy description for different kind of legal documents by the definition of an XML-DTDs (NIR-DTDs), an example of which can be seen in Figure 1. The standard establishes constraints in the hierarchy of the formal elements of a legislative text (collections of articles), and a specification of the meta data which can be applied to a legislative document or to parts of it [6]. The advantages of XML format for legal documents are briefly summarized as follows: standardized definition of the structure of the document; automated assistance for the creation of legal texts; regulatory impact assessment on sorting; improved navigation within the legal texts; extensive research in the legislative databases; increased uniformity [5].

4.2   XmLegesEditor

A tool was built in order to obtain an holistic approach to the drafting process: xmLegesEditor is a specific integrated legislative drafting environment developed at ITTIG/CNR for supporting the adoption of NIR XML standards. The effort

```
<intestazione><titoloDoc id="titolo">DISCIPLINARY ACTION 5/13</titoloDoc></intestazione>
<formulainiziale> Attendances <h:br/>
      Mayor        Bertrandi Paolo <h:br/>
      Councillors   Bianchi Maurizio  <h:br/> Tomasi Alberto  <h:br/>
      <preambolo>
            <h:p> WHEREAS, disciplinary charges were served on City employee
                  Mr Giovanni Pedrotti, born in Rovereto on 18 March 1985, national identity number
                  PDRGVN85P55L378O  and living in Mattei Street 73, Trento </h:p>
            <h:p> WHEREAS, the Councillor Bianchi Maurizio reminded the serious allegations
                  which were proffered against Mr Giovanni Pedrotti in note 46/13 </h:p>
      </preambolo>  <h:br/> NOW, THEREFORE, BE IT RESOLVED <h:br/>
</formulainiziale>
<articolato>
      <articolo id="art1">  <num>Art. 1.</num> <corpo>
            Giovanni Pedrotti's employment with the City of Trento shall be immediately terminated.
      </corpo></comma></articolo>
</articolato>
```

Fig. 1. Deliberation excerpt with NIR XML markup. Text with personal data is underlined.

made with the development of xmLegesEditor has been to establish a trade-off between a user-friendly approach to text authoring hiding the underlying XML structure, and the maximum flexibility and extensibility in the exploitation of the high potentiality of content expression offered by XML documents [1]. Typically, WYSISYG word-processors are mainly oriented to texts' style markup rather than structural and semantic markup. XmLegesEditor proposes an original approach to this problem: the basic idea is that the user should be constrained by the editor to perform only valid operations on the document in such a way that, starting from a valid document, only valid documents can be produced [1]. A fundamental feature of xmLegesEditor is that it is a free resource, distributed with an open source license (GNU-GPL v3): the idea is to offer a shared highly customizable and extensible platform to develop specific functions and easily integrate existing or new designed tools as external modules.

5   Opening Public Deliberations: Some Privacy Remarks

Public deliberations contain in many cases personal data that requires to be protected due the disclosure. This is the typical case of balancing between transparency and privacy rights (see European Data Protection Supervisor (EDPS) [8,9] and [14,15],[18],[21]). The Italian DPA has stated several times about this problem for cases of publication of personal data in deliberations and administrative acts (Dec. 26/10/1998 [doc. web n. 30951], Dec. 2/9/1999 [doc. web n. 1092322], Dec. 23/2/2012 [doc. web n. 1876679], Dec. 7/10/2009 [doc. web n.

1669620]), prescribing the adoption of all technical measures for protecting privacy and to respect the principles of necessity, proportionality and minimization. Despite these DPA decisions were focused on privacy concerns related only to publication and not on reuse, they were complying with the recommendations of Art.29 Working Party and of EDPS on the reuse of PSI.

5.1 Call for Anonymisation

In the Opinion 7/2003 on the re-use of public sector information [24], Art. 29 Working Party insisted on the role that anonymisation can play in this sector and made the same recommendation in his Opinion 3/2013 on purpose limitation [26] and in the Opinion 6/2013 on open data and public sector information (PSI) reuse [27], stressing - in a stronger way - the necessity of anonymising personal data for the disclosure as open data, having in mind the connection between the scenario of reuse of personal open data and the potentiality of big data and data analytics (see Annex 2: Big data and open data). It is important to note that according to the WP29 anonymisation is not the only measure that a PSB must adopt in order to publish open data protecting privacy rights: the PSB should necessary conduct a robust and detailed privacy impact assessment identifying the risks and the measures adopted, following a case by case approach. However, although anonymisation is not considered a sufficient tool, in many cases it is strongly recommended or imposed as necessary. This position was followed in February 2013 by the Italian DPA in his Opinion on the draft of the Transparency Decree [doc web. n. 2243168]: anonymisation is required as necessary measure to assure the privacy of citizens for the publication (as open data) of public information for which the publication is not mandatory. The Transparency Decree adopted the solution proposed. We experienced firsthand the need for anonymisation by discovering with a simple Google search an ordinance where a mayor imposed a mandatory medical treatment to a citizen suffering from psychiatric illness. Name, birthdate and residence address of the citizen were all explicitly written resulting in a clear privacy breach. At the time of our search the ordinance wasn't present on the communality website anymore, yet we managed to found a copy inside Google cache.

5.2 Anonymisation Level

Some doubts arise on what kind of anonymisation the European DPAs and Member States legislators are referred to [7]. In the Opinions mentioned before, Art. 29 Working Party refers to a strict concept of anonymisation, that is required to avoid the constraints imposed by privacy legislation, while, in other cases, the argumentation is open to different levels of anonymisation and different technical possibilities, in relation to the probability of re-identification, to its costs and to the context of processing ([9],[14,15],[25]).

5.3   What to Anonymise

Deliberations may contain personal information under the form of names, addresses, birthdates, sex. In Fig. 1 we may see an example of some word that must (and must not, like Council members) be anonymised. Additional personal information can be found in documents referenced by the deliberation, such as *note nr 46/13* in the example. Since these documents might contain identifying information about physical persons named in the deliberation, if they are publicly available in non-anonymised form, references to them must be cancelled out. Also, referenced documents such as addendums can be in any format, including images. Trying to aid anonymisation of images by automatic means is much more difficult than dealing with plain text.

6   Anonymisation Techniques for Open Data

During last years several clamorous cases of privacy breaches occurred after the publication of supposedly anonymised datasets [4],[13],[23]. In 1997, Sweeney showed it is possible in the US to find the identity of a person by just knowing his age, sex, ZIP code with 5 digits and crossing this data with voting records, which are public in the US [20].

6.1   Reference Guide

Since UK is spearheading open data movement in Europe, its citizens are increasingly worried about their personal data being published on the internet. To address their concerns, UK government released a valuable Code of practice for anonymisation [22]. It targets a broad audience, explaining in simple terms risks and methods related to anonymisation. Anonymising deliberations falls into the so-called case of qualitative data anonymisation, where identifying information such as names and addresses is either cancelled out before publication, or generalized by applying a method called banding. An example might be substituting the address *Mattei Street, 73, Trento 38122* with a generic *Trento, 38XXX*. Banding preserves more information and it is valuable for researchers in social sciences when studying anonymised transcripts of interviews with people. Another option could be to use a technique called pseudo-anonymisation to associate a unique key to each anonymised person and substitute names in the text with that key. This would allow to recognize that the same person is mentioned in different deliberations without disclosing the actual identity of that person. To validate the effectiveness of the anonymisation ICOs Code of practice recommends performing the so-called *motivated intruder* test before publishing anonymised data. The test prescribes to play the role of an individual who wants to identify people in the anonymised dataset *if* motivated for some reason (i.e. sell data, blackmail people, stalking, etc). The intruder is supposed to try crosslinking anonymised data to existing sources by only using legal means, like searching the internet, enquiring people, looking at public records and so on.

The Working Party Art.29 in the Opinion 06/2013 [27] cites the motivated intruder test but seems skeptical about its effectiveness: among other things, it stresses how not all possible motivations can always be foreseen. The Opinion recommends so-called re-identification tests, where attempts to re-identification are done regardless of the possible supposed gains. Recently the Working Party Art.29 also published a detailed guide on anonymisation techniques [28] cast in the EU legal framework, where it offers a much welcomed quantitative approach to the problem of anonymisation. Reviewed techniques range from the simplest k-anonymisation by generalization to the most advanced randomization method of differential privacy. The report concludes there is still no silver bullet, and a case by case analysis must be performed prior the publication of any dataset.

6.2   Solution for Deliberations

Pseudo-anonymisation can be discarded right away because deliberations are published both in original and later in anonymised form, allowing to easily associate a person name to its key. While Working Party report on anonymisation [28] is clearly of importance when publishing statistical datasets, unfortunately is less relevant in our case. Statistical data is usually provided under the form of a table where it is relatively easy to understand how persons could be grouped to protect their anonymity. On the other hand, personal information in deliberations is scattered all over the text, and people mentioned in them are mostly unrelated. References to other documents containing additional information about persons in the deliberation also offer lots of clues for cross-linking attacks. In order to make a quantitative assessment of the amount of disclosed information, it would be necessary to mark and collect all such data (in the case of disciplinary action of Figure 1 it could be the office where the employee was working, his position, etc). Over time, this would give a clear historical picture of what has been released and allow more precise choice of anonymisation to perform in new documents. Although interesting, conducting such an analysis at present seems too onerous for a public administration. For these reasons, our current choice is to adopt the approach of cancelling out identifiers (such as names, social security numbers) and main quasi-identifiers (such as gender, birth-dates and postal codes).

7   Use Case: Open Data Initiative of Trentino

The Autonomous Province of Trento (PAT) has promoted territorial development based on competitiveness and innovation through specific innovation programs and laws. In particular, the Provincial Development Plan (PSP Piano di Sviluppo Provinciale) aims to adopt the information society as the fundamental resource for its territorial development. This vision was confirmed by the Provincial law 16/2012 which foresees the adoption of the Open Source software and of the Open Data paradigm. Then, the PAT approved the provincial guidelines about the Open Data formats, metadata and licences (Del. 2858/2012).

Deliberations (from the Province and from other municipalities too, including the City of Trento) are public information that the local government open data initiative is planning to open following new transparency rules. As we have seen in the previous sections, opening these deliberations matters for privacy protection. Moreover, as deliberations are only a part of the data to be published in the open data catalog *dati.trentino.it* from the Province, the issue of ensuring privacy of information requires a more comprehensive solution than only anonymising data in the deliberations. However, the same techniques that we apply to tackle the problems in the deliberations, can be used to deal with the anonymisation on other types of text.

7.1 The Semantic Stack

In order to support anonymisation, the open data initiative of Trentino includes a semantic stack that encompasses tools to parse and understand content of the datasets. Considering the semantification of text, the semantic stack will include Natural Language Processing (NLP) [12] techniques to parse sentences, and also Word Sense Disambiguation (WSD) [3] and Named Entity Recognition (NER) and Disambiguation (NED) [12] techniques, among others. By relying on NLP tools, we can parse all the sentence to its components, such as subjects, predicates, verbs, tagging each word with its parts of speech (verb, noun, adjective, adverb, ...), then using WSD, we can disambiguate the meaning of each word in the text, which would allow us to recognize synonyms in the text, such as car and automobile, and differentiate homonyms, such as bank (of the river) and bank (the financial institution). Once we know the meaning of each word, this will simplify the task of identifying name references in the text, task that is performed using NER, and later to disambiguate the exact entity that is being referred, using NED. In the case of the deliberations we want to be able to automatically recognize person name references in the text, but it is also very important to know who is being referred by the name, because not all the names need to be anonymised, for example, public names, such as the President of the Council, or the signatories of the Deliberations do not need to be anonymised.

7.2 A Possible Solution: xmLeges Extension

We propose an extension to the xmLeges software and a possible *ex ante* procedure for the anonymisation of the deliberations problem, rather than an *ex post* solution. We suggest a workflow that includes the editing step of the deliberations, which would allow the authors to identify, with some automatic support, the parts that need privacy protection. The workflow is outlined in Figure 2 as follows:

1. the deliberations are edited inside xmLeges software, which will automatically suggest the common XML structure;
2. during the editing process, all the text is parsed and disambiguated using NLP, WSD, NER and NED tools, allowing the editor to automatically find the name references in the text;

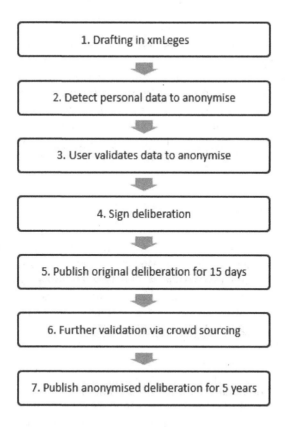

Fig. 2. *Ex ante* approach for anonymisation

3. the xmLeges allows the user to manually check all the text that was automatically marked to be potentially anonymised, to accept or reject these elements. The user should also be able to manually mark the names, addresses, and other text that s/he thinks needs to be anonymised;
4. the user sends the finished deliberation for signing and receives it back when this is done;
5. the deliberation is published for 15 days as is;
6. during these 15 days, the user can still further mark the parts of the deliberation that need to be anonymised. At this step, one can design a crowdsourcing-like approach for marking (tagging) the text that needs anonymisation;
7. the anonymised deliberations are published for 5 years, allowing the readers to ask for the original deliberations using the proper channels to obtain them.

In order to accomplish the above steps, we plan to extend the open source xmLeges with the semantic technologies available in the Trentino Open Data Project. The semantic xmLeges (S-xmLeges) will also be made available as an open source project.

In some cases it will be difficult to adopt the *ex ante* approach requiring the usage of the S-xmLeges tool for the creation and edition of the deliberations, as this would require training and switching editing tools that people in public administrations are already familiar with. When this is the case, we can adapt the *ex ante* approach to convert it into an *ex post* approach by allowing people to create the deliberations as they want, publish them for the required 15 days as is, and then, we adapt the steps described above as outlined in Figure 3:

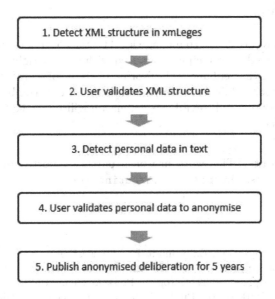

Fig. 3. *Ex post* approach for anonymisation

1. the deliberations are loaded into xmLeges software, which will parse them into the common XML structure;
2. user validates the XML structure, making sure the suggested tagging is appropriate
3. all the text is parsed and disambiguated using NLP, WSD, NER and NED tools;
4. the xmLeges allows the user (or via crowdsourcing-like approaches) to validate all the text that was automatically marked;
5. the anonymised deliberations are published for 5 years.

7.3 Possible Issues

The main technical issues that we foresee with this approach is the ability to fully automatically recognize rather technical terms that are part of the lexicon in the legal domain. This can be dealt with a vocabulary that can be built based on existing legal dictionaries, and creating crowdsourcing tasks whenever new terms are not found in the dictionary, asking the crowd of experts in the legal domain to define these terms. Also, given that the state of the art WSD, NER and NED tools are not perfect, a human would need to double check some of the annotations created by these tools, when the confidence in the disambiguation or recognitions is below a threshold.

8   Conclusions

In this paper we proposed how to manage legal text according to Italian transparency laws and open data principles, balanced with privacy rights. We suggested an *ex ante* solution that enhances the Norme in Rete software with a semantic open source stack for publishing anonymised deliberations, combined with an *ex post* solution. The proposed S-XmLeges extension will be tested in the Trentino Open Data project.

Acknowledgements. This work has been partly supported by the Trentino Open Data project (see `http://dati.trentino.it/`).

References

1. Agnoloni, T., Francesconi, E., Spinosa, P.: xmLegesEditor, an OpenSource visual XML editor for supporting Legal National Standards. In: Proc. of V Legislative XML Workshop, pp. 239–252. European Press Academic Publishing (2007)
2. Aichholzer, G., Burkert, H.: Public Sector Information in the Digital Age: Between Markets, Public Management and Citizens' Rights. Edward Elgar Publishing, Incorporated (2004), `http://books.google.it/books?id=a0AbDHMb5rAC`
3. Andrews, P., Pane, J.: Sense induction in folksonomies: a review. Artificial Intelligence Review, 1–28 (2013), `http://dx.doi.org/10.1007/s10462-012-9382-7`
4. Barbaro, M., Zeller, T.: A Face Is Exposed for AOL Searcher No. 4417749. The New York Times (August 2006),
`http://www.nytimes.com/2006/08/09/technology/09aol.html`
5. Biagioli, C.: Modelli funzionali delle leggi. Verso testi legislativi autoesplicativi. Series in legal information and communication technologies, EPAP (2009), `http://books.google.it/books?id=A6RqQgAACAAJ`
6. Biagioli, C., Francesconi, E., Spinosa, P.L., Taddei, M.: A legal drafting environment based on formal and semantic XML standards. In: ICAIL, pp. 244–245 (2005)
7. Dos Santos, C., Bassi, E., De Terwangne, C., Fernandez Salmeron, M., Tepina, P.: Policy Recommendation on Privacy and Personal Data Protection as Regards Re-Use of Public Sector Information (PSI). Masaryk University Journal of Law and Technology (MUJLT) 6(3) (2012)
8. EDPS: Public access to documents containing personal data after the Bavarian Lager ruling of 24 March 2011 (2011)
9. EDPS: Opinion on the open-data package (April 18, 2012) (2012)

10. European Parliament: Directive 2003/98/EC, PSI Directive (2003)
11. Giovannini, M.P., Palmirani, M., Francesconi, E.: Linee guida per la marcatura dei documenti normativi secondo gli standard NormeInRete, Agenzia per l'Italia digitale. Series in Legal Information and Communication Technologies, vol. 9. Agenzia per l'Italia digitale, vol. 9 (November 2012)
12. Margonar, S., Giunchiglia, F., Pane, J.: A Large Scale Name Matching and Search Framework. Technical report, DISI - University of Trento (March 2013), http://eprints.biblio.unitn.it/4161/1/DISI-13-026.pdf
13. Narayanan, A., Shmatikov, V.: Robust De-anonymization of Large Sparse Datasets. In: Proceedings of the 2008 IEEE Symposium on Security and Privacy, SP 2008, pp. 111–125. IEEE Computer Society, Washington, DC (2008), http://dx.doi.org/10.1109/SP.2008.33
14. Nissenbaum, H.: Privacy as Contextual Integrity. Washington Law Review 79(1) (2004)
15. O'Hara, K.: Transparent government, not transparent citizens: a report on privacy and transparency for the Cabinet Office. Technical report, Cabinet Office (September 2011), http://eprints.soton.ac.uk/272769/
16. Pagallo, U., Bassi, E.: Open Data Protection: Challenges, Perspectives and Tools for the Re-use of PSI. In: Hildebrandt, M., O'Hara, K., Waidner, M. (eds.) Digital Enlightenment Yearbook 2013, pp. 179–189. IOS Press (2013)
17. Ponti, B.: La trasparenza amministrativa dopo il d. lgs, 33 (marzo 14, 2013). Analisi normativa, aspetti organizzativi ed indicazioni operative. Series in legal information and communication technologies, Maggioli (2014)
18. Raab, C.: Privacy Issues as Limits to Access. In: Aichholzer, G., Burkert, H. (eds.) Public sector information in the digital age: between markets, public management and citizens' rights, pp. 23–46. Edward Elgar (2004)
19. Ricolfi, M., Sappa, C. (eds.): Extracting Value From Public Sector Information: Legal Framework and Regional Policies. Quaderni del Dipartimento di Giurisprudenza dell'Università di Torino, ESI (2013)
20. Sweeney, L.: Computational Disclosure Control - A Primer on Data Privacy Protection. Tech. rep. MIT (2001)
21. Turilli, M., Floridi, L.: The ethics of information transparency. Ethics and Inf. Technol. 11(2), 105–112 (2009), http://dx.doi.org/10.1007/s10676-009-9187-9
22. UK Information Commissioner's Office: Anonymisation: managing data protection risk code of practice (2012)
23. Wikimedia: What are readers looking for? Wikipedia search data now available. Wikimedia Blog (September 2012),
    http://blog.wikimedia.org/2012/09/19/what-are-readers-looking-for-wikipedia-search-data-now-available/
24. WP29: Opinion 7/2003 on the re-use of public sector information and the protection of personal data - Striking the balance (wp83) (July 2003)
25. WP29: Opinion 4/2007 on the concept of personal data (wp136) (April 2007)
26. WP29: Opinion 03/2013 on purpose limitation (wp203) (April 2013a)
27. WP29: Opinion 06/2013 on open data and public sector information ('PSI') reuse (wp207) (June 2013b)
28. WP29: Opinion 05/2014 on Anonymisation Techniques (wp216) (April 2014)

Online Dispute Resolution and Models of Relational Law and Justice: A Table of Ethical Principles

Pompeu Casanovas[1,2] and John Zeleznikow[3]

[1] Institute of Law and Technology, Autonomous University of Barcelona, Barcelona, Spain
[2] Center for Applied Social Research, Royal Melbourne Institute of Technology, Melbourne, Australia
pompeu.casanovas@uab.cat
[3] Laboratory of Decision Support and Dispute Management, College of Business, Victoria University, Victoria, Australia
John.Zeleznikow@vu.edu.au

Abstract. *Regulatory systems* constitute a set of coordinated complex behavior (individual and collective) which can be grasped through rules, values and principles that constitute the social framework of the law. *Relational law*, *relational justice* and the design of *regulatory models* can be linked to emergent agreement technologies and new versions of Online Dispute Resolution (ODR) and Negotiation Support Systems (NSS). We define the notions of *public space* and *information principles*, extending the concept of 'second order validity' to the fields of ODR and NSS.

Keywords: regulatory systems, Semantic Web (SW), Ethics, Normative Multiagent Systems (nMAS), ODR systems, Negotiation Support Systems, Fifth Party.

1 Introduction

The relational perspective to law emerged from the interplay between lawyering practices, contract studies, and socio-legal scholarship, alike. It stresses a view of contracts as relations rather than discrete transactions looking at the evolving dynamics of the different players and stakeholders within their living constructed shared contexts. The term "relational" emphasizes the complex patterns of human interaction and exchange. It means that relational regulatory models are complex, and that their strength certainly stems from sources other than just the normative power of positive law. We will call this set of coordinated individual and collective complex behavior which can be grasped through rules, values and principles that constitute the social framework of the law, *regulatory systems*.

How can relational law, relational justice and regulatory systems be linked to the newer versions of Online Dispute Resolution? And how Web 2.0 (the social web) and Web 3.0 (Web of Data) are related to this sociolegal approach?

In the Web 3.0 law turns out to be interactive, relational, deploying thorough multilayered governance regulatory systems. A *hybrid perspective* takes into account

P. Casanovas et al. (Eds.): AICOL IV/V 2013, LNAI 8929, pp. 54–68, 2014.

phenomena that are different in nature —e.g. linked open data; the conceptual structure of legal data, metadata and rules; the conceptual structure of networked governance; the so-called "fifth party" in Online Dispute Resolution (ODR) and Negotiation Support Systems (NSS) developments.

This paper contributes to the ongoing discussion by contending that ethical principles can bring the required perspective to draw and interpret the general design for such regulatory models. Ethics play a major role in this relational approach. Following some recent work on Data Protection and Privacy by Design, and some recent attempts to integrate fairness and transparency to frame ODR and NSS (dispute resolution technologies, negotiation support systems), we will show how regulatory models can integrate moral, political and legal principles to avoid the drawbacks that may come from a purely normative approach.

2   Relational Justice, ODR and Ethical Principles

The CAPER[1] regulatory model (CRM) stems from the area of Freedom, Security and Justice (FSJ) to manage police interoperability and to protect citizens' rights in the European space [5]. This appears to be a quite specific and overregulated domain, deserving much attention by legal drafters and actors in the political arena. Snowden revelations and the recent Bowden Report to EU Parliament in September 2013 have contributed to a greater awareness of the need for privacy protection, balancing safety and security [7].

It is our contention that, stemming from a relational approach to law and justice, distance from security to liberty can be shortened. There is a dynamic and ongoing relationship between both dimensions of human freedom. Properties such as validity can be applied to test the legal outcome of agreements; but issues of ethics and trust which are essential in mediation, ODR and SSN can be applied as well to regulatory designs of FSJ domain.

Accountability, asymmetrical network governance and responsible data protection are some of the aspects to be pointed out. The CAPER regulatory model encompasses legal boundaries and empowerment capabilities alike. The evolutionary context created by criminal threats to the open society must be taken into account here, because it sets a bottom-up permanent and dynamic landscape of changing scenarios. The common resilience of governments, companies and citizens is essential when dealing with such a landscape, and therefore, the suggested standards assume that citizens, and not only governments, are entitled to cooperate with police organizations and with the justice to fight organized crime. But do-it-yourself-justice situations must be bounded and ruled through democratic means of governance and legal controls: this is why it is so important to define a global public space in which cooperation and collaborative ways of citizens' participation can find a legal place to develop safely. Crisis mapping and new forms of crowdsourced constitutional law are among the successful forms of what it has been already called *digital neighborhood*.

---

[1] CAPER stands for "Collaborative information, Acquisition, Processing, Exploitation and Reporting for the prevention of organised crime", see http://www.fp7-caper.eu/

Examples such as those of the Vancouver riots, warn against the unintended consequences of mob behavior that may follow from the indiscriminate use of social media to help local authorities to identify rioters [47].

Relational justice is a bottom-up justice produced through cooperative behavior, agreement, negotiation or dialogue [12, 13]. The standard typology of ODR systems lists automated negotiation, computer assisted negotiation, online mediation and online arbitration [50]. Such systems are conceived to operate in a transnational and global space, and usually designed to reach agreements independently of any specific legal domain (family law, private international law, e-commerce, consumer law…). ODR systems incorporate (and actually operate) through argumentative means, between both persuasion and deception [23].

However, in spite of many attempts to implement them into the market and as a private or e-government regular service, ODR tools have not been *so widely used and developed* as it was expected only five years ago [55].

The reasons for such a slow development as Web Services are manifold. As it happened in the early times of ADR developments, big companies have already developed dispute solving devices as a normal service being offered at their website. E-Bay and Wikipedia systems are among the well-known examples. It is currently referred as example Colin Rule's assertion about the 60 million cases solved by e-Bay in a single year. However, there is another important aspect to be taken into account. Colin Rule also asserts that "costs have an impact on not only access to but also perceptions of distributive justice. If ODR is less expensive than other alternatives, it enhances access. Outside big marketplaces, however, there are few business models for sustainable ODR systems" [39]. The acceptance of ODR is dependent on a country's legal culture and its institutional acceptance (in national commerce courts for example): not all countries have had an equal degree of reception of ODR [1].

Moreover, ODR entails more complex procedures than ADR: the so-called "fourth party" refers to the technology component, but the notion of "fifth party", the provider of technology, is most needed to understand practical and legal consequences [lodder]. Accordingly, Carneiro, Novais and Neves [23] are suggesting technical reasons for the slowness in constructing ODR technology: a lack of multi-domain tools that can address more than one legal field leads to currently available tools only being available for only a single domain, drastically diminishing its application. The "fifth-party" is still under development. "Template-based" Negotiation Systems, in which no solution is proposed by the system, might be complemented with the aid of more proactive technologies, i.e. systems based on game and bargaining theory [36] [37, 38].

We would like to advance two arguments to foster ODR and legally valid negotiated agreements.

First, the idea of open social intelligence (OSI) can help to constitute a new framework [14][42]. Castelfranchi [32] asserts that the social mind cannot be conceived as a mere aggregate of individual abilities, but a set of social affordances. Therefore, social interactions organize, coordinate, and specialize as artifacts, tools, to achieve some outcomes for a collective work. OSI elements and Artificial Intelligence (AI) components should be enhanced and combined into ODR toolkits (web services, platforms, mobile applications…) to facilitate citizens' and consumers' participation, and an open use and reuse of the accumulated knowledge. Achieving this, it does not necessarily means Crowdsourced Online Dispute Resolution (CODR), as advanced in [28].

Second, ethical components deserve a closer attention, and once incorporated into ODR, they turn out to be essential for its broader implementation and acceptance because the notion of *validity* or *legality* is transformed as well through networked regulatory models in ODR scenarios.

AI-oriented ODR can help, indeed, to overcome some of the traditional barriers pointed out by inner and external criticisms. A few of them rely on the limitations over the communication process. It is true that compared to face-to-face settings, nonverbal cues (facial gestures, voice inflection, intonation, facial reddening…) are usually absent in ODR settings. But at the same time the flexibility, mobility and fastness of proactive technologies can be enhanced through Multi-agent systems (MAS) and emotion-sensitive sensors. Virtual institutions developing agreement technologies, and face-recognition imaging, e.g., are already mature enough to be used in real settings [43]. COGNICOR, the automated conflict resolution company that won the 2012 European start-up award, constitutes an example of such a successful innovative ODR strategy.[2] In addition, this approach contributes to uncovering new conflicts and legal issues, e.g. disputes about reputation rights in social networks and across the web [57]. MODRIA is another example of an innovative company dealing with reputation conflicts.[3]

Standards and regulations provide another side of the problem. Empirical studies on consumers' behavior, strongly show that most e-buyers ignore national consumer laws. E.g. The findings by Ha and Coghill [26] in an Australian survey on online shoppers suggest that most respondents are not aware of the following issues: (i) which organizations are involved in e-consumer protection; (ii) government regulations and guidelines; (iii) industry codes of conduct; (iv) self-regulatory approaches adopted by business; and (v) the activities of consumer associations to protect consumers in the online marketplace. After harvesting all available P3P Policies (Platforms for Privacy Preferences Protocol) —the 100,000 most popular Web sites (over 3,000 full policies, and another 3,000 compact policies) — Reay, Dick and Miller [46] concluded that privacy provisions are largely ignored by consumer web sites. New strategies, such as providing structured legal information directly on mobile applications, seem to be appropriate for using ODR systems more efficiently and bringing mediation to consumers and citizens.[4]

There are several proposals for drafting legal standards for mandatory ODR in Europe [20]. Quite recently the United Nations Commission on International Trade Law (UNCITRAL) set up a Working Group to develop: (i) procedural rules, (ii)

---

[2] http://www.cognicor.com/, http://thenextweb.com/eu/2012/06/22/smart-complaint-resolution-service-cognicor-wins-the-european-commissions-new-grand-startup-prize/

[3] https://www.modria.com/

[4] Cf. See GEOCONSUM, a mobile application to provide consumer legal information https://play.google.com/store/apps/details?id=com.idt.ontomedia.geoconsum&hl=en

operational guidelines for providers and neutrals, (iii) minimum requirements for providers and neutrals, including accreditation and quality control, (iv) creation of equitable principles for the resolution of disputes, (v) and enforcement mechanisms.[5]

Rule and Rogers [49] observe that a cross-border resolution system requires "all participating entities to exchange information around the world, in real time, in multiple languages". Therefore, the challenge is constituted by data standards application and "a public, comprehensive set of rules to govern the inter-operation of all of the organizations participating in the global system".

All of this has a strong flavor of *déjà vu*: it is similar to the Uniform Domain Name Dispute Resolution Procedure (UDRP) adopted by International Corporation for Assigned Names and Numbers (ICANN) [29]. Such problems are also similar to the experiences in the Freedom, Security and Justice Area (FSJ), where the patchwork of local, national, international, European and international norms might be reorganized through interoperable regulatory models.

At the kernel of these trends is applying XML standard, LOD, and Data Protection policies to the management, classification, communication and organization of ODR global knowledge. It implies a change in the understanding of ODR *valid* outcomes.

Again, what is meant by a "legal" or "valid" agreement cannot be only conceptualized stemming from the field of international private law.[6] As it will be shown in the next section, agreements and negotiations through ODR and NSS can be better understood as *legal components of a global public space* which has to be *anchored* in some notion of what global law is or should be. This is properly the field of computational and informational ethics.

3   ODR, Ethical Principles and the Redefinition of the Global Public Space

Negotiation, conflict and dispute resolution studies have been always focused on political and ethical grounds. In these approaches, justice is at the center of discussions. Sometimes, when dealing with ethical issues, *trust*, over other possible moral issues, has been considered as the main ODR procedural value. Therefore, computer models applying argumentation schemes theory are trust-centered schemes [letia], and *building trust* is also the focus of other studies on predictors of disputants' intentions to use ODR services [57] or on intermediation and consumer market inefficiencies [21]. Rule and Friedberg [48] consider ODR as just one tool in a

---

[5] See [29] for a comparison between EU ADR/ODR regime and UNCITRAL's Draft Rules. "The UNCITRAL draft Procedural Rules envisage a three-stage procedure: (1) automated/assisted negotiation between the parties without a human neutral, which may include blind-bidding techniques; (2) mediation/conciliation; and (3) arbitration leading to a decision which can be enforced".

[6] After analyzing UNCITRAL's draft Rules for ODR, Cortés and Esteban de la Rosa contend [19]: "low-value e-commerce cross-border transactions, the most effective consumer protection policy cannot be based on national laws and domestic courts, but on effective and monitored ODR processes with swift out-of-court enforceable decisions".

broader toolbox (amongst techniques coming from marketing, education, trust seals and transparency). From this point of view, trust is not analyzed as a moral value, but as carrying on social and economic values in the market, depending upon *reputation*. This is why trust is so time-consuming and hard to build.

Focusing on trust is a result of applying to the Internet the traditional ADR perspective in which interests and private gains and losses prevailed over other public aspects [2]. Thus, trust and confidence, meant efficiency as well. The role of lawyers, arbitrators and mediators in balancing attitudes (neutrality, impartiality) are supposed to induce confidence and to bring efficiency to the system.

Nevertheless, under the "fifth party" perspective, the structural framework comes to play. *Fairness*, and not only trust, matter.

"Is it a violation of neutrality if eBay runs the overall dispute resolution system while also deciding individual case outcomes? The company strives to build fair and open dispute resolution processes, but the fact remains that eBay will not offer a system it believes operates contrary to the overall objectives of the marketplace. Should the standard for process impartiality be changed in ODR? Perhaps we should worry more about the overall appearance of partiality (the "kangaroo court" phenomenon) than obsessively trying to wring every last drop of bias that might exist at every stage in the process. In one possible solution, ODR systems could substitute a mediator requirement to "serve in a balanced capacity" rather than an impartial capacity. Rather than just protecting one party, this protects everyone, including the system, *thus upholding the notion of fairness.*" [39].

However, marketplaces take place in an open society that is becoming global very fast. This is not only an economic issue, but a social and political one. ODR procedures and outcomes call for democratic legal forms. The three-step model for ODR systems proposed by Lodder and Zeleznikow [37] [38][7] can be harmonized within a legal framework encompassing *fairness and transparency*. But as some reviewers point out sharply, "it is not clear however, in the ODR context, how to achieve transparency, in what areas and how to cope with its implications" [31].

Answering this criticism is far from simple, because the intersection between both values reflects the tension between the public and the private that is transforming the national version of the rule of law into a global set of *legitimated* governance mechanisms (in absence of some version of a global state).

"*Transnationalism* – law beyond the state – may be the key to predictability, and thus to the sort of justice, or fairness, that is central to the rule of law" [52] [51]. *Systemic fairness*, "developing and applying a set of predictable transnational rules" (ibid.), or *meta-justice*, developed by Alex Mills intending "the justice of the principles governing the global ordering of legal systems that private law embodies"

---

[7] The first step involves finding out the BATNA (best alternatives to the negotiated agreement), i.e. what happens if the negotiation were to fail. Next stage would involve facilitating conflict resolution by means of argumentation. In case not all of the issues are resolved, the third step would employ analytical techniques to complete the resolution process.

(ibid.), are some of the notions that have been proposed to grasp this shifting turn of the law becoming global. What meta-justice principles are, and what do consist of? How could they be applied to computer systems?

Philosophers, legal theorists and computer scientists have been cooperating to give a reasonable answer to the questions raised by global justice.[8] It is our contention that bringing together *fairness and transparency* requires a more complex conceptualization of the tensions produced within the hybrid field of transnational regulations. i.e., adopting a relational justice perspective and working out the notions of complex regulatory systems and complex regulatory models can shed some light to this changing legal world. Table 1 summarizes the *Principles of fair information practices* (FIPs) following the tradition of Alan F. Westin (1967) [56]:

Table 1. FIPs. Source: [33]

1. *Openness and transparency*	There should be no secret record keeping. This includes both the publication of the existence of such collections, as well as their contents.
2. *Individual participation*	The subject of a record should be able to see and correct the record.
3. *Collection limitation*	Data collection should be proportional and not excessive compared to the purpose of the collection.
4. *Data quality*	Data should be relevant to the purposes for which they are collected and should be kept up to date.
5. *Use limitation*	Data should only be used for their specific purpose by authorized personnel.
6. *Reasonable security*	Adequate security safeguards should be put in place, according to the sensitivity of the data collected.
7. *Accountability*	Record keepers must be accountable for compliance with the other principles.

These foundational principles have been embedded into EU Directives and regulations, and have fostered academic, theoretical and practical discussions during the last twenty years.

Leaning on the first comparative tables by Cavoukian [18] on Privacy by Design Principles, we have completed them with the Principles of the Semantic Web Linked Open Data, Legal Information Institutes Principles, ODR, Crowdsourcing and Crisis Mapping (Table 2).

---

[8] "Nowadays, a system designer must have a deep understanding not only of the social and legal implications of what he is designing but also of the ethical nature of the systems he is conceptualising. These artefacts not only behave autonomously in their environments, embedding themselves into the functional tissue or our society but also 're-ontologise' part of our social environment, shaping new spaces in which people operate." [54]

Table 2. Comparison between Fair Informational Practices (FIPs), Privacy by Design (PbD), Linked Open Data principles (LOD), Principles of Legal Information Institutes (LIIP), and ODR Crowdsourcing, and Crisis Mapping Principles

Privacy by Design Foundatio nal Princi ples [10] [18]	Fair Informa tion Practice Principles (GPS) [33]	Cavoukian Extended Principles [18]	Semantic Web LOD Principles [3]	Legal Information Institutes Principles [8][25]	ODR Principles [2] [58] [59]	Crowd-sourcing Principles [4] [6] [27] [40]	Crisis Mapping Princi ples [41] [44] [45]
1. Proactive not reactive; Preventative not Remedia		Established methods to **recognize** poor privacy designs, to anticipate poor privacy practices and outcomes, and to correct the negative impacts	URIs to denote things, HTTP Dereference Serialization formats Proactive modeling: XML, RDF, SPARQL, OWL Interconnecte dness	Technological investment, information, free access to law an legal information	Willingness to enter into a negotiation and be fair	Participatio n Collabora tive work, governance and decision making	Informing Reporting Proactive participa tion Conflict prevention and crisis manage ment
2. Privacy as the Default Setting	3.Purpose Specification 4.Collection limitation, Data minimizatio n 5.User Retention, Disclosure Limitation	Privacy becomes the prevailing condition - without the data subject ever having to ask for it -no action required.	Dereferencin g Accessibility, Secure data exchange, protection, Storage, Metadata, Ontologies, Alarm Systems, Trust	Republication Anonymization	Fairness-Enabling Discovery (Disclosure Limitation)	Trust: disclosure limitation	Harmless Digital neighbor-hood Causing no harm
3. Privacy Embedded into Design		Systemic program or methodolog y in place to ensure that privacy is thoroughly integrated into operations standard-based and validable).	Dereferencin g Looking up data, structured data, Data protection, Storage, Metadata, Enrichment, Core Ontologies, Domain Ontologies, Rules, Principles, Trust, Validation	Republication Reusing Authentication (Authoritative versions) Integrity	Fairness-Bargaining in the shadow of the law and the use of BATNAs	Trust: Empower ing people	GIS monitoring Implemen-ting Digital Neighbor hood
4. Full Functionality Positive-Sum, Not Zero-Sum		Multifunctio nal solutions: legitimate non-privacy interests and objectives, early, desired functions articulated, agreed metrics applied.	Web Science, Universality, Linked Data, Human Giant Graph, Accessibility, Data protection, Metadata, Core Ontologies, Domain Ontologies, Rules, Principles, Trust, Validation,	Balanced interests (publisher/ state/ user)	Fairness-Enabling Discovery (Privacy Limitation)	Trust: self-interest; monitorizati on, metrics applied	Trust: aggregated interests and values; monitored processes; metrics applied

Table 2. (*Continued*)

5. End-to-End Security, Full Lifecycle Protection	7. Security		Secure user participation, Ontology sustainability, folksonomies	Integrity, Security, Maintenance	Secure environment	Integrity: secure environment and participation	Volunteers' Security
6. Visibility, Transparency Keep It Open	2.Accountability 8. Openness 10.Compliance		Transparency Accountability Content value, tagging and semantic enrichment	Accountability Distributed Authority of republished materials	Developing transparency	Trust: Transparency, work quality	Validation Transparency
7. Respect for User Privacy Keep it User-Centric	1. Consent 6. Accuracy 9. Access		End user-centered systems, personalization,	Personalization. End user-centered systems	Accuracy	Aggregated value	Truthful and accurate information

There is a coincidence on objectives, structure and number of principles. What is worthwhile highlighting is that the main focus of their discourse lies in a deeper level, disclosing the ethical ground on which principles are based. Privacy by Design (and Privacy by Default) principles tend to stress the respect for user privacy and informed *consent*. Linked Open Data principles highlight the *accountability* of the protocols settled on data use and reuse by companies, administrations and governments. The principles lied down by Legal Information Institutes to rule the free reproduction and dissemination of legal content are focused on the *republication* of targeted legal materials.

Principles for crowdsourcing are less centered, as they are depending upon the field in which they apply and they are intertwined with remuneration for work — labor micro-tasks (Mechanical Turk e.g.) or research challenges. *Trust* seems to be crucial for self-interested participation. But when the task to be carried out is entirely voluntary and people do not seek economic compensation, the situation changes. In the domains of crisis mapping (emergencies, natural disasters, humanitarian crisis...) and election monitoring what is sought is reliable information on local events. *Truth* constitutes the main focus.

These focal points have their counterpart —consent/ publicity; accountability/ public security; reputation/ intellectual property, compensation/ quality, validation/ causing no harm— in a non-homogeneous *continuum* of rights and duties. PbD are *user-centered*, LOD are *data/protocol-centered*, LIIP are platform or *service-centered*, crowdsourcing principles are *task/centered*, crisis mapping principles are *reporting/centered*. It is noteworthy that from PbD to crisis mapping monitoring the focus shifts from private to more public concerns.

This leads to a different definition of the private-public space *continuum*, in which rights and duties to be complied with are almost the same (as showed by the similarity of principles) but have different *weights*. Therefore, *public consciousness, public space, public domain, public community* can be distinguished, stemming from the different models of relational law that principles allow, and the different kinds of citizens' rights than can be put in place (*civil rights, global rights, added-value rights, common rights*).

We think that ODR principles fit into this broad landscape in a particular way. As shown in table 3. On the one side fairness must be protected as a general condition of dispute settlement. On the other hand transparency is a condition for enabling discovery in order to not to alter the outcome of the negotiation. Thus, ODR principles are *process-centered*. They can be enacted and applied in a *public global space*, in which what has to be protected is not only the specific outcome of a negotiation, but the system as a whole: it is important that trust can be enhanced through fairness and the *legality* of the final outcome.

Table 3. Fairness ODR Principles. Source: [59]

Fairness Principle 1 – developing transparency	For a negotiation to be fair, it is essential to be able to understand and if necessary replicate the process in which decisions are made. In this way unfair negotiated decisions can be examined, and if necessary, be altered.
Fairness Principle 2 – enabling discovery	Even when the negotiation process is transparent, it can still be flawed if there is a failure to disclose vital information. Such knowledge might greatly alter the outcome of a negotiation.
Fairness Principle 3 – bargaining in the shadow of the law and the use of *BATNAs*[9]	Most negotiations in law are conducted in the shadow of the law. These probable outcomes of litigation provide beacons or norms for the commencement of any negotiations (in effect BATNAs). Bargaining in the shadow of the law thus provides standards for adhering to *legally just and fair norms*. Providing disputants with advice about BATNAs and bargaining in the shadow of the law and incorporating such advice in negotiation support systems can help support fairness in such systems.

But to understand what "legally just and fair norms" mean in the application of the third Fairness Principle, that is to say, calculating BATNA while negotiating at the same time "in the shadow of the law", the evaluative test of the CRM can be performed in each specific mediation process, or can be embedded within the Negotiation Support Systems (NSS).[10]

---

[9] BATNA stands for "Best Alternative to a Negotiated Agreement".

[10] "For example, in the AssetDivider system, interest-based negotiation is constrained by incorporating the paramount interests of the child. By using bargaining in the shadow of the law, one can use evaluative mediation (as in a family mediator) to ensure that the process is fair. The Split-Up system models how Australian family court judges make decisions about the distribution of Australian marital property following divorce. By providing BATNAs it gives suitable advice for commencing fair negotiations. The BEST-project (BATNA establishment using semantic web technology), based at the Free University of Amsterdam, aims to explore the intelligent disclosure of Dutch case law using semantic web technology.It uses ontology-based search and navigation. The goal is to support negotiation by developing each party's BATNA" [59].

Doing so, *the validity of the system triggers the legality of the negotiation process and possible upcoming agreements that might follow. Therefore, legality is a by-product of the enforceability, effectiveness, efficiency and justice of the normative system.* The ODR principles are *anchored* into complex regulatory models that grasp the real values and properties of the functioning of the whole system (the 4th and 5th Parties pointed out by Lodder and Zeleznikow). Fig. 1 plots this dynamic process, in which justice plays a major role as inner component of the model.

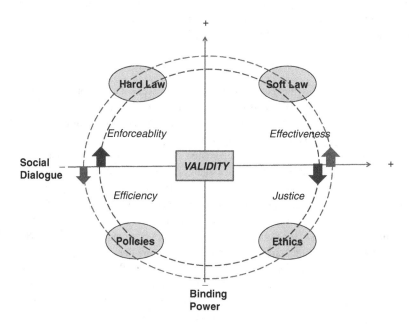

Fig. 1. Three axes, four first order properties, and one second order property to ˙model regulatory systems. Source: [26]

4   Conclusions: Models of Relational Law

In this paper we have outlined a way to conceptually model from a descriptive and empirical approach some elements that refine and slightly modify the normative notion of law, stemming from its implementation in SWRM and complex regulatory systems. We have contended that the validity of norms, rules and principles cannot be directly applied as an identification property to single out their legality. The design of regulatory systems, either in nMAS or embedded into Web Services, ODR platforms and NSS devices, entails a complex framework. Ethical principles are more important than ever in this global space in which the power of nation-states is not the only source of law. Contexts and fields of application are shaping the final scope of regulatory outcomes.

We have compared broadly some of these principles, adding Semantic Web LOD, LII, ODR, CR and Crisis Mapping to the originally tables plotted by Ann Cavoukian.

Technology is being used to the extent that fits the users' needs, and not the other way around. This is still an unfinished and ongoing work. As more fields are added, privacy and data protection analysis becomes a problem of aggregation, and the idea of privacy becomes situated within a global space in which latent and explicit conflicts can be classified into stable structural frameworks.

PbD principles are equally important, then, but ethics and technology can play other kinds of roles, centered on individual rights too, but having a collective dimension able of being organized into structured and coordinated political actions. Disclosing government information, denouncing corruption, managing emergencies in natural disasters, and monitoring elections means organizing *crowd, collective intelligence*. This implies a new challenge for democratization, fostering the construction of relational law models adapted to different problems, frameworks and coordinated tasks to design regulatory programs for specific, emerging transnational fields and actions.

We have shown that Semantic Web technologies and SWRM open up new ways for implementing, handling and performing legal rights and duties in these fields. But it is our contention that they must be built up and anchored in the perspective of what relational law means. Law is becoming at the same time more and less dependent on legal texts. More dependent because Legal Open Data will allow a fast and cheap accessibility to a great bulk of accumulated, stored texts in connected repositories. Less dependent because people will be using its content in many ways, not only interpreting it canonically, seeking from authoritative opinions. Law is being linked, dereferenced, crowdsourced, reinterpreted in a way that intertwines legal norms with ethical and political issues and principles.

Using Floridi's metaphor of third-order technologies, SW and LOD are certainly situated in a kind of autonomous and self-consuming contained "in-betweeness" [43]. But conflicts and law have always had a high degree of open *heteronomy*. Humanity-in-the-loop [58] very likely will lead to a situation in which agents (whether artificial or humans) interact through regulations and conflicts. Applying national constitutional norms, or even private or public international law only, to harness SWRM *hybrid* models of regulation it is not realistic. It does not close the gap between legal theory and the new developments of the Web.

Acknowledgements. CAPER, EU Grant Agreement 261712; Crowdsourcing DER DER2012-39492-C02-0; SINTELNET FP7-ICT-2009-C-286380; Consumedia IPT-2011-1015-430000; Crowdcrisiscontrol IPT-2012-0968-390000.

References

1. Al-Adwan, M.K.M.: The Legitimacy of Online Alternative Dispute Resolution (ODR). International Journal of Business and Social Science 2(19), 167–169 (2011)
2. Bernard, P., Garth, B.: Dispute Resolution Ethics: A Comprehensive Guide. ABA, Section of Dispute Resolution, Washington D.C. (2002)
3. Berners-Lee, T.: Linked Data (updated 2009),
   http://www.w3.org/DesignIssues/LinkedData.html
4. http://www.ted.com/talks/tim_berners_lee_on_the_next_web.html

5. Bodriagov, O., Buchegger, S.: Crowdsourcing and Ethics: The Employment of Crowdsourcing Workers for Tasks that Violate Privacy and Ethics. In: Elovici, Y., Altshuler, Y. (eds.) Security and Privacy in Social Networks, pp. 47–66. Springer, Dordrecht (2009)
6. Boehm, F.: Information Sharing and Data Protection in the Area of Freedom, Security and Justice. Springer, Heidelberg (2012)
7. Bott, M., Young, G.: The Role of Crowdsourcing for Better Governance in International Development. PRAXIS The Fletcher Journal of Human Security 27, 47–70 (2012)
8. Bowden, C.: The US surveillance programmes and their impact on EU citizens' fundamental rights. In: Presented to the European Parliament's Committee on Civil Liberties, Justice and Home Affairs (LIBE Committee), PE 474.405 (September 05, 2013) , http://www.europarl.europa.eu/meetdocs/2009_2014/documents/li be/dv/briefingnote_/briefingnote_en.pdf
9. Bruce, T.R.: Foundings on the Cathedral Steps. In: Peruginelli, G., Ragona, M. (eds.) Law via the Internet. Free Access, Quality of Information, Effectiveness of Rights, pp. 411–422. European Press Academic Publishing, Florence (2009)
10. Burns, R.: Connecting Grassroots to Government for Disaster Management. The Wilson Center (2012), http://burnsr77.github.io/assets/uploads/ Workshop_BackgroundReading.pdf
11. Cameron, K.: The Laws of Identity...as of November 5, 2005. Microsoft Corporation (2005), http://www.identityblog.com/stories/2005/05/13/ TheLawsOfIdentity.pdf
12. Carneiro, D., Novais, P., Neves, J.: Towards Domain-Independent Conflict Resolution Tools. In: Proceedings of the 2011 IEEE/WIC/ACM International Conferences on Web Intelligence and Intelligent Agent Technology, vol. 2, pp. 145–148. IEEE Computer Society (2011)
13. Casanovas, P., Poblet, M.: Concepts and fields of Relational Justice. In: Casanovas, P., Sartor, G., Casellas, N., Rubino, R. (eds.) Computable Models of the Law. LNCS (LNAI), vol. 4884, pp. 323–339. Springer, Heidelberg (2008)
14. Casanovas, P.: Agreement and Relational Justice: A Perspective from Philosophy and Sociology of Law. In: Ossowski, S. (ed.) Agreement Technologies. LGTS, vol. 8, pp. 19–42. Springer, Heidelberg (2013)
15. Casanovas, P.: Social Intelligence: A New Perspective on Relational Law. In: Schweighofer, E., Meinrad Handstanger, H., Hoffmann, F.K., Primosch, E., Schefbeck, G., Withalm, G. (eds.) Festchrift für Friedrich Lachmayer. Zeichen und Zauber des Rechts, pp. 493–510. Editions Weblab, Bern (2014)
16. Casanovas, P.: Philosophy and Technology. Special Issue on Information Society and Ethical Inquiries (2014), doi:10.1007/s13347-014-0170-y
17. Casanovas, P., Barral, I. (Guest Editors): Special Section on Legal XML and Online Dispute Resolution. Democracia Digital e Governo Eletrônico 10, 1–432 (2014)
18. Castelfranchi, C.: Minds as Social Institutions. Phenomenology and the Cognitive Science (2013), doi:10.1007/s11097-013-9324-0
19. Cavoukian, A.: Privacy by Design. The 7 Foundational Principles. Implementation and Mapping of Fair information Practices. Information an Privacy Commissioner, Ontario, Canada (2010)
20. Cortés, P., de la Rosa, F.E.: Building a Global Redress System for Low Value Crossborder Disputes. International and Comparative Law Quarterly 62(2), 407–440 (2013)

21. Cortés, P.: Developing Online Dispute Resolution for Consumers in the EU: A Proposal for the Regulation of Accredited Providers. International Journal of Law and Information Technology 19(1), 1–28 (2012)
22. Datta, P., Chatterjee, S.: Online Consumer Market Inefficiencies and Intermediation. The DATA BASE for Advances in Information Systems 42(2), 55–75 (2011)
23. D'Inverno, M., Luck, M., Noriega, P., Rodríguez-Aguilar, J.A., Sierra, C.: Communicating open systems. Artificial Intelligence 186, 38–94 (2011)
24. Egger, F.N.: Deceptive Technologies: Cash, Ethics & HCI, SIGCHI Bulletin, p. 11 (May/June 2003)
25. Floridi, L.: Technology's In-Betweeness. Philosophy and Technology 26(2), 111–115 (2013)
26. Greenleaf, G., Mowbray, A., Chung, P.: The meaning of 'free access to legal information':A twenty year evolution' (on LSN) Law via Internet Conference, Cornell University, Ithica, USA (October 2012)
27. Ha, H., Coghill, K.: Online shoppers in Australia: dealing with problems. International Journal of Consumer Studies 32, 5–17 (2008)
28. Harris, C.G., Srinivasan, P.: Ethics and Crowdsourcing. In: Altshuler, Y. (ed.) Security and Privacy in Social Networks, pp. 67–83. Springer, Dordrecht (2013)
29. van der Herik, J., Dimov, D.: Towards Crowdsourced Online Dispute Resolution. Journal of International Commercial Law and Technology 7(2), 99–111 (2012)
30. Hörnle, J.: Encouraging Online Alternative Dispute Resolution in the EU and Beyond. European Law Review 38, 187–208 (2013)
31. Kannai, R., Schild, U., Zeleznikow, J.: Modeling the evolution of legal discretion – an Artificial Intelligence Approach. Ratio Juris 20(4), 530–558 (2007)
32. Kersten, G.E., Vahidov, R., Arno, R.: Lodder and John Zeleznikow: Enhanced Dispute Resolution Through the Use of Information Technology. Group Decision and Negotiation 2, 525–530 (2011)
33. Kittur, A., Nickerson, J.V., Bernstein, M.S., et al.: The Future of Crowd Work. In: Proceedings of the 2013 Conference on Computer Supported Cooperative Work, CSWC 2013, pp. 1301–1318. ACM (2013)
34. Langheinrich, M.: Privacy by Design - Principles of Privacy-Aware Ubiquitous Systems. In: Abowd, G.D., Brumitt, B., Shafer, S. (eds.) UbiComp 2001. LNCS, vol. 2201, pp. 273–291. Springer, Heidelberg (2001)
35. Larson, M.: Crowdsourcing: From Theory to Practice and Long-Term Perspectives. Dagstuhl, September 1-4 (2013)
36. Letia, L., Groza, A.: Planning with argumentation schemes in online dispute resolution. In: IEEE International Conference on Intelligent Computer Communication and Processing, pp. 17–24 (2007)
37. Lodder, A.: The Third Party and Beyond. An Analysis of the Different Parties, in particular The Fifth, Involved in Online Dispute Resolution. Information & Communications Technology Law 15(2), 143–155 (2013)
38. Lodder, A., Zeleznikow, J.: Developing an Online Dispute Resolution Environment: Dialogue Tools and Negotiation Systems in a Three Step Model. The Harvard Negotiation Law Review 10, 287–338 (2005)
39. Lodder, A., Zeleznikow, J.: Enhanced Dispute Resolution Through the Use of Information Technology. Cambridge University Press (2010)
40. Mars de, J., Exon, S.N., Kovach, K., Rule, C.: Virtual Virtues: Ethical Considerations for an Online Dispute Resolution Practice. Dispute Resolution Magazine 17(1), 6–10 (2010)

41. Marshall, C.C., Shipman, F.M.: Experiences Surveying the Crowd: Reflections on Methods, Participation, and Reliability. In: WebSci 2013, Paris, France, May 2-4, pp. 234–243. ACM (2013)
42. Meier, P.: On Crowdsourcing, Crisis Mapping and Data Protection Standards (2012), http://irevolution.net/2012/02/05/iom-data-protection/
43. Noriega, P., Chopra, A.K., Fornara, N., Lopes Cardoso, H., Singh, M.: Regulated MAS: Social Perspective. Normative Multi-Agent Systems, pp. 93-134 (2013)
44. Ossowksi, S. (ed.): Agreement Technologies. LGTS. Springer, Heidelberg (2013)
45. Poblet, M.: Spread the word: the value of local information in disaster response. The Conversation (January 17, 2013), http://theconversation.com/spread-the-word-the-value-of-local-information-in-disaster-response-11626
46. Poblet, M., Leshinsky, R., Zeleznikow, J.: Digital neighbours: Even Good Samaritan crisis mappers need strategies for legal liability. Planning News. 38(11), 20–21 (2012), http://search.informit.com.au/documentSummary;dn=002163980680827;res=IELBUS
47. Reay, I., Dick, S., Miller, J.: A Large-Scale Empirical Study of P3P Privacy Policies: Stated Actions vs. Legal Obligations. ACM Transactions on The Web 3(2), Art. 6 (2009)
48. Rizza, C., Guimarães, A., Chiaramello, M., Curvelo, P.: Do-it-yourself justice considerations of social media use in a crisis situation: the case of the 2011 Vancouver riots. In: IEEE/ACM International Conference on Advances in Social Networks Analysis and Mining, pp. 720–721. IEE Computer Society (2012)
49. Rule, C., Friedberg, L.: The appropriate role of dispute resolution in building trust online. Artificial Intelligence and Law 13, 193–205 (2005)
50. Rule, C., Rogers, V.: Building a Global System for Resolving High-Volume, Low-Value Cases. Alternatives to the High Cost of Litigation 29(7), 135–136 (2011)
51. Schultz, T.: Private Legal Systems: What Cyberspace Might Teach Legal Theorists. Yale Journal of Law & Technology 10, 151–193 (2007)
52. Schultz, T.: The Roles of Dispute Settlement and ODR. In: Arnold Ingen-Housz, K. (ed.) ADR In Business: Practice and Issues Across Countries and Cultures, vol. 2, pp. 135–155. Kluwer, Amsterdam (2011)
53. Schultz, T.: Internet Disputes, Fairness in Arbitration and Transnationalism: A Reply to Julia Hörnle. International Journal of Law and Information Technology 19(2), 153–163 (2011)
54. Turel, O.: Predictors of disputants' intentions to use online dispute resolution services: the roles of justice and trust. Doctoral Dissertation. Canada, McMaster University (2006)
55. Turilli, M.: Ethical Protocols Design. Ethics and Information Technology 9, 49–62 (2007)
56. Wahab, M.S.A., Katsh, E., Rainey, D. (eds.): Online Dispute Resolution: Theory and Practice A Treatise on Technology and ODR. Eleven International Publishing, The Netherlands (2012)
57. Westin, A.F.: Privacy and Freedom Atheneum, N.Y (1967)
58. Ye, Q.: Research on Disputes about the Reputation Right in Networks, E-Business and Information System Security. In: EBISS 2009. IEEE (2009)
59. Zeleznikow, J.: Methods for incorporating fairness into the development of an online family dispute resolution environment. Alternative Dispute Resolution Journal 22, 16–21 (2011)
60. Zeleznikow, J., Bellucci, E.: Legal fairness in ADR processes – implications for research and teaching. Alternative Dispute Resolution Journal 23, 265–273 (2012)

Drafting a Composite Indicator of Validity for Regulatory Models and Legal Systems

Andrea Ciambra and Pompeu Casanovas

UAB Institute of Law and Technology, Universitat Autònoma de Barcelona, Barcelona, Spain
andrea.ciambra@gmail.com,
pompeu.casanovas@uab.cat

Abstract. The aim of this paper is to lay the groundwork for the creation of a composite indicator of the validity of regulatory systems. The composite nature of the indicator implies a) that its construction is embedded in the long-standing theoretical debate and framework of legal validity; b) that it formally contains other sub-indicators whose occurrence is essential to the determination of validity. The paper suggests, in other words, that validity is a second-degree property, i.e., one that occurs only once the justice, efficiency, effectiveness, and enforceability of the system have been checked.

Keywords: Validity, indicators, regulatory models, regulatory systems, Privacy Impact Assessment.

1 Introduction

The aim of this preliminary and exploratory study is to lay the groundwork for the creation of a *composite indicator* of the legal validity of norms. The composite nature of the indicator implies *a)* that its construction is embedded in the long-standing theoretical debate and framework of legal validity; *b)* that it formally contains other sub-indicators whose occurrence is essential to the determination of validity. The study suggests, in other words, that validity is a secondary property of a legal norm, i.e., one that occurs *only once* the norm's justice (J), efficiency (E_y), effectiveness (E_s), and enforceability (E_c) have been proved. This basic hypothesis can be rewritten in a plainer fashion as:

$$V_n \leq (J + E_y + E_s + E_c) \tag{1}$$

where the norm's legal validity V_n is less than or equal to the sum of the four sub-indicators. This also suggests that the norm's compliance with all four sub-indicators is a necessary and sufficient condition for its legal validity, i.e., that the occurrence of all four sub-indicators *implies* the norm's validity in a regulatory system previously defined:

$$(J + E_y + E_s + E_c) \Leftrightarrow V_n \tag{2}$$

These are not to be understood as formulae, but as simple way to convey a first intuition about validity. There are two tasks to comply with. The first one is theoretical: we should set a sound conceptual framework. The second one is technical: once concepts are cleared up, we can proceed to construct the composite indicator.

P. Casanovas et al. (Eds.): AICOL IV/V 2013, LNAI 8929, pp. 69–81, 2014.

The hypothesis does not assert the validity of all types of norms and normative systems, but only that in a regulatory model it becomes possible to assign the specific validity of the regulations, i.e., the validity of the system as a whole according to the previous occurrence of these sub-indicators. In this way, very likely, what the composite indicator really measures is the *institutional strengthening* of the whole system. That is to say, the emergent pragmatic aspect of regulatory systems that we can equate with their legal existence.

The analysis of the state of the art of the theoretical debate on legal validity highlights an unresolved issue[1] in the determination of *a*) a generalisable threshold for the existence of a norm's validity and, *b*) the assessment of the concept of legal validity as a *continuum* or gradient rather than a discrete quality of a norm (i.e., a *yes/no* dichotomy). This paper starts briefly from the study of this debate (Section 2) to locate its working hypothesis into the theoretical framework of legal validity. It is carried out in tight connection with the objectives of CAPER,[2] a large-scale collaborative project within the 7th Framework Programme of the European Union (EU) that aims to build an information-sharing Internet platform for the detection and prevention of organised crime.

It also analyses structural issues related to the construction of composite indicators (especially in the social sciences as well as in non-quantitative, discursive contexts) to advance a tentative indicator for the benchmarking or 'measurement' of the CAPER Regulatory Model (CRM), i.e. the specific set of rules laid down to run the government of the platform and its compliance with European and National regulations, including ethical principles, Data Protection Impact Assessments, and Best Practices mentioned in the new draft of the Regulation (Section 3). Section 4 identifies a number of research paths that, even besides the advances of the CAPER project, unfold thanks to the development of a technically-reliable indicator of validity for regulatory models.

2   The Theoretical Debate: Legal Validity

The definition of under what conditions law and norms can be considered *valid* is one of the most disputed debates in legal theory and the philosophy of law, a "major jurisprudential battleground" [6], the "pineal gland of law" [5], as well as a litmus test to identify the field's main theoretical fracture lines. The contrast between different schools of thought on legal validity lies essentially in the inevitable relation between and potential overlap of legality and morality, i.e., the middle ground between what law is and what it ought to be. Historically, the debate has polarised across a continuum that spans from natural lawyers' morality thresholds on one extreme (an *unjust law* is certainly *not law*) to positivist law's

---

[1]   See Hage and von der Pfordten [1] and, in particular, Posher [2] and Spaak [3], as well as Grabowski [4]. For a general view and the general shape of the debate, see Pattaro [5].

[2]   CAPER is the acronym for «Collaborative information Acquisition, Processing, Exploitation, and Reporting for the prevention of organised crime». Curiously enough, a valuable work on indicator validation in the context of environmental social impact assessment by Bockstaller and Girardin, mentioned later in this paper, also refers to the development of another 'CAPER' project, namely, the «Concerted Action of Pesticide Environmental Risk indicators». The two projects should not, of course, be confused and all mentions in this paper refer to the former.

formalist tests on the other end (insofar as a norm abides by the formal requirements and conventions overtly accepted by a given political and social community, it is valid law). These two extremes diverge significantly also in epistemological terms.

Broadly speaking, natural law considers, on the one hand, law as a consequence or a subsequent derivation of the fundamental moral standards, principles, and values embedded in a community. This assumption allows natural lawyers to perform the validity assessment *ex ante*, i.e., as soon as the norm stems from the moral endowment of the community, it is *inevitably just*. On the other hand, positivist lawyers test validity *once the norm is established*, since they are concerned with the respect of the procedures and process that led to the ultimate formulation of the norm. The legal validity of a norm, therefore, "is established not by arguments concerning its value and justification but rather by showing that it conforms to tests of validity laid down by some other rules of the system" [7]. Positivists move the validity test more and more backwards up to a core of fundamental norms—i.e., "those ultimate rules of recognition" that are a "matter of social fact" [6]— that cannot be contested lest the whole legal system be questioned. Positivists, in other words, hold an idea of law as "that which is" rather than "that which ought to be" [8]. This *ex post* approach exposes the positivist understanding of just or valid law to a historical vulnerability and a recurrent criticism, emphasising the attempt of positivist lawyers to *justify* as being valid (to the extent that they are formally correct) certain norms, laws, and policies that would generally raise moral concerns when cast against the background of (potentially) universal or majoritarian principles and values.

Despite the theoretical conundrums and the need to locate each current of thought at a given point on the 'validity debate' continuum, most readings of the validity problem imply a controversial issue of subjectivity and relativism and emphasise the lack of a defined, generalisable, adaptable and context-free *measurement of legal validity*, i.e., the lack of a reliable indicator that—whatever the legal context, juridical structure, and constitutional/institutional order—may signal a norm's legal validity or invalidity into the regulatory system. The theoretical debate on legal validity underscores, moreover, the importance of language, meaning, and semantic contextualisation in the attempt to abstract a general concept of validity. This emphasis has two main analytical implications.

First, any advance in this field needs to avoid the risk of *trivialising* the issue as of linguistic or cultural misunderstanding, i.e., the assumption that, since many scholars have analysed the validity issue from the semantic perspective of either certain languages[3] or certain specific fields of application (e.g., the practice of law in court or the normative underpinnings of law- and policy-making), the different contributions to the debate may after all be agreeing on essential concepts and (more or less inadvertently) mystifying or baffling their mutual dialogue by means of ambiguous, relative or unsettled discursive vehicles and semantic structures.

Second —and consequently—, any advance in the validity debate *should* try to overcome these persistent definitional uncertainty and endemic 'relativism', strive for a notion of the validity concept which may reliably and flexibly used in diverse contexts and under varying conditions, and elicit the immediate and unequivocal understanding of the recipients and the users of norms —be they citizens, lawyers, lawmakers or scholars.

---

[3]   See, for instance, the detailed analysis of the validity debate in German philosophy of law carried out by Grabowski on Kelsen, Radbruch, Dreier, Alexy and Habermas [4].

There are at least two ways of tackling this problem. From a logical point of view, validity can be faced as an emergent semantic property of inferential processes and then linked to the argumentative discourse on normative semantics. This is the way lately chosen, for instance, by Prakken and Sartor [9]: arguments about norms are modelled as the application of argument schemes to knowledge bases of facts and norms. But, from an empirical point of view, this normative approach does not help to know how the system works.

From an empirical approach, the problem can be described as a controlled induction process. The assumption is that validity is a second order property of a regulatory model that applies to the evaluation of regulatory systems. A regulatory system can be defined as a set of functionally interacting elements (not, or not only, as a set of logically consistent norms). A regulatory model tests how well the system is working—a process of assessing performance against some stated criteria or a known measure (i.e., a benchmark). This is why it makes sense to construct indicators to *validate* the system [10].

3   The Construction of the Indicator

This section of the paper lays the groundwork for further research on and assessment of a new composite indicator of validity for regulatory systems. We are not the first to suggest composite indicators for the legal field. Vallbé and Casellas [28] are proposing a model for the costs of discovery of legal information, the relationship between governmental online presence and legal publication, and the quality of regulation. Vallbé [29] just constructed a composite indicator for judicial performance (a judicial regional authority index, related to the degree of decentralization of states).

But very likely ours is one of the first attempts to model in this way some concepts stemming from legal theory. It should be noted too that this paper does not start from a normative point of view. *Regulatory spaces* [11] or *meta-regulatory strategies* [12] have been already proposed from a socio-legal perspective to cope with the transnational plurality of normative sources. It is our contention that we can take a different and simpler starting point. The process of construction and validation of the indicator suggested here draws from several examples and methodological notes in current literature and focuses in particular on a field that has been developed significantly over the last two decades and may serve, by all means, as a *lesson learned* or good practice in this regard: privacy impact assessments (PIAs). .

3.1   Preliminary Lessons: The Case of Privacy Impact Assessments

Impact Assessments (IA) consist of all sorts of studies, measurements and reflections about the social, ethical and legal effects and consequences of certain policies, regulations and practices. From the past twenty years on it has become commonplace to apply IA to privacy (Privacy Impact Assessments, PIAs), regulations (RIAs), surveillance (SIAs) and data protection (DPIAs). Implementing a PIA or a DPIA means a sort of monitoring audit that goes along the process of creating, testing, reviewing and eventually enforcing a regulatory tool (including technological projects and economic planning). They have been adopted mainly to evaluate intended legislation and public policies in PIAs have been currently adopted by Common Law countries like USA, Canada, UK, Australia, and New Zealand for the protection of civil (human) rights regarding personal data [13]. A PIA is conceived as a methodology and a process for identifying and evaluating risks to privacy, checking for compliance with legislation and aiming at avoiding or mitigating those risks [14].

PIAs are the immediate precedent for Data Protection Impact Assessments (DPIAs), as foreseen by the EU Directive proposal. The IA Document defines DPIAs as a PIA: "A process whereby a conscious and systematic effort is made to assess privacy risks to individuals in the collection, use and disclosure of their personal data. DPIAs help identify privacy risks, foresee problems and bring forward solutions". Constructing an empirical notion of *validity* is key to evaluate the functioning of regulatory systems after a PIA (or DPIA) has been carried out.[4]

3.2   A Composite Indicator of Legal Validity

Composite indicators are increasingly valued in the social sciences because of their "ability to integrate large amounts of information into easily understood formats for a general audience" [15]. A composite indicator is a synthetic index of several individual indicators, a quality that allows analysts to present complex content more rapidly, compare different contexts or timeframes more intuitively, and reduce the amount of data or graphic content that needs to be used to deliver the necessary information. Their convenience and growing systematic use in a number of policy fields and academic sectors demand, however, a degree of methodological consciousness that may add up to the indicator's credibility and reliability. Rather than one able to convey as much information as possible with the most compact index available, this study suggests the creation of an indicator whose composite nature is mostly qualitative by the moment. At least at its present stage.

Legal validity as a composite indicator implies that the object of the assessment is *not (legally) valid unless* all four sub-indicators reach a certain threshold. This indicator, in other words, to be applied as an evaluative tool, is not meant to compile synthetic information from a dataset; rather, it depends on its sub-indicators to show a certain value for it to be an actual measure of validity. The creation of the indicator is still at an embryonic stage of development. We tend to adhere to the general definitions and guidelines issued by the Organisation for Economic Cooperation and Development (OECD) about the construction of composite indicators. Even though the OECD has consistently increased its reliance on (and, proportionally, its careful methodological improvement of) composite indicators of economic and sustainability performance, especially in cross-country comparisons, its guidance for the process of creating one are extremely useful even outside the scope of economics. The OECD's recommended 'checklist' [16] suggests a few fundamental steps

---

[4]   According to the EU Commission Staff Working Paper SEC(2012) 72 final, a Data Protection Impact Assessment (DPIA) is a process whereby a conscious and systematic effort is made to assess privacy risks to individuals in the collection, use and disclosure of their personal data. DPIAs help identify privacy risks, foresee problems and bring forward solutions. The definition of these general concepts stems from D7.1, EU Commission Staff Working Paper SEC(2012) 72 final, the Joint Proposal for a Draft of International Standards on the Protection of Privacy with regard to the processing of Personal Data (Madrid, 2009), and Directives 95/46/EC, 2002/58/EC, and 2009/136/EC.

towards the creation of a valid composite indicator: 1) defining a consistent theoretical framework for the selection of relevant variables, objectives, and potential recipients of the tool; 2) selecting adequate data according to "analytical soundness, measurability, ... and relevance of the indicators to the phenomenon being measured and the relationship to each other" [16]; and 3) normalisation of all indicators by weighing them to a ground coefficient, in order to make all variables comparable and the composite result homogeneous. At the current stage of work, the composite indicator of legal validity is at phase 2: the theoretical framework is already well enshrined in the long-standing debate on validity, normativity, legality, and morality; the following sub-sections start defining the sub-indicators by attaching them to a given variable and suggesting a suitable set of existing data to assess its occurrence and/or intensity. Fig. 1 shows the general structure of the model that we are fleshing out.

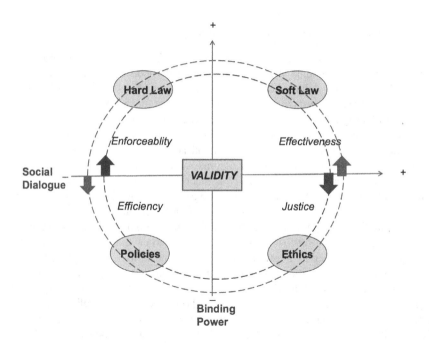

Fig. 1. Structure of the CAPER Regulatory Model. Source: [26, 27]

Sub-indicator No. 1: Efficiency. Efficiency is an indicator of governance that refers to the quality of the outcome produced (regulations, services or products): in this case, the relationship between regulatory systems and Agencies and Administration policies. In order to build this indicator for the EU Freedom, Security and Justice area, we are focusing only on one small part of the Rule of Law World Bank indicators, i.e., how well the regulatory system is able to perform within a multilevel governance organisation. Therefore, it refers primarily to its *institutional strengthening* (IS) dimension. (Do note that validity refers to IS as a second order property). IS points to the collective property that emerges from the process of implementing a model seeking a certain balance between the binding power of the rule of law and the dialogue among all the stakeholders, including the different polices, web service providers, and citizens. Two more related dimensions are at stake: the interoperability between databases and technological languages [17], and the multi-level inter- and intra-organisational dynamics.

Sub-indicator No. 2: Effectiveness. Effectiveness is an indicator of governance that refers to the relationship between the results achieved and the resources used (cost in relation to the outcomes achieved). We propose to measure the effectiveness of soft law mechanisms regarding non-binding regulations, directives and guidelines of the UN and the EU Commission. The performance of statements, principles, codes of conduct, and codes of practice can be summarised for Security Information Governance and Data Protection combining the COBIT maturity model and ISO 17799—as suggested by von Solms [18]. COBIT can be seen as being used on a strategic level, indicating the 'what' as far as governance is concerned. On the other hand, ISO 17799 can be seen more as being used on a lower level, specifying the 'how', as far as information security management is concerned [19].

Sub-indicator No. 3: Enforceability. Enforceability entails the possibility to be argued in court to ground a judicial ruling. It belongs to the adjudication legal system, in which certainty of law matters. In terms of measurement, enforceability presents a meaningful semantic challenge, since historically analysts and organisations have indulged in elaborate indicators of *enforcement* rather than enforceability, i.e., *ex post* analyses of the actual degree of compliance with an established norm rather than an analysis of a norm's potential for compulsoriness. A valuable example comes from studies on the rule of law and the performance of law enforcement and justice, especially on a global scale, at which comparable and normalised results are most needed [20]. Indicators such as the administrative processes that lead to a norm's enforcement, measurements of political influence or intervention in the enforcement process, and the respect of due process guarantees throughout all procedures are common indices of enforcement and compliance in this kind of studies.

The composite indicator suggested in this paper, however, looks more at a characterisation of enforceability at an earlier stage of the policy- and law-making processes, i.e., we are more interested in the possibility for practitioners and administrators to evaluate *preliminarily* whether a norm presents any issues when it comes to actual enforcement and prospective compliance by the act's recipients. Under common circumstances, of course, compliance is highly correlated to the hierarchical value of the legal vehicle used to implement it: in other words, it can be expected that a norm or act be more easily enforced if it is carried out through hard law or strictly mandatory provisions. A gap between a norm's legal vehicle—e.g., hard law in the form of a regulation—and the norm's capability to be implemented and enforced—e.g., bottlenecks in the administrative procedure, misled targeting, excessive costs—undermines the norm's validity, as it increases uncertainty and lowers effectiveness. Standard compliance metrics are generally drawn from corporate performance[5] or administrative auditing [21]. This paper aims to promote further academic debate on a middleground indicator of enforceability that would measure the potential for a smooth implementation process and the lack of any ethics- and performance-related shortcomings in the enforcement of a certain law or policy act.

Sub-indicator No. 4: Justice. The ethical criterion of *justice*, needed to deem whether a norm or law is valid or invalid, is perhaps the most controversial or questionable point in the methodological argument that designs the composite legal validity indicator suggested in this paper. The assessment of this indicator tends inevitably to a subjective appraisal of qualitative, non-tangible, and/or discursive data such as perception, opinions or 'feelings' about a subject's experience of justice in its relationship with society or authority. A more technical and quantitative analytical vantage point has focused, conversely, on justice as it is usually 'materialised' in court and procedure: part of this literature has developed, accordingly, a number of indicators to measure the performance of justice systems, especially in terms of effectiveness and social cost.[6] Measurement and 'quantification' is therefore the most challenging issue raised by an 'ethical' indicator of justice.

The lack of data—especially when "a certain behaviour cannot be measured or no one has attempted to measure it" [15]—affects the reliability of the indicator. This is all the more true when dealing with a variable, the perception of justice, which can be parameterised only through discursive and content-related analysis of language 'vehicles', i.e., all those linguistic and semantic units that constitute communication, meaning, and ideas and whose cataloguing and typology may render a conceptual map of what is conceived as *just* in a growingly objective and socially-accepted way. The construction of such methodology—calling for an attempt to perfect certain techniques that are commonly adopted in discourse and content analysis in the social sciences and *normalise* an indicator of justice against a comparable and replicable minimum standard—presents perhaps the richest opportunity for further debate and research development. Metrics, typologies, data mining, bibliometrics, and content analysis all contribute to the potential toolkit that may provide the first-hand raw data needed to develop and validate the justice indicator envisioned and suggested in this paper.

3.3 Methodological Caveats on the Indicator's Validation

The applications of a consistent and reliable indicator of legal validity are manifold. The possibility to assess—regardless of context and time—whether a suggested

---

[5] Corporate services—e.g., the Compliance Week (http://www.complianceweek.com/)— information service are nowadays a full-fledged industry providing additional assistance and expertise to corporations interested in ethics and compliance audits.

[6] Harvard University is developing a tailored project on indicators of safety, justice, and the rule of law, involving a number of academic and civil society institutions in six partner countries:
http://www.hks.harvard.edu/programs/criminaljustice/research-publications/measuring-the-performance-of-criminal-justice-systems/indicators-in-development-safety-and-justice.

regulation, norm or law passes a test of legal validity is necessary guidance for policy- and lawmakers. The composite nature of this indicator, moreover, implies that, insofar as the norm passes the validity test, it is also just, enforceable, effective in reaching its goals, and efficient in terms of resources or time needed. The indicator proposed in this paper, in other words, complies with the basic function of any indicator, i.e., "to reduce the volume and complexity of information which is required by decision makers" [22]. Such an indicator provides the analyst or the lawyer with a threshold after which validity is identified straightforwardly and relays, at the same time, "a complex message in a simplified manner" [23]. A composite indicator on validity, ideally, would condense convolute information on a norm's qualities in just one single measurement.

There are a number of methodological caveats that need to be taken into consideration when *validating* an indicator, especially if its design derives—as it is the case with legal validity—from the need to fill a theoretical vacuum with significant concrete implications in the routine activities of practitioners and professionals. We concur that the scientific and practical value of an indicator is intrinsically connected to its compliance with fundamental criteria of acceptability. Within the closed epistemic community of scholars, students, and practitioners of a given discipline, general consensus and acceptance validate an indicator "if it is scientifically designed, if the information it supplies is relevant, if it is useful and used by the end users" [24].

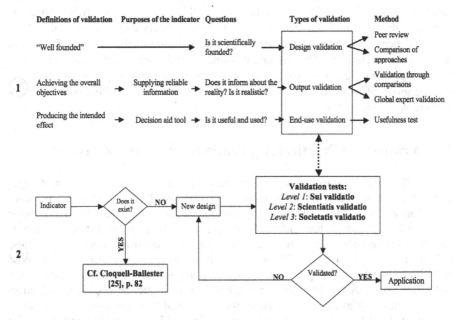

Fig. 2. Re-elaboration of the process flows of indicator validation in (1) Bockstaller and Girardin, [24], and (2) Cloquell-Ballester et al., [25]. The dashed line emphasises the convergence of both models on a tri-partite validation test.

An indicator's *design* must respect generally-accepted rules and prescriptions; the *feedback* of relevant scientific peers must confirm the viability of the indicator as an analytical instrument; and the *output* of the indicator must be intelligible, accessible, and useful to the target recipients of the tool. This tri-partite scheme is commonly adopted in the literature about social composite indicators (see Fig. 2), especially in well-developed fields such as environmental impact assessment or sustainability studies. The design-feedback-output model can also be interpreted hierarchically, as with the "3S methodology" and the three progressive stages of "*sui validatio*", "*scienciatis validatio*", and "*societatis validatio*": this scholarship argues that the indicator's "credibility" grows proportionally to its ability to pass this cumulative test [25].

Both tri-partite models of validation presented above are useful to effectively increase the scientific reliability of a new indicator and respond to the requirements of acceptability established as standard in a given community. This position paper, therefore, after outlining the main characteristics of its proposed indicator of legal validity, also recommends that this design and prospective analytical tool be subject to the scrutiny of peer researchers and practitioners in order to gather valuable feedback and responses as regards: *a*) the scientific adequacy of the model proposed in this paper; *b*) positive comparison between this indicator and analogous or comparable tools already validated by its recipients; and *c*) positive reception from potential end-users as far as the necessity, the appropriateness, and the practical potential of this indicator in its day-to-day, professional or 'routine' uses are concerned. Besides suggesting a new research agenda on this topic for the close future and invite all interested contributors to engage in the debate outlined above, this paper calls explicitly upon the users that this indicator has been tailored to: practitioners and professionals whose choices depend—to a varying degree—on the recognition of the validity of a certain norm, law or regulation. This empirical quandary and concrete objective have been the lynchpin and the true raison d'être of this work in the first place, and it is this specific group of recipients and potential users that the whole CAPER project and its deliverables are aimed at.

4   Prospective Practical Applications: the CAPER Project

The CAPER project aims to create a common platform for the prevention of organised crime through sharing, exploitation and analysis of both open and private information sources.[7] One of the main objectives of the project is to establish a common platform through which law enforcement agencies (LEAs) from different countries can share information to pool resources and improve mutual interoperability in their fight against organised crime. The development of the project envisages the analysis and collection of data not only from openly available sources such as televised, radio, and visual broadcasts or Internet content, but also from internal resources and information exclusively available to LEAs in the exercise of their functions. The sensitive content of the data and materials made available by LEAs to design and create the platform makes it all the more important for all actors involved in the project to test

---

[7] http://www.fp7-caper.eu/fr.html.

all proposed action and objectives against an indicator of validity, in order to clarify since the earliest stage of development that all planned measures meet a generally-accepted standard of legitimacy.

The creation and validation of a reliable and context-free indicator of legal validity is, therefore, crucial for the development of the CAPER platform and the usability of its instruments. CAPER is also a valuable measurement of the complexity of cooperation, information sharing, and interoperability in such a sensitive field, in which LEAs manage significant amounts of delicate information and implement a number of actions that affect—one way or the other—different societal groups as well as the populace at large. There is an ethical red line lingering over the blurred boundary between the information that LEAs need to perform their duties and the information whose management requires additional regulation and caution as it enters the sphere of privacy of citizens and other subjects of law. The model of legal validity indicator suggested in this paper addresses this issue by 'quadrupling' the dimensions implied by the validity of a norm, measure or decision. An action set out by LEAs in the framework of the CAPER project, therefore, will be asked to pass a validity test that, *per se*, also confirms that this measure is efficient in terms of its practical implementation, effective in meeting strategic objectives and carrying out the necessary tasks, enforceable through the deployment of the available instruments and resources, and, most importantly, that this measure is *just* to the extent that it complies with privacy requirements and is not detrimental to the recipients' individual rights only for the sake of its application. For this set of reasons privacy impact assessments have been a relevant source of practices, examples, and information for the definition of this composite indicator. The CAPER project is a valuable starting point for the refinement of this kind of 'ethical' indicators and assessment protocols, even though issues of subjectivity, qualitative appraisal, and discursive/non-neutral techniques remain open to further public scrutiny and debate in the scientific community.

5   Conclusions and Future Work

In this paper we have drawn the main lines to build up a regulatory model for the monitoring and evaluation of regulatory systems. We have suggested that validity is not a first-order property of the system, but a second-order property *a*) along the axis compulsoriness/social dialogue; *b*) the linear function four-tuple [enforceability, effectiveness, efficacy, justice]; and *c*) the resulting institutional strengthening. The Caper Regulatory Model (CRM) provides the benchmark with which this model will be tested to evaluate the functioning of the European platform for police interoperability to fight organised crime.

References

1. Hage, J.C., van der Pfordten, D.: Concepts in Law. Springer (2009)
2. Poscher, R.: The Hand of Midas: When Concepts Turn Legal, or Deflating the Hart-Dworkin Debate. In: Hage, P.J.C., von der Pfordten, P.D. (eds.) Concepts in Law, pp. 99–115. Springer, Netherlands (2009)

3. Spaak, T.: Explicating the Concept of Legal Competence. In: Hage, P.J.C., van der Pfordten, P.D. (eds.) Concepts in Law, pp. 67–80. Springer, Netherlands (2009)
4. Grabowski, A.: Juristic Concept of the Validity of Statutory Law. A Critique of Contemporary Legal Nonpositivism. Springer, Heidelberg (2013)
5. Pattaro, E. (ed.): A Treatise of Legal Philosophy and General Jurisprudence - Volume 1: The Law and The Right. Springer (2012)
6. Sartor, G.: Legal Validity as Doxastic Obligation: From Definition to Normativity. Law Philos 19, 585–625 (2000)
7. Raz, J.: The authority of law: essays on law and morality. Oxford University Press, Oxford (2009)
8. Gray, J.C.: The Nature and Sources of the Law. The Columbia University Press (1909)
9. Prakken, H., Sartor, G.: Formalising arguments about norms. In: Ashley, K.D. (ed.) JURIX 2013: The Twenty-sixth Annual Conference on Legal Knowledge and Information Systems, pp. 121–130. IOS Press, Amsterdam (2013)
10. Kaufmann, D., Kraay, A., Mastruzzi, M.: The Worldwide Governance Indicators: Methodology and Analytical Issues. Social Science Research Network, Rochester (2010)
11. Lange, B.: Regulatory Spaces and Interactions: An Introduction. Soc. Leg. Stud. 12, 411–423 (2003)
12. Bomhoff, J., Meuwese, A.: The Meta-regulation of Transnational Private Regulation. J. Law Soc. 38, 138–162 (2011)
13. Wright, D., Wadhwa, K., de Hert, P., Kloza, D.: A Privacy Impact Assessment Frameworks for Data Protection and Privacy Rights. DJLS/2009 -2010/DAP/AG (2011), http://www.piafproject.eu/ref/PIAF_D1_21_Sept_2011.pdf
14. Wright, D., de Hert, P.: Introduction to Privacy Impact Assessment. In: Wright, D., de Hert, P. (eds.) Privacy Impact Assessment. LGTS, pp. 3–32. Springer, Berlin (2012)
15. Freudenberg, M.: Composite Indicators of Country Performance. Organisation for Economic Co-operation and Development, Paris (2003)
16. Handbook on constructing composite indicators: methodology and user guide. OECD, Paris (2008)
17. Wallwork, A., Baptista, J.: Understanding interoperability. In: Backhouse, J. (ed.) D4.1: Structured account of approaches on interoperability, pp. 19–28. FIDIS - Future of IDentity in the Information Society (2005)
18. Von Solms, B.: Information Security governance: COBIT or ISO 17799 or both? Comput. Secur. 24, 99–104 (2005)
19. Pretorius, E., von Solms, S.-H.: Information Security Governance using ISO 17799 and COBIT. In: Jajodia, S., Strous, L. (eds.) Integrity and Internal Control in Information Systems VI. IFIP s, vol. 140, pp. 107–113. Springer, Heidelberg (2004)
20. Agrast, M., Botero, J., Martínez, J., Ponce, A., Pratt, C.: The World Justice Project Rule of Law Index. The World Justice Project, Washington, DC (2012)
21. Measuring compliance effectiveness - our methodology. Australian Taxation Office, Camberra (2012)
22. Donnelly, A., Jones, M., O'Mahony, T., Byrne, G.: Selecting environmental indicator for use in strategic environmental assessment. Environ. Impact Assess. Rev. 27, 161–175 (2007)
23. Fisher, W.S.: Development and Validation of Ecological Indicators: an ORD Approach. Environ. Monit. Assess. 51, 23–28 (1998)
24. Bockstaller, C., Girardin, P.: How to validate environmental indicators. Agric. Syst. 76, 639–653 (2003)

25. Cloquell-Ballester, V.-A., Cloquell-Ballester, V.-A., Monterde-Díaz, R., Santamarina-Siurana, M.-C.: Indicators validation for the improvement of environmental and social impact quantitative assessment. Environ. Impact Assess. Rev. 26, 79–105 (2006)
26. Casanovas, P.: A Note on Validity in Law and Regulatory Systems (Position Paper). Quaderns de filosofia i ciència 42, 29–40 (2012)
27. Casanovas, P., Zeleznikow, J.: Online Dispute Resolution and Models of Relational Law and Justice. Presented at the Joint Workshop AICOL-SINTELNET at JURIX, 26th International Conference on Legal Knowledge and Information Systems, Bologna, Italy, (December 11, 2013)
28. Vallbé, J.J., Casellas, N.: What's the cost of e-Access to Legal Information? A composite indicator (2014), http://goo.gl/yNn0xz
29. Vallbé, J.J.: Measuring the Judicial Power of Regions: A Judicial Regional Authority Index. European Consortium for Political Research (2014), http://www.ecpr.eu/Events/PaperDetails.aspx?PaperID=16452&EventID=12

Measuring the Complexity
of the Legal Order over Time

Monica Palmirani and Luca Cervone

CIRSFID, University of Bologna, Bologna, Italy
{monica.palmirani,luca.cervone}@unibo.it

Abstract. One of the main problems in a legal order is how to manage the complexity of its changes over time. These modifications produce a very intricate network of citations in the legal order, so experts and citizens alike have serious difficulties accessing the normative content. Without countermeasures (e.g., simplification policy, codification, consolidation), the evolution of the legal order over time increases the uncertainty of the normative system and of the knowledge-acquisition process. This paper provides a theoretical model based on a set of indexes for measuring the complexity of each modificatory act using explicit modifications provisions. The global measurement provides an understanding of the complexity of the legal order over time. Secondly, we produce a diagram system for visualizing these indexes of the legal order per year and per document. The model was tested on an annotated corpus the Piedmont region has recently released that contains all its legislation as open data using the XML NormeInRete standard.

Keywords: Lifecycle Modelling, NormeInRete, Legal XML, Complex System.

Introduction

One of the main problems in a *legal order* (LO),[1] especially in civil law systems, at the national and local levels alike, is to how to manage the complexity of its changes over time. These modifications produce a very intricate network of citations in the legal order, so experts and the citizens alike wind up having serious difficulties accessing normative content. This is especially true in those countries, such as Italy, where codification is not a mandatory practice or where consolidation is not a lawful technique. Unless countermeasures are taken (e.g., simplification policy, codification, consolidation, and deregulation), the evolution of the legal order over time increases the uncertainty of the normative system and of the knowledge-acquisition process.

---

[1] We will be adopting the definition of *legal order* offered by Alchourron and Bulygin [1] as a set of legal systems over time (a *diachronic* normative system). A *legal system* is a set of legal sources that are *valid* at a given time *t* (a *static* legal system). The concept of validity has been a subject of debate in legal theory for decades. In this paper, because we are analysing bibliographic sources of law (i.e., acts published in an official gazette with a document representation), we consider the legal system the set of legal provisions that are effective (in operation) [13] at a given time *t*.

P. Casanovas et al. (Eds.): AICOL IV/V 2013, LNAI 8929, pp. 82–99, 2014.

A *legal order* is a *complex dynamic system*[2], in that the individual behaviors of its nodes (e.g., acts), coupled with local changes (e.g., modifications), produce side effects across the entire system (the legal order itself) in a nonlinear way over time. The propagation of the modification effects between the legal system documents can be represented with a *directed acyclic graph*[3], not with a simple sequence of nodes. Moreover modifications, as a set of temporal events, produce a network of relationships among the different acts that follow a non linear-time model[4], and it is difficult for citizens, as well as for legal experts, to navigate through an updated legal system, especially if retroactive modifications produce bifurcations in the legal order.

It is quite easy to monitor the lifecycle of a single act, especially in a given time, but to have a global vision of the entire legal order of relationships is quite a task. Moreover, normative citations usually create semantic references (e.g., reference to definitions or the introduction of exceptions) among fragments of legal sources, and this semantic is difficult to detect and properly represent.

This paper provides a theoretical model for measuring the *dynamic complexity* of the legal order using explicit modifications, where for *complexity*[5] we define the amount of information needed in order to fully describe a phenomenon, capturing all its properties. Because the modifications increase the information necessary for describing the lifecycle of a textual provision, and also of the legal norms connected, we provide a mathematical formula for calculating the side effects produced by a modificatory act in the legal order. To this end we define three main criteria: (i) quantitative modifications (e.g., any modification applied to other documents, articles, paragraphs, etc.); (ii) qualitative modifications (e.g., textual modification vs. temporal suspension); and (iii) textual unit affected by a modification (e.g., a fragment or a chapter, an annex, an entire document). The final parameter we call the *index of dynamic complexity*, considering that any modification in the legal system produces a knowledge-acquisition side effect that often spreads to all the other connected acts or to the implicitly associated secondary regulations. So much is this the case that a bad article is usually the most frequent part of the text amended and affected by modifications: the legislator is inclined to find an immediate new good textual formulation without considering side effects on other acts in the legal order.

---

[2] For a full definition of *complex dynamic system* and *non-linear* system see [7].

[3] For a complete definition of *directed acyclic graph* see the Reinhard D., Graph Theory, Graduate Texts in Mathematics, Vol. 173, Springer, 2010.

[4] The linear time is the most classical theory in physics, introduced by Newton. The linear time model regulates the current life of every individual (e.g. present, past, future). However the concept that time is tied to the reference system is the foundation of *special relativity* theory of Einstein. When we have a bifurcation of the linear timeline inside of a legal order, we have created two different perspectives of the same legal order, equally legal valid, where the point of view of the subject (e.g. judge) is fundamental for defining the correct reference systems (e.g. legal system).

[5] In this paper the term *complexity* is not linked to the linguistic complexity of the text, or the simplicity of the normative provision, but with the mathematical definition provided in the *complex theory*. However it is possible here to provide an intuitive definition applied in the social science: "Complexity theory is the study of complex, nonlinear, dynamic systems with feedback effects" [19].

Another phenomenon that strongly undermines the linearity of the legal system is the modification of a modification. Considering the lawmaking process, it is often simpler to modify a modificatory act than to modify the original act. This legal-drafting technique is forbidden in most legal traditions, and it is especially a problem in the civil law system, where the lawmaking workflow can be simplified by making it possible for a legislative assembly to discuss and accordingly amend a short act instead of a long one.

In this paper we also provide a visualization of the measurement of *dynamic complexity* using a mathematical model based on indicators. We have produced four different visualizations: (i) a bubbles diagram per year where the size of each bubble is proportional to the index of the year's *dynamic complexity*; (ii) a bubbles diagram per modificatory document; (iii) a timeline diagram for navigating the dynamic legal order over time; and (iv) a Sankey diagram for connecting each modificatory act with the modified acts in order to track and highlight the so-called "modification of a modification" phenomenon.

In order to apply the above-mentioned method, we needed to have a legal textual corpus annotated with modifications, making it possible to calculate the index of *dynamic complexity*. The Piedmont region has recently released all its annotated legislation database (called Arianna)[6] in open data form under the CC-by license. The corpus (at April 2014) contains 2,144 regional laws (a total of 18,244 articles for an average of 8.5 article per document); 408 documents are modificatory acts, with a total of 1,233 modifications applied to the Piedmont region's legal system from 1971 to 2014[7].

The paper is divided into three parts, illustrating (i) the theoretical model for creating the measurement indicators; (ii) the method for applying the model to the legal corpus; and (iii) the visualization model.

1   The Complexity of the Legal Order

The *legal order* is defined in Bulygin [1], Bobbio [2], and Guastini [8] as a sequence of legal system as it changes over time. On this definition, a legal order offers a diachronic[8] and dynamic view of the legislative system. The *legal system* is the set of the norms that are *valid* at a given time,[9] and so it offers a *static* vision of the legal order at a given time t. This theoretical model, sourced from the theory of law, shows

---

[6]   http://arianna.consiglioregionale.piemonte.it/

[7]   Statistic web portal of the project is visited for this paper at April 27, 2014:
http://sinatra.cirsfid.unibo.it:8080/exist/rest/db/piemonte_q ueries/stats.xql

[8]   *diachronic*: it is a qualification of a phenomenon that changes through time and it includes the dynamic characteristic. The opposite is *synchronic* that is a qualification of a phenomenon that is static and so frozen in a given time t.

[9]   In determining what set of norms is *valid* at time t, we look at the set of norms produced in accordance with the rules set out in the constitution for a norm to be effective at time t.

how difficult it is to achieve a correct view of an updated legislative system at a fixed instant t.

In this scenario it is quite difficult to manage the *dynamic complexity* of the legal order, considering the multiple modifications that affect the normative system as a whole. These modifications are effective at a given time t_i, producing a virtual new version of the target document. The modifications could overlap in the timeline (e.g., modifications may have different temporal intervals of efficacy), and when they are applied in different sequences, they may yield completely different results (e.g., deletion of paragraph 5 may yield different normative contents depending on whether it is done before or after inserting a new paragraph). This *dynamic complexity* is compounded by the large number of documents that usually make up the normative system; thus, for example, in Italy the database of the High Court of Cassation includes 1 million documents, and today nobody can list the law in force—such is the complexity of the Italian legal order, riddled with a welter of cross-references among modifications.

Another important observable fact can strongly undercut the certainty of the normative system: modifications applied to a future or past time (so-called future or retroactive modifications). The following figure shows all the versions of document A and how the acts are intertwined with one another by modifications, among other factors. Document M_1 modifies two documents: A_0 and Q_1. Q_1 modifies D_k that is later modified by M_j. Document M_j produces also two sets of modifications on A: one effective at t_j and the second effective at t_m (in the future). In meantime, document C_p retroactively modifies document M_j at $t_p > t_m$, thus affecting the entire previously produced versioning chain. This event indubitably impairs our ability to have a grasp of the legal system over time, creating a dynamic complexity whereby the system is different depending on the time at which the end-user looks at the collection of legal documents.

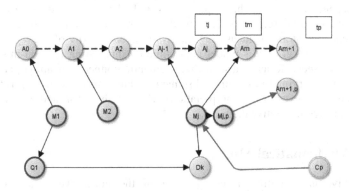

Fig. 1. Modification of a modification (C_p modifies M_j) and multiple modifications over time affecting the same document. M_j produces two sets of modifications: one at t_j and the other at t_m.

A third fact needs to be considered: modifications applied to the past (retroactive modifications at given t_p) that create a discontinuity at an earlier point in the timeline (an event that leads forking, for example in t_2). In this case the legal order needs to be bifurcated if it is to maintain both of the legal systems generated by the retroactive artefact event p. In figure 2 the event p provokes a forking of the temporal line in the t_2 producing two parallel timelines.

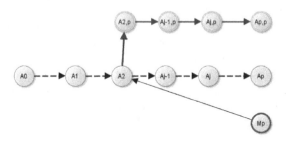

Fig. 2. Retroactive modification and forking of the timeline of the normative system

Finally, it is important to also mention incomputable reasons that make the legal order extremely complex to manage: (i) bad legislative drafting techniques may include implicit modifications (e.g., "All the norms incompatible with the present act are repealed"); (ii) incomplete and vague references (e.g., "The norms in the financial domain are suspended for the entire the fiscal year"); and (iii) general formulations in the text that make it impossible to precisely identify the target of the modifications (e.g., "the third article of the law is suspended").

Other semantic (e.g., linguistic) reasons or syntactic ones (e.g., the length of the document) could aggravate the complexity of the legal order. This is supported by other studies presenting similar findings in regard to complexity in the legal domain by looking at different characteristics of the legal text such as its structure, language, and interdependence [3]. In this paper, we would like to focus on measurable and objective parameters (e.g., explicit modifications over time) so as to build measurement indicators for measuring the *dynamic complexity* of the *legal order*. A potential improvement of this model is possible by integrating the non-modificatory normative references so as to also include semantic connections among acts. In the future, we will be trying to use minimal semantic qualifications (e.g., positive citation, negative citation, exception, interpretation) as a method for adding associations expressed through normative citations.

2   A Mathematical Model

Our objective is to define a measurement of the normative system over time (diachronic), in such a way that we can then design a measurement indicator of the complexity of the legal order. Complexity is measured with a formula dependent on modifications over time. In the worst case, modifications fork the system into different branches, sometimes even in such a way that they overlap (e.g., retroactive modification or annulment by the constitutional court). Let us begin by introducing some terminology [13]:

- Modifications M_{j-1} are applied to document D_{j-1}, yielding an updated document D_j where α is the content of the document, j is a point in time when the set of modifications is applied, and f is the function of the transformation that applies M_{j-1} to D_{j-1} so as to produce the new version of the document, namely, the updated document D_j:

$$f(D_{j-1}(\alpha), M_{j-1}) = D_j(\alpha) \tag{1}$$

- M_{j-1} is a vector with all the modifications. It is possible to associate with each type of modification a weight w in order to produce a formula for calculating the impact of the modification on D_j and so on the entire normative system:

$$M_{j-1}(m_1{*}w_1, m_2{*}w_2, m_3{*}w_3, m_4{*}w_4, ..., m_x{*}w_x) \tag{2}$$

- A *versioning chain* is a set of versioned text linked to the abstract concept of the legislative document D (e.g., Italian Act no. 256 of 2005 is legislative document D, in the FRBR[10] model is the *Work* [6]). *A legal document's versioning chain* is the set of all the documents virtually versioned:

$$versioning\ chain\ vc(D) = \sum D_{1,n} \qquad n = [0, \infty[\tag{3}$$

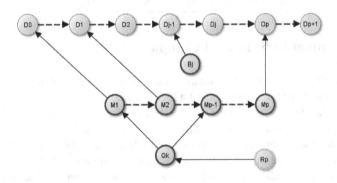

Fig. 3. Example of extended versioning chain

- The *extended versioning chain* of a document D is the set of all the versioning chains of D and the versioning chains of the modificatory documents. This indicator is particularly important in understanding the impact a document has in the legal order over time, including the whole constellation of modifier documents that are connected using the modificatory relationship. This means

---

[10] Functional Requirements for Bibliographic Records, FRBR, http://www.ifla.org/publications/functional-requirements-for-bibliographic-records

that all the modification of modifications are calculated in this parameter. For example, we consider the document D, it is modified by M and B, M is modified by Q that is modified by R.

The $evc(D)$ indicator includes all the *versioning chain* of D, B, M, R, Q considering that the D is influenced by all those documents).

extended versioning chain $evc(D) = vc(\text{D}) + (vc(M_j))$	(4)

- A *legal system* is the set of all the legal document D effective[11] in a given time t. It is defined as follows:

$LS(t) = D_t(\alpha) , D_t(\beta), D_t(\delta), \ldots D_t(\omega)$ where $\alpha, \beta, \delta, \ldots, \delta$ = legislative unit (document, article, etc.) $LS(t) = \{$a system of documents D that are *effective* at a given time $t\}$. Legislative system is the synchronic view of norms. t is a fixed time in a discrete representation.	(5)

- A *legal order* is defined as a sequence of legislative systems over time where time follows a discrete model:[12]

$LO = \{LS(t_1) , LS(t_2), LS(t_3), \ldots LS(t_j)\} \; j \in \text{N}$	(6)

3   Measurement Indicators of Complexity

On the basis of the abovementioned mathematical model, we have defined three measurement indicators of the complexity of a legal order.

The first indicator (*active impact indicator - AII*) is based on Formula 2. We have assigned the following weights to the modifications considering that the legal corpus coming from the Piedmont region was not enriched with the NormeInRete [13] [10] qualifications (e.g., integration, substitution, repeal, temporal modifications [12] [5]). For this reason the indicators number 6 and 7 of the following table were not applied.

---

[11] The concept of *effectiveness* depends the legal system that we are considering. In Italy the effectiveness, usually, starts after 15 days after the publication in the official gazette. The same rule is applied for the Regional Law.

[12] A discrete model of time is a mathematical model where the continuity of the timeline is simulated as a conjunction of discrete points in time using integers.

[13] NormeInRete is the legal XML standard approved by the Italian Government in 2001.

Table 1. Table of the weights assigned to different types of modifications

1.	Type of situation	Weight
2.	Each document modified	1
3.	Each modification of any fragment of the document	1
4.	Each modification of a modification	2*n. of reiteration
5.	Each modification of a citation	2
6.	Each retroactive modification	4
7.	Repeal of the entire document	$Z(D)^{14}$

This table could be extended and enriched with other type of modifications on the basis of the taxonomy of modifications we have previously presented in other papers [12][15] [16].

Now we can define the *active impact indicator* of document D ($z(D)$) as the sum of the product of m_i; the modification detected in the text; and w_i, the weight assigned on the basis of the abovementioned Table 1:

$$z(D) = \sum m_i * w_i \qquad i \in N \tag{7}$$

We can also calculate the *impact indicator* of the legal order at a given time y ($n(LO,y)$) as the sum of the z_j (D_j, t_j):

$$n(LO,y) = \sum z_j (D_j, t_j) \qquad j \in N \text{ and } t_j \le y \tag{8}$$

If we limit the y to one year, we can obtain a measurement of the *active impact indicator* for the given year and so we can evaluate the extent to which the legal order has been affected during that year.

Finally, we would also like to measure the *passive impact indicator*, based on how many modifications an act has received during its lifecycle, where pm_i designates the passive (i.e., inbound) modifications received:

$$p(D) = \sum pm_i * w_i \qquad i \in N \tag{9}$$

These indicators need a mathematical normalization process making it possible to compare the result values in different temporal periods as well as among different normative systems. Normalization can be achieved by several methods (e.g., max, max-min, standard deviation, etc.). We have decided to use the *max-min* method that returns values in the interval [0,1], so the $Z(D)$ is the normalized indicator:

---

[14] See the next paragraph for the $Z(D)$ normalized indicator definition.

$$Z(D) = \frac{z(D) - z(D)_{min}}{z(D)_{max} - z(D)_{min}} \tag{10}$$

4   A Method for Applying the Model

We are going to apply this model to the body of all the legislative acts the Piedmont region has recently released as open data[15]. Because at the time we conducted this study the Piedmont region had not released the entire NormeInRete[16] (now NIR [10]) XML collection in one bulk, but rather released one XML document for each HTML page, we created a crawler that would extract all the XML files for each of the legislative acts from the official portal. Every night the crawler scrapes the NIR XML files. Because the XML markup of the files released by the Piedmont region wound up being stratified over the time, the files were not homogenously annotated: they were also incomplete, and the modifications were not marked up in depth. The NIR standard evolved over the time with three major releases (res. 1.0, 1.1, 2.0), so it was applied in different manners during the last decade by the Piedmont, generating a mixed XML corpus. For this reason the first task was to harmonize the corpus, to repair the incompleteness and refine the errors, and secondly to apply the measuring formula.

The incompleteness affected, mostly, the relevant meta information: the unique identifier of the legal source (URN[17]) was present only in the more recent documents (starting from 2008), and where it was present, the date of the act was incomplete. For instance in the following example the date part of the URN is limited to the year (2008), when the URN grammar strongly recommends the full date canonical format (2008-01-14):

<urn>**urn:nir:regione.piemonte:legge:2008;1**</urn>

This incompleteness affected all the normative references in the XML documents, making it more difficult to detect the interconnection among the documents in the legal corpus network. The URNs of the same legal source have different formats in the references (<rif>), so detection of links was not easy. The following box provides an example of a link to the above mentioned legal source, where it is possible to see a discrepancy of URN:

---

[15]   http://www.dati.piemonte.it/catalogodati/dato/100646-arianna-leggi-regionali-storiche-e-vigenti-regolamenti-regionali.html

[16] http://www.digitpa.gov.it/sites/default/files/DigitPA_Linee_Guida_NIR_V_1.0_0.pdf

[17] URN means Uniform Resource Name; it is a persistent logical name of the resource that not depend on the physical location in the server. For NormeInRete the technical committee developed a special grammar called URN:LEX. See the http://datatracker.ietf.org/doc/draft-spinosa-urn-lex/?include_text=1

```
<rif xlink:href="urn:nir:regione.piemonte:legge:2008-01-14;1">legge regionale
14 gennaio 2008, n. 1</rif>
```

The second problem was the completeness of normative references in the Piedmont corpora. The tool used by the Piedmont for detecting the URN in the text is also tasked with building the URN, but sometimes the text was incomplete, and an adjustment was necessary so as to permit a correct and complete markup. In the following example, the URN is incomplete because the date is partial (giving us only the year 1995), but in some other parts there is the complete reference with the complete date, so the two links are interpreted by the program as two different navigation targets:

```
<rif xlink:href="urn:nir:regione.piemonte:legge:1995;93#art3-
com3">articolo 3, comma 3 della l.r. 93/1995</rif>
<rif xlink:href="urn:nir:regione.piemonte:legge:1978-03-15;13#art9">articolo 9
della legge regionale 15 marzo 1978, n. 13</rif>
```

The third problem is that all documents before 2008 had incomplete modification markups. The markup was complete only for structural and atomic changes (e.g., article, paragraph, letter) and not for partial textual amendments. The following example shows a partial modificatory provision that was not properly marked up with the tag <mod> (modification) around the text:

```
<articolo id="art2">
    <num>Art. 2.</num>
      <rubrica/>
      <comma id="art2-com1">
      <num>1.</num>
        <corpo> Al <rif xlink:href="urn:nir:regione.piemonte:legge:2000-01-
24;4#art7-com1">comma 1 dell'articolo 7 della legge regionale 24 gennaio 2000,
n. 4</rif>, sono soppresse le parole <<società miste o consorzi a prevalente
partecipazione pubblica>>.
        </corpo>
      </comma>
</articolo>
```

The method we have adopted to address these problems is to enrich and refine the original XML sources by:
- building the URN where it was not present in the markup, using the other tags present in the original document (<dataDoc> and <numDoc>). So we built the following new metadata:
  <urn norm="urn:nir:regione.piemonte:legge:1974-09-02;28"/>;
- refining the entire URN of the original sources including the complete date of the document detected by the tag <dataDoc norm="14012008">
  <urn norm="urn:nir:regione.piemonte:legge:2008-01-14;1"/>;
- refining the @href attribute in all the <rif> elements so as to complete date;

- refining the <mod> in the partial modifications whenever possible [5][11]; and
- refining the old documents' metadata block.

The enriched XML files, created using *xQuery* techniques, are stored in an *eXist*[18] [14] database and published on the portal[19] as a dataset under a CC-by license held by CIRSFID. In order to avoid introducing errors in the refinement process abovementioned, we foster the *eXist* database [14] features for implementing some checking mechanisms (e.g. univocal URN checking, navigation of all the links, data checking, etc.).

5   Calculating the Indicators

After polishing the dataset we applied several *xQuery* queries to the *eXist* database and we detected all the <mod> and <ref> elements so as to create a map of the modificatory acts and the modified acts. All the modifications were qualified following the previous table, and the corresponding weights were assigned. During the calculation, we have also compared the number of <mod> elements with the number of <ref> elements (target of the modification), and where discrepancies came up we used the greater of the two values (max criterion). In order to detect modifications of modifications, we assigned weight 2 to the <ref> included in the tag <virgolette>.

On the basis of that information, we created an *active impact indicator –Z(D)–*for each modificatory document and for each year.

The normalization process produced a table of values (see the annex) that we have represented in the regression graph below:

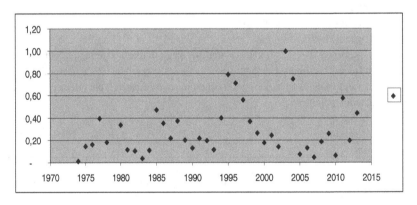

Fig. 4. Regression graph for detecting a phenomenon

The graph shows that dynamic complexity peaks in 2003, but that behaviour is otherwise generally regular: every ten years or so the dynamic complexity falls steeply, only to rise to a new peak. This data registers an interesting phenomenon that should

---

[18] eXist is a no-SQL database, native XML http://exist-db.org/exist
[19] http://sinatra.cirsfid.unibo.it/lod/piemonte/

be analysed with the experts, with additional information about the political activities (e.g. during the election periods 2000, 2005; 2010 the legislative activity were minimized), the sociological (e.g. immigration) and historical events (e.g. economic crisis), the natural disaster (e.g. flooding, 1977, 1994, 2000) and, last but not least, constitutional normative modifications that impacted on the regional regulation (e.g. constitutional law No. 3/2001 modified the competences between state and regions). The national modifications at the constitutional level, which happened in 2001, produced an huge amount of modifications in the Piedmont legal system during the years 2003 and 2004, and this is manifestly visible view in the graph.

6   Visualization Technique

Since the indicators are too technical and odd to afford a good grasp by end-users, we have tried out several libraries of document visualization tools[20]. We have used three libraries to create graphs using the *json* technology and so to provide the graphical tool with the necessary input data: (i) bubble graphs for presenting the *active impact indicator* −Z(D)−; (ii) Sankey diagrams for presenting the relationships between modifiers and modified acts; and (iii) timeline graphs.

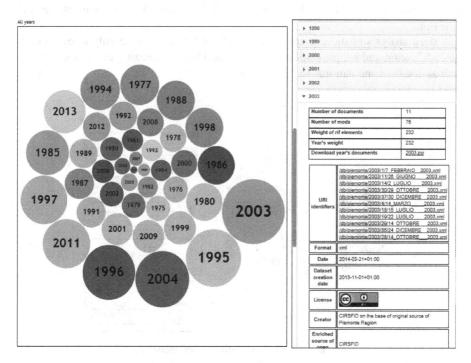

Fig. 5. Active Impact Indicator per year

---

[20] http://d3js.org/

This is the result of the graphic tool using the bubble graph, which in the window on the right visually renders the information on the left. This information makes it possible to navigate between the graph and the text, and to verify the correctness of the analysis using the XML NIR file. All the statistical data are visualized in order to permit a validation of the analysis. The dataset is also released with RDF metadata so as to enable machine-readable reuse.

The document with the higher *active impact indicator* is the regional Act No. 19, of 22 July 2003. It includes 26 modifications and 92 references of modifications to the regional Act No. 16/1999, that is the Code of the Mountain Law. Piedmont region is the second region in Italy for number of municipalities in the mountain area (51,8% of the region is mountain, 15,4% of the residents live in mountain land and there are 530 municipalities in the mountain area[21]), so this code is the most important act for the regulation of the country. The competences about the environment changed thanks to the constitution law No. 3/2001, so this high indicator is the consequence of the relevant change at the national level. A second example is the Act No. 65, of 1995 about the river natural reserves. Piedmont region includes the longest river of Italy (Po river), with a high risk of flooding[22]. In November 1994 Piedmont had the worst flooding disaster with a relevant number of victims and significant damages. The act No. 65/1995 was intended to modify all the regional regulation about the river natural reserves and so it modified 13 regional acts. This produced a strong impact in the legal order.

A deep analysis of these graphs could lead the legislative assembly to plan better the codification and simplification processes and also to managing exceptional events such as disasters and national legal changes.

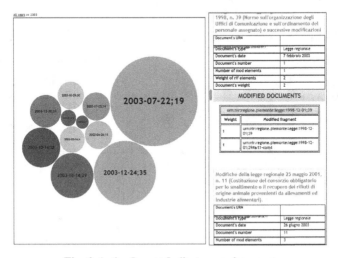

Fig. 6. Active Impact Indicator per document

---

[21] http://noi-italia2010.istat.it/index.php

[22] Piedmont is the region with the higher number of victims for flooding in Italy with 73 flooding events between 1950-2014.

The visualization of the relationships between the modifiers and the modified acts was simplified using a Sankey graph that can very intuitively render modifications of modifications. In the figure below, we can see that Act No. 18 of 6 June 2001 modifies an Act No. 4 of 2001, which in turn modifies a third document, Act No. 50 of 2000. This last document modifies yet another document issued in 1986 (see the figure n. 7). This graph permits the legal expert to detect immediately the modifications of modifications, to understand the origin of the changes and to favor also the correct legal interpretation of the norms.

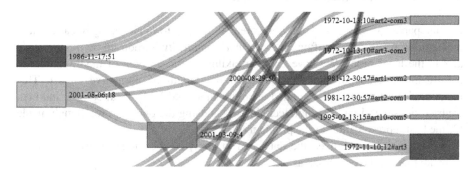

Fig. 7. Sankey diagram

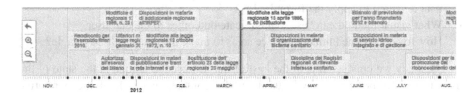

Fig. 8. Timeline

Finally, we also drew a timeline of the legal order so as to present the document collection in a dynamic manner. This visualization makes it possible to grasp the diachronic characteristic of the normative system over time and the connection with the modifier acts.

7   Validation

The system and the methodology described in this paper attracted the interest of the "Quality of Legislation" office of the Piedmont region legislative assembly. For this reason there is the intention to install this tool in the official web portal of the Arianna database for starting an evaluation phase with the cooperation of the legal civil servants and of the citizenry.

8   Related Work

Other researchers ([3][4][11][18]) have previously addressed this multifactor topic of the complexity of the legal system, but no one who has analysed the problem of a legal order's dynamic complexity over time has taken a diachronic approach based on modifications and references.

9   Conclusion and Future Work

In this paper we have set out a theoretical model for calculating measurement indicators capable of quantifying the *dynamic complexity* introduced by modificatory acts in the legal order. We have done so taking special account of modifications of modifications, types of modifications, and the length of normative chains. This method of measurement was applied using an annotated XML legal corpus, and then we tried out several visualization techniques, choosing the ones that we thought made for the best representation of the problem.

We have also normalized the data, and in applying the model, some difficulties came up that can be expressed as follows:

- How to find the necessary information in the XML annotated legal corpus?
- How to refine the original information so as to obtain an objective source for the data to which to apply the formulas?
- What visual method can best communicate outcomes to the end-user?
- How to analyze the human interface so as to make the data easily navigable?

The study we conducted is to the best of our knowledge unique in the state of the art, since other projects have focused on the complex network of the normative system [4][17]. The paper demonstrates that legal order is a *complex dynamic system* and it is possible to calculate the corresponding complexity parameters fostering the related theory. In the future, we intend to develop the timeline graphic for each document, while also bringing out the temporal level of the impact indicators by using a dynamic parameter for time. Secondly, we would like to design a network graph illustrating all the maps of the relationships detected using a *directional acyclic graph*.

Acknowledgements. This study was awarded second place in the visual category at the Piedmont Visual Contest. We would like to thank the Open Data Piedmont team for cooperating with our research group and providing us with the dataset.

References

[1] Alchourrón, C., Bulygin, E.: Normative Systems. Pringer-Verlag, Wien (1971)
[2] Bobbio, N.: Teoria generale del diritto, Giappichelli, Turin (1993)

[3] Bommarito II, M.J., Katz, D.M.: A mathematical approach to the study of the united states code. Physica A: Statistical Mechanics and its Applications 389(19), 4195–4200 (2010)

[4] Boulet, R., Mazzega, P., Bourcier, D.: Network Analysis of the French Environmental Code (2012)

[5] Brighi, R., Lesmo, L., Mazzei, A., Palmirani, M., Radicioni, D.: Towards Semantic Interpretation of Legal Modifications through Deep Syntactic Analysis. In: JURIX 2008, pp. 202–206 (2008)

[6] de Oliveira Lima, J.A., Palmirani, M., Vitali, F.: Moving in the Time: An Ontology for Identifying Legal Resources. Computable Models of the Law, Languages, Dialogues, Games, Ontologies, pp. 71-85 (2008)

[7] Fuchs, A.: Nonlinear Dynamics in Complex Systems. Springer (2013)

[8] Guastini, R.: Teoria e Dogmatica delle fonti. Giuffrè, Milano (1998)

[9] Levy, D.L.: Applications and Limitations of Complexity Theory in Organization Theory and Strategy. In: Rabin, J., Miller, G.J., Bartley Hildreth, W. (eds.) Handbook of Strategic Management, 2nd edn., Marcel Dekker, New York (2000)

[10] Lupo, C., Vitali, F., Francesconi, E., Palmirani, M., Winkels, R., de Maat, E., Boer, A., Mascellani, P.: General xml format(s) for legal sources - Estrella European Project IST-2004-027655. Deliverable 3.1, Faculty of Law, University of Amsterdam, Amsterdam, The Netherlands (2007)

[11] Mazzei, A., Radicioni, D., Brighi, R.: NLP-based extraction of modificatory provisions semantics. In: ICAIL 2009, pp. 50–57. ACM (2009)

[12] Palmirani, M., Brighi, R.: Model Regularity of Legal Language in Active Modifications. In: Casanovas, P., Pagallo, U., Sartor, G., Ajani, G. (eds.) AICOL-II/JURIX 2009. LNCS, vol. 6237, pp. 54–73. Springer, Heidelberg (2010)

[13] Palmirani, M., Brighi, R.: Time Model for Managing the Dynamic of Normative System. In: Wimmer, M.A., Scholl, H.J., Grönlund, Å., Andersen, K.V. (eds.) EGOV 2006. LNCS, vol. 4084, pp. 207–218. Springer, Heidelberg (2006)

[14] Palmirani, M., Cervone, L.: Legal Change Management with a Native XML Repository. In: Governatori, G. (ed.) The Twenty-Second Annual Conference on Legal Knowledge and Information Systems, JURIX 2009, 16-18 December, pp. 146–156. ISO Press, Amsterdam (2009)

[15] Palmirani, M.: Legislative Change Management with Akoma-Ntoso. Legislative XML for the Semantic Web, Law, Governance and Technology 4, 101–130 (2011)

[16] Palmirani, M.: Time Model in Normative Information Systems. In: The Proceeding of Workshop The Role of the Legal Knowledge in the eGovernment, ICAIL 2005, Wolf, Tilburg (2005)

[17] Winkels, R., de Ruyter, J.: Survival of the fittest: Network analysis of dutch supreme court cases. In: Palmirani, M., Pagallo, U., Casanovas, P., Sartor, G. (eds.) AICOL-III 2011. LNCS, vol. 7639, pp. 106–115. Springer, Heidelberg (2012)

[18] Winkels, R.G.F., Boer, A., Plantevin, I.: Creating Context Networks in Dutch Legislation. In: The Twenty-Sixth International Conference on Legal Knowledge and Information Systems, JURIX 2013. Frontiers in Artificial Intelligence and Applications, vol. 259, pp. 155–164. IOS Press, Amsterdam (2013)

Appendix

Modificatory acts with the corresponding value for building the *AII*.

YEAR	#doc	#mod	#ref	#All	normalization	normalization	mod/doc
1974	2	2	3	3	0,01	0	1
1975	13	16	34	34	0,15	0,14	1,23
1976	9	16	37	37	0,16	0,15	1,78
1977	12	53	91	91	0,39	0,38	4,42
1978	11	26	42	42	0,18	0,17	2,36
1980	19	78	57	78	0,34	0,33	4,11
1981	13	5	27	27	0,12	0,10	0,38
1982	6	14	24	24	0,10	0,09	2,33
1983	4	4	9	9	0,04	0,03	1,00
1984	11	11	25	25	0,11	0,10	1,00
1985	15	68	110	110	0,47	0,47	4,53
1986	22	42	82	82	0,35	0,34	1,91
1987	10	19	51	51	0,22	0,21	1,90
1988	11	55	87	87	0,38	0,37	5,00
1989	15	43	47	47	0,20	0,19	2,87
1990	6	7	30	30	0,13	0,12	1,17
1991	16	51	41	51	0,22	0,21	3,19
1992	9	20	46	46	0,20	0,19	2,22
1993	9	12	26	26	0,11	0,10	1,33
1994	14	43	93	93	0,40	0,39	3,07
1995	14	63	183	183	0,79	0,79	4,50
1996	25	98	165	165	0,71	0,71	3,92
1997	18	63	130	130	0,56	0,55	3,50
1998	13	21	86	86	0,37	0,36	1,62
1999	5	39	62	62	0,27	0,26	7,80
2000	11	21	41	41	0,18	0,17	1,91
2001	7	13	57	57	0,25	0,24	1,86
2002	5	14	33	33	0,14	0,13	2,80
2003	11	76	232	232	1,00	1,00	6,91
2004	11	62	174	174	0,75	0,75	5,64
2005	3	7	17	17	0,07	0,06	2,33
2006	6	12	30	30	0,13	0,12	2,00
2007	2	5	11	11	0,05	0,03	2,50
2008	10	14	43	43	0,19	0,17	1,40
2009	4	10	60	60	0,26	0,25	2,50

2010	4	4	15	15	0,06	0,05	1,00
2011	13	75	134	134	0,58	· 0,57	5,77
2012	4	6	46	46	0,20	0,19	1,50
2013	7	28	103	103	0,44	0,44	4,00

Time, Trust and Normative Change. On Certain Sources of Complexity in Judicial Decision-Making

Michał Araszkiewicz

Jagiellonian University, Faculty of Law and Administration, Department of Legal Theory,
Bracka 12, 31-005 Kraków, Poland
michal.araszkiewicz@uj.edu.pl

Abstract. The aim of this paper is to outline a structure of legal knowledge that is involved in resolution of complex legal cases comprising intertemporal issues and constitutional problems. Although the topics of dynamics of legal systems are already well-elaborated in the AI and Law literature, the problem of constitutional admissibility of certain types of changes to the legal systems remains an underexplored issue. The model developed in this paper is designed to fill in this gap. The meta-information concerning admissibility of certain changes to legal systems (with regard to relevant constitutional principles) should become a standard element of any well-developed database of statutory legal knowledge.

Keywords: case-based reasoning, constitutional review, principles, rule-based reasoning, time trust.

1 Introduction

As a matter of course, time is an important factor as regards judicial application of rules. Statutory provisions may be modified or repealed by the legislator and new provisions may enter into force. A special type of legal rules, that is, intertemporal rules are a tool used to resolve potential and actual conflicts between (older and newer) legal rules. If explicit intertemporal rules are absent, the collisions may still be dealt with by means of *lex posterior* argumentative mechanism (the so called implicit derogation). Intertemporal rules may themselves be modified and repealed and there may be conflicts between them, too. This leads to creation of a multi-level framework of rules. The existence of such multi-level framework is a common feature of contemporary legal systems. Provided that the intertemporal legal rules are encompassed in well-drafted and clearly defined provisions, the reasoning with those collision rules and meta-rules does not lead to particularly complex questions of law. The more problematic situations arise, however, when certain intertemporal rules prescribe for retroactivity of object level rules, thereby allowing their application to the states of affairs which obtained before these object rules came into force. Although the application of retroactivity technique is not entirely forbidden in contemporary legal systems (as a matter of fact it is used fairly often as regards procedural matters, including the context of tax proceedings), it should be used rather

P. Casanovas et al. (Eds.): AICOL IV/V 2013, LNAI 8929, pp. 100–114, 2014.
© Springer-Verlag Berlin Heidelberg 2014

carefully due to the risk of potential conflict with important constitutional principles, such as protection of acquired rights and legal certainty.

The context outlined above poses a challenge to AI and Law research on models of legislation and in particular to the dynamics of legal systems. In our opinion, such models should not only be able to indicate which rules are to be applied to certain states of affairs (taking the temporal and dynamic dimensions of legal systems into consideration), but they should also signalize potential constitutional problems stemming, *inter alia*, from the identified retroactivity of certain legal rules. An important qualification is that imposing a task of resolution of these constitutional problems on legal knowledge systems would perhaps be too ambitious in the present state of research. However, there are no fatal obstacles that would preclude the developers of the systems to include the signalization function in their work.

The investigations are based on an actual Polish legal case concerning the so called right of perpetual usufruct. The Section 2 of the paper is devoted to detailed description of this case as well as to the explanation of the legal provisions that are applicable to it. In Section 3 the structure of legal rules applicable to the problem is outlined. Section 4 deals with the constitutional problem that was identified in connection with the *prima facie* answer stemming from the analysis of legal rules alone. The focus is on the concept of trust of the citizen who may rightly believe that the state vested him certain rights that should not be taken away by means of (especially retroactive) normative change. Section 5 discusses selected topics from the related work. Section 6 concludes.

2   The Case of Perpetual Usufruct Annual Payment

Perpetual usufruct is a kind of real property right present in the Polish legal system. It is possible to establish it on real property owned by the State or by municipalities. The economic justification of perpetual usufruct is as follows. Neither the State nor the municipalities are interested in transferring real property to private parties, because this type of property is a very convenient source of income (especially important one for the budgets of municipalities). On the other hand, private parties are interested in investing in real property not owned by them only if it is warranted that they will be able to control it for a sufficiently long period of time in order to obtain revenue from the investment.

The right of perpetual usufruct satisfies both criteria. The right is typically established by means of a contract concluded between a municipality and a private party (referred to as perpetual usufructuary) for the period of 99 years. The scope of rights of the perpetual usufructuary is very similar to the one that is assigned to owner, with two important qualifications. First, the perpetual usufructuary should act in accordance with the contract concluded with the municipality (for instance, he or she may be obligated to construct and maintain certain objects on the land). Second, the perpetual usufructuary is obligated to pay certain fees to the municipality, including the annual fee.

The amount of the annual fee may be updated by both parties. The municipality is obviously interested in increasing the amount of the fee. Typically, the increase of this amount stems from certain economic indicators. The perpetual usufructuary is interested in decreasing this amount. This may take place if he or she incurred expenditures that lead to increase of value of the real property. In such case the amount of the annual fee is decreased in proportion to the amount of the incurred expenditures.

Let us now present a case study concerning the application of the abovementioned rules in a setting that became complex due to legislative changes. The perpetual usufructuary (hereafter referred to under a fictitious name of Mr. Kowalski) filed a motion for update of the annual fee in December 2009, where the update was expected to become effective form the 1st January 2010. Before, the amount of this fee had been updated on the 1st January 2009. He demanded the fee be decreased due to the expenditures incurred by him in 2009. In the course of the proceedings before the municipal authorities, the applicable provisions were changed by means of the Act of 28 July 2011 amending the Real Estate Management Act (REMA)[1]. The amending law contained the following intertemporal provision:

[INTERTEMPORAL PROVISION] In cases initiated and not completed before the entry into force of this Act, concerning the update of the fees for perpetual usufruct, the provisions of the Real Estate Management Act, as amended by this Act, shall be applicable.

The amending act modified the crucial Article 77.1 (the first sentence) of the REMA. Here below we present the former and the new version of this provision:

[REMA 77 FORMER] The annual fee for perpetual usufruct of the land may be updated no more frequently than once a year, if the value of the property changes.

[REMA 77 NEW] The annual fee for perpetual usufruct of the land may be updated no more frequently than once every three years, if the value of the property changes.

Hence, the legislative modification concerned the minimal amount of time interval between the updates of the amount of the annual fee. This time interval has been changed from one year to three years. Interestingly, this change was caused by the lobby of the perpetual usufructuaries, who were economically pressed by frequent updates of the annual fees by the municipal authorities. However, the reform precluded also the perpetual usufructuaries from demanding the update more frequently than one time every three years.

The new regulation came into force on 8th October 2011 in the course of the proceedings in Mr. Kowalski's case. Taking into account that the amount of the

---

[1] The REMA place of publication is the Journal of Laws 2010.102.651 (consolidated text as amended). The amending law's is the Journal of Laws 2011.187.1110.

annual fee had been updated in 2008, the municipal authorities refused to update the fee again on the basis of the motion filed in 2009 (the acceptance of such motion would cause the effective decrease of the annual fee from 2010 on). The authorities did not question any statements of facts, but they contended that due to the legislative change and the content of the INTERTEMPORAL PROVISION, Mr. Kowalski's case should be decided on the basis of the REMA 77 NEW. Due to the obligatory interval of three years, the motion filed by Mr. Kowalski could be effective from the beginning of 2012 on, hence he would be obligated to pay the high amount of the annual fee for the years 2009-2011. The perpetual usufructuary took recourse to the court.

Before the court, he argued that the decision of the municipal authorities violated his constitutional rights stemming from the principle of democratic state governed by the rule of law, especially the principle of protection of trust of citizens in legal stability and the principle of protection of acquired rights. Moreover, he claimed that the principle of equal protection of pecuniary interests was violated. The court did not feel competent to assess this complex issue on its own and filed a motion for preliminary judgment to the Polish Constitutional Tribunal.

3   The Rule-Based Framework for the Case-Study

Before entering into the discussion of constitutional issues involved in the case let us begin with a contention that on the basic level, the case comprises quite basic reasoning patterns using legal rules. This section is devoted to an outline of a semi-formal framework that captures well the problems concerning the application of rules in this cases. The elaboration presented here is comparable to standard accounts present in the literature (see Section 5 for the discussion).

We adopt a perspective on representation of legal rules which is more fine-grained than the most foundational account of rules as normative conditionals [1] :

$$\text{IF } A_1, A_2, A_3, \ldots, A_n \text{ THEN } B,$$

because it is our intention to analyze certain possible types of conflicts between rules and the role of meta-level intertemporal rules. In consequence, we adopt the following definition of a legal rule.

Definition 1. Legal Rule. A legal rule is a tuple <A, C, R, LINK, S, F, E, T>, where:
1) A is the Addressee of the rule, where $A \in \mathbf{A}$, the set of Addresses of any legal rule in question. The Addressee of the legal rule may be indicated by means of an indefinite expression (for instance, "who") or by means of a description (for instance, "municipal authority").
2) C is the set is the (possibly empty) set of Conditions of application of a legal rule in question to a case. C is a subset of **C**, the set of all conditions of application of legal rules in the legal system in question. The structure of rule

condition should be given in the form of features ascribable to individuals (in the language of logic, as predicates assigned to variables).

3) R is the set of legal Results produced by the rule in question. R is a subset of **R**, the set of all types of legal results known in legal system in question. As for the types of legal results, we adopt a standard distinction, we adopt a standard distinction between prescriptive rules (accounting for a certain type of behavior as obligatory or permissible) and constitutive rules (or counts-as rules; assigning certain statuses to persons, objects and states of affairs).[2]

4) LINK is a relation between the Conditions and Results of a given rule. Each LINK belongs to the set **LINK**: the set of all possible relations between Conditions and Results of any legal rule. In consequence, this relation does not have to be identified with material or defeasible implication; bi-implications and certain other types of relations are to be considered.[3]

5) S is the source parameter of the rule in question. The set of all sources **S** is based on the set of sources of binding law in a given jurisdiction. For instance, the set S based on the Polish Constitution would encompass the following elements: <Constitution, Ratified International Treaty, Statute, Regulation, Local Act>.

6) F is the force parameter[4] of a rule in question. It is the time interval from the date of entry of a legal rule into force to the date of its formal derogation. In case of rules in force, this parameter is monadic: only the date of entry into force is indicated. If the version of a given rule is changed, the force parameter encompasses all dates of entry into force and derogations of particular versions of a rule.

7) E is the parameter of efficacy of a rule in question. Efficacy is understood here as the time interval designating states of affairs to which the rule in question may be (in principle) applied. Note that the F and E do not have to indicate the same intervals (although in standard situation they should). A legal rule may be efficacious as regards states of affairs that took place before entering of a rule into force. Also, in certain settings, a rule may still be used to assess certain states of affairs even if it is formally derogated. The Efficacy parameter encompasses data concerning the efficacy of particular versions of a rule.

8) T is the parameter of territorial range of applicability of a rule in question. In typical situation it will cover the territory of a given state, but as local regulations are concerned, only the territory of a given province or municipality will be indicated by this parameter.

---

[2] It should be noted, however, that this distinction is criticized in recent literature, cf. [2]. As for the distinction between three categories A, C and R it is introduced for the sake of transparence of the representation of legal rules. If standard predicate logic is adopted for representation of legal rules, then each of these categories is ultimately represented by means of atomic formulas. Also, the distinction between these three categories is useful as regards practical needs of legislative technique. For instance it is more convenient

[3] See [3] for the discussion of different types of this relation.

[4] We use the term „force" instead of „validity" for a reason. The concept of legal validity is ambiguous and theoretically controversial. Cf. [4].

The definition described above enables us to discuss reasoning with a fine-grained approach. It also encompasses not only the data about the content of the rule (parameters 1-4) but also certain metadata important for construction of argumentative patterns (parameters 5-8).

Let us present an example of application of this framework to representation of a legal rule (REMA 77). First we represent its FORMER version, that is, before the reform from the year 2011:

A: perpetual usufructuary, municipal authority

C: [at_least_1_year_from_the_previous_update_of_annual_fee]

R: [update_of_annual_fee_possible]

LINK: similar to bi-implication

S: Statute

F: 1 January 1998 – present

E: 1 January 1998 – present

T: Poland.

Let us now present a version of this rule stemming from the reform. Let us note that some information will be replaced and some will be added to the content of the rule.

A: perpetual usufructuary, municipal authority

C: [at_least_3_years_from_the_previous_update_of_annual_fee]

R: [update_of_annual_fee_possible]

LINK: similar to bi-implication

S: Statute

F:

    1)   (FORMER) 1 January 1998 – initiation of NEW force interval

    2)   (NEW) 8 October 2011 – present

E:

    1)   (FORMER) 1 January 1998 – initiation of NEW efficacy interval

    2)   (NEW)[All_cases_initiated_before_8Oct2011_and_not_completed_by_
       8Oct2011] OR 8 October 2011 - present

T: Poland.

The foregoing presentation is based on the following assumptions that are in accordance with the linguistic conventions accepted in legal communication community.

1. Modifications of legal rules do not change their identity. The rule REMA 77 remains one and the same object although its parameters change. This assumption is in accordance with the very concept of modification of a rule: it is an object (a rule) which is modified (its features are changed).

2. Although a certain version of a rule may be formally eliminated from the legal system by means of derogation, it still may be efficacious and used to assess certain factual situations from the past. Vice versa, new versions of rules may be applied retroactively. The latter phenomenon is made visible through the complex representation of efficacy interval of REMA 77 in the new version.

As rules are applied to reality, let us now define a factual situation.

Definition 2. Factual Situation. A Factual Situation is a set P of persons (p_1, p_2, ..., p_n), a set O of objects (o_1, o_2, ..., o_n), a set A of actions (a_1, a_2, ..., a_n) and a set PRED n-ary predicates defined over sets of persons, objects and actions that represents a given legal case. Moreover, the Time and Territory parameters are ascribable of legal cases. Not all types of entities have to be present in every case.

For instance, a partial description of Mr. Kowalski's case would encompass the following set:

1) Persons: Mr. Kowalski; predicate: [perpetual_usufructuary]
2) Objects: -.
3) Actions: update_of_annual_fee; predicate:
   [1_year_from_the_previous_update_of_annual_fee].
   Filed a motion [before_8_October_2011_not
   completed_before_8_October_2011].

As we do not assume here any kind of deductive character of rules, the rules bring their results about by means of application of argument schemes. The basic account of a rule-based argument scheme is as follows.

Definition 3. Rule-Based Argument Scheme. A Rule Based Argument Scheme is a reasoning pattern comprising the following sentences.

Premise 1. There is a Legal Rule <A, C, R, LINK, S, F, E, T>.

Premise 2. There is a Factual Situation <P, O, A, PRED, Time, Territory>.

Premise 3. There is a Subsumption Relation[5] between the Factual Situation and the Legal Rule, that is, elements of the Factual Situation are qualified either is elements of sets given by the description of a Legal Rule.

Conclusion. The legal result R as prescribed by the Legal Rule should follow.

Definition 4. Subsumption Relation. There is a Subsumption Relation between the ranges of predicates P_m and P_n if and only if $P_m \subset P_n$.

The Subsumption Relation is, therefore, defined by means of set-theoretical relation of inclusion.[6] It should not be conflated with applicability of rules, for it covers also different types of conceptual relations that stem from constitutive or "counts-as" rules. Let us also note that the existence of the Subsumption Relation does not lead automatically to the acceptance of conclusions that follow from Rule-Based Argument Schemes.

The account of reasoning-with-rules patterns as argument schemes enables us to construct legal rules on the basis of legal provisions or even its parts (like in case of REMA 77 rule) without any necessity to reconstruct "complete" legal rules from different parts of statutory text. The phenomenon of defeasibility of such accounted legal rules is captured by the possibility of asking critical questions to the argument based on rules.

---

[5] The Subsumption Relation between the (ranges of) two predicates may be accounted for by means of set-theoretical inclusion.

[6] See [5] for the detailed discussion of different types of set-theoretical relations between legal predicates.

As regards the application of the Rule-Based Argument Scheme, it is straightforward. Obviously, his case has been initiated before the 8th of October and not completed by this date. Due to this fact, it should be decided on the basis of REMA 77 NEW legal rule, according to which the minimal interval between updates of the amount of annual fee for perpetual usufruct is 3 years. As a consequence of this, in the factual situation given by Mr. Kowalski's case, and contrary to his interest, the update of annual fee should be declared inadmissible before the beginning of 2012.

The rule-based framework presented above is able to represent different problems concerning reasoning with legal rules in a fine-grained manner and consistently with the actual argumentative practice of lawyers. Moreover, the framework is also applicable for representing the process of amending and repealing of statutory provisions. It should be noted that the process of legislation is itself a rule-based process regulated by law. In the example discussed above it was simply asserted that the content (the C parameter) and the efficacy of REMA 77 was changed as a result of amendment of the statute. The rule-based framework presented here is able to represent this process through the introduction of the concept of meta-rules.

Definition 5. Meta-Rule. A Legal Rule is a Meta-Rule if an only if at least one of the legal Results prescribed by it concerns:

1) addition, or
2) modification, or
3) repeal

of any parameter <A, C, R, LINK, S, F, E, T> of any legal rule.

In this contribution it is assumed that Meta-Rule should explicitly refer to other rules. In consequence, intertemporal rules become paradigmatic examples of Meta-Rules.[7]

The change of structure of REMA 77 rule was a result of application of two Meta-Rules: the modifying rule changing the time interval between the updates from one to three years and the rule reconstructed from the INTERTEMPORAL PROVISION, modifying the Efficacy parameter of REMA 77.

---

[7] However, the problem of influence of one rules on another ones is much broader than presented in Definition 5. For instance, in case of implicit derogation stemming from lex posterior argumentation it is necessary to analyze the semantic content of rules is order to identify any modifications in the legal system in question. These issues are beyond the scope of the present contribution. Also, the problem of hierarchy of legal norms leads to serious complications as regards the discussion of Meta-Rules. In particular, the role of constitutional norms is apparently unclear in this context: do they automatically restrict the scope of application of lower-level rules? The answer to this question depends on the institutional setting concerning resolution of constitutional conflicts in the jurisdiction in question. In Polish legal system the common courts are not allowed to decide such questions on their own, therefore the constitutional layer in the present model is accounted for in separation from the basic multi-level rule-based framework. I am grateful to one of the anonymous reviewers for indicating these problems, which are worth further exploration.

The degree of complexity involved in the structures of legal knowledge described above is significant, although not fatal from the perspective of formal modeling and computational features. However, it would not be satisfactory if a legal knowledge base system simply yielded a legal answer to Mr. Kowalski's case that he is not entitled to the update of the amount of the annual fee for perpetual usufruct. This is because for any lawyer, the presence of important constitutional issues in this case is evident. The next section is devoted to the outline of these issues and for a proposal of their operationalization.

4   Enter Constitutional Problems

It was already pointed out that Mr. Kowalski satisfied all formal and substantial conditions to obtain the update (decrease) of the perpetual usufruct annual fee. However, in the course of the proceedings before the municipal authorities the normative change took place to the detriment of this interest. A question arises, whether the change of REMA 77 rule is acceptable with respect to constitutional standards.

The following exposition has preliminary character and it points out certain problems that should became the subject of interest of AI and Law community rather than formulates more concrete suggestions concerning their elaboration in legal knowledge based systems. However, as it is apparent from the discussion below, it would be difficult to assume that there is a predetermined right answer to questions of this sort; in consequence, the developed legal knowledge systems should focus on signalizing potential constitutional problems (for instance, in legislation support systems) and outline their structure rather than indicate potential anticipated solutions. The latter task should be the object of research in more distant future due to too high level of complexity.

The Polish Constitutional Tribunal (the PCT) is the polish constitutional court authorized to perform judicial review: it may invalidate statutory provisions on the basis of their incompatibility with the Polish Constitution. The PCT is also responsible for development of the Polish constitutional jurisprudence and its decisions form an important part of the landscape of the constitutional law, although they do not have strictly binding character. In particular, the PCT interprets and develops important constitutional principles, many of which are drawn from the very general clause contained in the Art. 2 of the Polish Constitution[8]:

The Republic of Poland is a democratic state of law, realizing the principles of social justice [the State of Law Principle, hereafter the SLP].

Among the sub-principles that were formulated by the PCT as "normatively entailed" by the SLP, the principle of protection of citizens' trust in the state and its

---

[8] The Polish Constitution of April 2nd, 1997 was published in the Journal of Laws 1997.78.483 (as amended).

laws (the Trust Principle). The Trust Principle itself implies (according to the PCT) a number of more specific principles, among which the following should be listed:

1) the principle of legal certainty,
2) the principle of protection of acquired rights and on-going interests,
3) the prohibition of retroactivity of law (see the Judgment of the PCT of April 13, 1999 K 36/98 and the Judgment of the PCT of April 10, 2006, SK 30/04).

Hence, each of legal rules introduced to the Polish legal system can be scrutinized with respect to the Principle of Trust and more specific principles that are, according to the PCT, entailed by it. As constitutional principles they should be accounted as goal-norms rather than action norms [6] or as optimization commands rather than definitive commands [7, 8]. The circumstances of any case should be assessed with respect to the criteria that are implied by the Principle of Trust and the procedure of weighing should be performed in order to conclude whether the regulation in question violates the constitutional principles indicated above.

In order to operationalize the Principle of Trust, the introduction of more specific criteria of assessment of cases in the light of this principle is needed. Let us present three instructive passages from the jurisprudence of the PCT that offer very useful clarification of the subprinciples of the Principle of Trust:

[Legal Certainty] *"(…) [T]he principle of trust in the state and its laws is based on the requirement of legal certainty, that is, the features of the law that provide legal security for individuals, allowing them to decide on their actions on the basis of full knowledge of the conditions of operation of the state and the legal consequences that may be entailed by the individual"* (the Judgment of the PCT of January 20, 2011, Kp 6/09).

[The Protection of Acquired Rights] *"(…) [T]he principle of the protection of acquired rights, provides for protection of both public and private rights, acquired either through a decision issued by a state authority or on the basis of law alone, as well as the protection of the so-called maximally developed expectations of rights, that is, the situation in which all essential requirements of being vested with a right are fulfilled by an individual"* (the Judgment of the PCT of March 30, 2005, K 19/02).

[The Relative Prohibition of Retroactivity] *"(…) [I]t is unacceptable to enact retroactive norms, if the entities, to which these standards apply, could not reasonably have foreseen this kind of legislative decision and if there are no extraordinary circumstances or constitutionally protected values that could justify such decision"* (the Judgment of the PCT of February 27, 2002, K 47/01).

In consequence, the Principle of Trust is in fact a bundle of sub-principles that possess quite complicated structure themselves and that are interconnected with each other. Their relative openness makes it possible for the PCT to apply them to different factual situations (assessed legal regulations). On the other hand, it is not possible to indicate any concrete criteria that would serve as a set of sufficient or necessary conditions or fulfillment or violation of these principles. Hence, their application to any regulation is ultimately grounded in the result of the process of balancing of values [6, 7, 8].

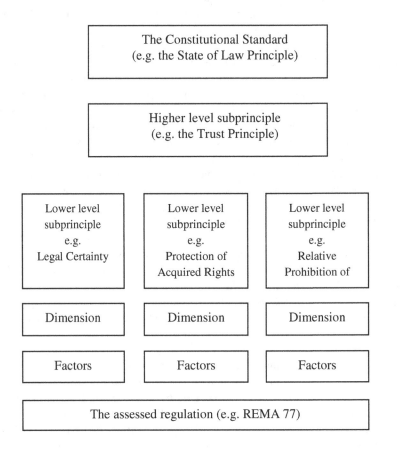

Fig. 1. The multilayered structure of knowledge involved in assessment of constitutionality of regulation

The case-based reasoning techniques such as the use of dimensions and factors [9, 10, 11] to characterize prior cases and then to analogize them to the case at hand can be potentially fruitfully used for the operationalization of reasoning with the sub-principles defined above. Two qualifications are in place here, however. First, due to the fact of lack of presence of formal precedential constraint in the jurisprudence of the PCT, the role of analogous reasoning based on factors will be lesser in comparison to their significance in common law legal cultures. The PCT will refer to its previous decisions for the sake of increasing persuasive power of its argumentation, but it will rather focus on the structure of the case at hand. Second, due to the fact that the PCT will assess the regulation in question against concrete constitutional standards (that is, against concrete provisions of the Constitution), as a result we will obtain a multilayered argumentative structure encompassing a number of intermediate steps between the constitutional standard and the regulation in question, assessed against the background of the circumstances of the analyzed case. This multilayered structure may be visualized on the following figure.

It is not possible to comment in detail on the possible problems resulting from reasoning with the structures outlined in the figure here above. Briefly, it would encompass the following stages: 1) identifying the (apparently) applicable Constitutional Standard together with the list of its subprinciples; 2) identifying factors abstracted from the factual situation that would support certain decision (such as the reasonable expectation of Mr. Kowalski that he would be entitled to the update of the annual fee); 3) developing dimensions in order to systematize different factors and to show their relative strength; 4) arguing that reaching certain point on a dimension counts as an argument for infringement of certain lower-level subprinciple; 5) arguing that the violations of lower level subprinciples are of such character that they lead to the conclusion that the higher level subprinciple has been violated; 6) arguing that there are (no) overriding constitutional arguments to the contrary, in order to conclude that the Constitutional Standard in question has (not) been violated. Obviously, this analysis cannot lead to any definitive conclusions before the actual PCT judgment is not issued. The degree of complexity involved in this domain of reasoning makes it implausible to assume that a legal knowledge base could accurately foresee the actual decision of the constitutional court.

5   Discussion and Related Work

The framework presented in this paper is a contribution to the theory of hybrid legal knowledge systems encompassing both rule-based and case-based elements, that traces its roots back to the work of Rissland and Skalak [12]. As for the rule-based part, the main focus in on the structure of rules. Instead of reconstructing rules as conditional sentences, a more complex approach is adopted to encompass several different parameters of rules. The topic of parameters of rules is deeply analyzed nowadays in the framework of RuleML research and in application of NLP tools to analysis of legislation [3, 13]. The present contribution encompasses many of the distinctions discussed in this literature and outlines briefly how these different parameters contribute to the argumentative practice in legal domain. The idea of argument schemes developed by Gordon and Walton [14] is employed in this context. The set of parameters used to characterize legal rules is also inspired by the work of Jaap Hage on Reason-Based Logic [15]. Another context that is relevant for the discussion of the problems tackled in this paper is the formal treatment of changes in normative systems, especially as regards the application of defeasible logics [16, 17, 18].

Although the issues of complexity of contemporary legal systems that stem from the legislative changes and intertemporal relations are a subject of intensive research nowadays, the intention of this paper is to show that this is just a tip of the iceberg. The degree of complexity of legal reasoning grows significantly if the constitutional context is projected on the rule-based framework. As it was argued, the concept of citizens' trust in the state and its law plays a particularly important role in this context, leading to the development of a very complex case-based jurisprudence concerning constitutional (sub)principles. Analysis of these issues requires the use of all case-based knowledge representation techniques, to begin with balancing of values, through development of dimensions and classification of cases with respect to them, to conclude with abstracting factors from factual situations' descriptions. Interestingly, due to the great emphasis on the process of balancing in statutory law culture [7], the considerations begin from the identification of high level principles to end up in identification of relevant factors. This order is reversed when compared to the development of case-based reasoning structures in American theory: there the research on factor and dimensions came first to be complemented by teleological considerations [19, 20].

6   Conclusions and Further Research

Contemporary research on AI and Law makes significant progress as regards modeling of temporal aspects of legal rules, their modification, the scope of their efficacy etc. This research domain is of utmost importance, however, for the sake of completeness of elaborated models and legal knowledge bases, the constitutional context should be taken into account. As it was shown on the basis of the actual case study (the case of Mr. Kowalski), even an elaborated rule-based framework may lead to confusing answers to legal questions, were the constitutional context not present. The Principle of Trust is one of the most important criteria for constitutional assessment of statutory regulations. The future research continuing this project will be devoted to operationalization of case-based reasoning concerning this principle. The obtained results should be integrated with the already existing systems employing temporal defeasible logics.

Acknowledgements. I would like to thank the organizers and participants of the AICOL 2013 – AI Approaches to the Complexity of Legal Systems workshop which took place in July 2013 in Belo Horizonte as one of the events at XXVI World Congress of Philosophy of Law and Social Philosophy where I presented a talk related to the topics discussed in this paper. I am grateful to Fernando Galindo for stimulating questions. I thank Pompeu Casanovas for his encouraging and very supportive attitude, as well as for the discussions on topics related to the frontiers of legal theory and information technology. Also, my thanks are due to three anonymous reviewers whose suggestions led to improvement of certain parts of this papers. All remaining deficiencies are responsibility of the author.

References

1. Sartor, G.: Legal Reasoning. A Cognitive Approach to Law, A Treatise of Legal Philosophy and General Jurisprudence, vol. 5. Springer (2005)
2. Hage, J.C.: Separating Rules from Normativity. In: Araszkiewicz, M., Banaś, P., Gizbert-Studnicki, T., Płeszka, K. (eds.) Problems of Normativity. Rules and Rule-following, Law and Philosophy Library Series (2014)
3. Athan, T., Boley, H., Governatori, G., Palmirani, M., Paschke, A., Wyner, A.: OASIS LegalRuleML. In: Verheij, B., Francesconi, E., Gardner, A. (eds.) ICAIL 2013: Fourteenth Intrenational Conference on Artificial Intelligence and Law, pp. 3–12. ACM, New York (2013)
4. Grabowski, A.: Juristic Concept of the Validity of Statutory Law: A Critique of Contemporary Legal Nonpositivism. Springer (2013)
5. Araszkiewicz, M.: Towards Systematic Research on Statutory Interpretation in AI and Law. In: Ashley, K. (ed.) Proceedings of the Twenty-Sixth Annual Conference on Legal Knowledge and Information Systems (JURIX), pp. 15–24. IOS, Amsterdam (2013)
6. Sartor, G.: Doing justice to rights and values: teleological reasoning and proportionality. Artificial Intelligence and Law 18, 175–215 (2010)
7. Alexy, R.: A Theory of Constitutional Rights, transl. J. Rivers, Oxford UP, Oxford (2009)
8. Araszkiewicz, M.: Balancing of Legal Principles and Constraint Satisfaction. In: Winkels, R.C. (ed.) Proceedings of the Twenty-Third Annual Conference on Legal Knowledge and Information Systems (JURIX), pp. 7–16. IOS, Amsterdam (2010)
9. Ashley, K.: Modeling legal argument: Reasoning with cases and hypotheticals. MIT Press, Cambridge (1990)
10. Aleven, V.: Teaching case-based argumentation through a model and examples (Unpublished doctoral dissertation). University of Pittsburgh Graduate Program in Intelligent Systems (1997)
11. Atkinson, K., Bench-Capon, T., Prakken, H., Wyner, A.: Argumentation Schemes for Reasoning about Factors with Dimensions. In: Ashley, K. (ed.) Legal Knowledge and Information Systems. JURIX 2013: The Twenty-Sixth Annual Conference, pp. 39–48. IOS Press, Amsterdam (2013)
12. Rissland, E., Skalak, D.: CABARET: Statutory Interpretation in a Hybrid Architecture. International Journal of Man-Machine Studies (IJMMS) 34(6), 839–887 (1991)
13. Gianfelice, D., Lesmo, L., Palmirani, M., Perlo, D., Radicioni, D.: Modificatory Provisions Detection: a Hybrid NLP Approach. In: Verheij, B., Francesconi, E., Gardner, A. (eds.) ICAIL 2013: Fourteenth International Conference on Artificial Intelligence and Law, pp. 43–52. ACM, New York (2013)
14. Gordon, T., Walton, D.: Legal reasoning with argumentation schemes. In: ICAIL 2009: Proceedings of the 10th International Conference on Artificial Intelligence and Law, pp. 137–146. ACM, New York (2009)
15. Hage, J.C.: Studies in Legal Logic. Springer, Berlin (2005)
16. Governatori, G., Palmirani, M., Riveret, R., Rotolo, A., Sartor, G.: Norm Modifications in Defeasible Logic. In: Moens, M.-F., Spyns, P. (eds.) Legal Knowledge and Information Systems - JURIX 2005: The Eighteenth Annual Conference on Legal Knowledge and Information Systems, pp. 13–22. IOS Press, Frontiers in Artificial Intelligence and Applications (2005)
17. Governatori, G., Rotolo, A., Riveret, R., Palmirani, M., Sartor, G.: Variants of temporal defeasible logics for modelling norm modifications. In: ICAIL 2007: Proceedings of the Eleventh International Conference on Artificial Intelligence and Law, pp. 155–159. ACM, New York (2007)

18. Governatori, G., Rotolo, A.: Changing legal systems: legal abrogations and annulments in Defeasible Logic. Logic Journal of the IGPL 18(1), 157–194 (2010)
19. Berman, D., Hafner, C.: Representing teleological structure in case-based legal reasoning: the missing link. In: ICAIL 1993: Proceedings of the 4th International Conference on Artificial Intelligence and Law, pp. 50–59. ACM, New York (1993)
20. Bench-Capon, T., Sartor, G.: A model of legal reasoning with cases incorporating theories and values. Artificial Intelligence 150, 97–143 (2003)

The Construction of Models and Roles in Normative Systems*

Alessio Antonini[1], Cecilia Blengino[2], Guido Boella[1], and Leendert van der Torre[3]

[1] Department of Informatics, Università degli Studi di Torino, Torino, Italy
{antonini,boella}@di.unito.it
[2] Department of Law, Università degli Studi di Torino, Torino, Italy
cecilia.blengino@unito.it
[3] ICR, University of Luxembourg, Luxembourg
leon.vandertorre@unilu.lu

Abstract. Roles are widely addressed in multi-agents systems with social norms but roles in legal systems are quite different. The relation between legal norms and roles have specific features that when comes to applications create a distance with the expectations from law practitioners. This paper analyse roles in legal systems with legal norms and present the extension of [1] about representing norms as social objects consenting the representation of the assignment of roles and the chain between principles, norms and roles.

Keywords: social ontology, legal reasoning, normative system, roles.

1 Introduction

Roles are basic bricks for the construction of social and normative system. Roles are widely addressed in a general perspectives without taking in account how roles and norms are created in real systems. This proposal addresses roles in legal normative systems focusing on the relation between roles and legal norms. The relation between legal norms and roles have specific features in particular about how roles are defined and assigned. In particular, role assignment is considered as a "normative act" defining the scope and the rules for acting as role holder. In this perspective a role is firstly being "hold" and only secondary "played" by agents. Furthermore, the focus on legal norms requires a strong distinction between the social expectation and the juridical function of roles. Considering the norm dynamics as perspective, this contribution addresses the following mechanisms:

1. the social characterization of agents and other entities,
2. the creation of models,
3. the assignment of roles to entities and
4. the connection between principles and norms and agents' actions playing roles

The main goal is to represent the dynamics of norms though exposing the hidden relations between the different information sources (laws, contracts, judgements, etc.)

---

* "European Legal Culture" funded by Compagnia di San Paolo.

P. Casanovas et al. (Eds.): AICOL IV/V 2013, LNAI 8929, pp. 115–129, 2014.
© Springer-Verlag Berlin Heidelberg 2014

The overall methodology involves the use of social ontologies to rebuild incrementally the state of affairs as the social objects within normative systems. We present an extended version of a social ontology[1] and its use to represent chains of norms and roles: norms implementing principles, roles defined in norms, principles implicated by norms and role assignments. Furthermore considering a semiotic perspective, we show how the general mechanism behind the creations of social concepts and social artefacts, for instance for the creation and use of new roles. The proposal focus at abstract level, entities (norms, principles, roles, concepts, agents) and relations are represented as graphs of resources from different data sources. The presented proposal is meant to represent the state of affairs of legal systems from different perspectives enabling different kind of legal reasoning. In other words, the presented framework can be applied to interconnect legal databases in order to rebuild the evolution of the legal system enabling many different analysis using custom interpretation theories implementing a specific perspective of the legal system.

The rest of the paper is structured as follows. In section 2 is discussed the state of the art about roles. Following in section 3 is discussed the concept of roles in normative systems and in section 4 is presented the extended social ontology for roles and models. In section 5 are presented and modelled two scenarios about chain of norms and role goals. Finally in section 6 are presented some final remarks and future works.

2   State of the Art, Methodology and Aims

Roles are been widely addressed from different perspectives in the multi-agent systems (MAS) and the normative multi-agent systems (NorMAS) communities. In general roles are used to abstract behaviour, position within organizations and, in normative systems, normative status (obligations, powers, permissions, etc.)

A role is a set of activities that were delegated by a social institution to agents (role holders). Roles connect powers to and goals consenting to reason about it abstracting from single agents. In MAS roles are described in many ways: in terms of rights, permissions and obligations[2], expectations, standardised patterns of behaviour[3], social commitments[4,5], goals and planning rules[6].

The overall metaphor behind the model of roles is the "agent play a role"[7]. This metaphor has several consequences: someone can or cannot play a role, a role can be player for a certain amount of time, an agent can switch roles, roles are played in a specific context, roles are related each other implicating games or protocols of behaviour. Furthermore, roles have a scope: there is a relation between acting in a role and organizations, roles playing roles and roles as pseudo-agents with their own mind set-up where discussed in [8,9]. As far as we know the works about roles are actually about social roles, it is still missing a study of roles within legal systems (considering the differences between legal norms and social norms), and following a theory about the evaluation of agents' playing roles considering legal norms.

2.1   The Relevance of Principles in Role Evaluation

The problem with models of role is related to the very metaphor behind them: "agents playing roles". The reasons are the dynamics of playing and the consequences on the mechanisms to handle roles. For instance, role conflict is reduced to a selection problem (which role to play) avoiding to deal with the conflicts of role purposes and the correctness of the use of the role powers . Moreover, the current metaphor is even less appropriate for roles in legal systems where roles are "owned" by agents.

In legal systems, a role cannot be not recognised by agents. The ownership of roles give much more freedom than the acting in a role: to remove an agent from a role it is not just a matter of fail the social expectation but it needs a legal support, such as a contract breaking, that can be quite difficult to build. This set up is much more rigid, the results is to give to agents the chance to establish their own interpretation of roles: combining goals, the use of powers, the interpretation of obligations, building new strategies, etc.

In order to evaluate a role acting it is required to consider norms in a broader sense. An overall evaluation of an agent acting requires a goal, but the goals of roles cannot be founded in the prescriptive content of norms with their definitions. The goal of norms and so the goal of roles can be founded in what is called the principles of norms. Legal norms are implementation of principles, following their goals do not always correspond to their effects or with the interpretations of the legal texts. Legal norms do not have only one meaning, it is always need to make interpretations: there is an intrinsic vagueness in law that is actually used by legislators to avoid arbitrary decision [10]. Norms contain open concepts connected to society and language. In legal systems roles are defined though legal norms: the descriptions of roles involves vagueness, open concepts like norms.

When comes to representations, legal norms are usually treated as set of rules that should be extracted from legal texts. Usually it is possible to find in legal texts scenarios, actions associate to positive and/or negative sanctions. Moreover, the text structure allow to extract a context, entities and rules (considering references and definitions). That gives the impression that a conversion of a legal system in a knowledge base is possible. The construction of knowledge base from legal contents is indeed possible but only considering one interpretation of norms at time and resolving vagueness. Rules represent only one of the possible interpretation at time.

The content of norms are far more than their legal text. Their meaning is grounded in existing social norms, principles and shared beliefs. Principles are part of the norms such as the prescriptive content uses to formulate rules[11]: rules indicate a specific behaviour that can be or cannot be followed but principles are generally considered what norms should maximize. Different theories [12] about principles agree on their quantitative nature. Differently from rules, principles do not allow a crisp evaluations it is not possible to be compliant to a principle. Principles require to consider the contingency and the material possibility to archive a desired effect [13]. Those aspects need to be considered because principles plays an important role in the use and interpretation of norm like norm scope and efficacy in society. For instance the efficacy of norms can be evaluated confronting the archived results with the desired effects. Norms depends on principles and so roles depends on both principles and norms, following agents' actions playing roles are bounded to principles and are part of the effect of norms.

2.2 Perspective on Roles

There are two aspects of roles corresponding to two different perspective of legal norms:

1. roles are prescriptive description of agents behaviour: capabilities, protocols, scope, goals, etc.
2. roles are symbols of social expectations about agents' behaviour in specific contexts (cognitive and socio/cultural artefacts): context-aware interpretations, pragmatics of powers, conflict resolution, principles, etc.

In artificial systems only the first aspect of roles is involved, formal systems belong to this category. Human and hybrid systems involves both aspects of roles, for instance human agents can follow formal procedures but they can also change, reshape or ignore them, change the rule of game or change the very meaning of the rules. The main assumption of this approach is that the state of affairs does not imply a specific interpretation, the collection of the social facts covers heterogeneous aspects about legal systems, that is a common ground for different kind of reasoning.

3 Roles as Social Objects

Roles and norms are both social artefacts, from now on we refer to them as "social objects". Social objects are created through a *communicative act* that became is some way *public and independent* from who performed the action and it can be *shared* through media. For instance documents and promises are social objects, when created they become independent and part of the society.

Social objects need to be interpreted, a contract without interpretation is just ink on paper and without the common knowledge it do not result in obligations: the difference between a real and a fake contract is not in their shape or content but in the circumstances they are been created. Agents rely on shared experience, models of objects, concepts and social objects to create new ones in an efficient way.

Example 1 (making contracts). For instance to make a new contract it is not required an long and extensively explaination about its meaning but only to indicate its specific parts such as objects and terms. This is possible because we rely on the shared knowledge about contracts, everyone will understand just recalling the term contract and reproducing the right circumstances (witnesses, signatures, etc.)

The representation of norms as social objects is been addressed in [1]. They presented an ontology to build abstract representations of social facts about norms and the mechanism to extract specific interpretations (historical, teleological, etc.) Now, we briefly recall the theory of social objects and following we provide an extension for principles and roles.

3.1 Social Objects

Social objects are a category of entities between ideal (abstract) and physical objects[14] sharing some features with both of them. Social objects are created trough the rule "Object = Inscribed Action". An action is constitutive and communicative, it says something about the social reality, and it is fixed (inscribed) in one or more objects (media used to spread the action). The contexts of social objects are part of them as the communication content, their use or creation if related to the common ground of agents (shared concepts) are used to build their interpretations by agents. For instance, legal concepts are used to complete an object meaning (recall example 3). Social objects are accessible through their inscriptions like papers, drawings, digital or human memory.

Social objects represent an incomplete knowledge with multiple possible interpretations. One possible meaning of a social object is the result of a reconstruction process including: the interpretation of its inscription, its context, the interpretation on the referred concepts and all the involved other social objects, and their integration. Social objects are composed and asynchronous[1,15] speech acts[16]. They are asynchronous because the communications are performed again and again when agents access to the objects. Social object are composed because made by several sources (e.g. pages or documents) and several speech acts. An agent loses the control of the context of use: when, who, how and why a social object is used. Therefore there are multiple sources of a social object interpretation and the different results comes from how those sources are handled. What is not at the stake is the general representation but how to use them during the reasoning processes. For instance a contract leads to trials not because it is not recognised as authentic but because the two parties do not agree on the consequences of the contract (obligations, etc.)

The social ontology we start from define three types of entities:

Agents called subjects that can act, communicate and create social objects.

Concepts include "ideal objects" and "physical objects". "Ideal objects" are entities like numbers that do not have a body, a unique definition and that exist outside time. "Physical objects" are all the entities with a physical body and a life cycle. For the purpose of speech acts both categories are considered concepts that can be used in a message.

Social objects that we discussed so far.

Social objects can be composed, agents can act as groups and concepts are part of conceptual or physical structures. Among entities of the same class it is defined a generic relation "part of" that stands for "is-a", semantic and other ontological relations. The ontology is meant to build abstract representations so "part-of" is an abstraction of all those relations that can be defined for lower level representations of the same entities.

Part of is a relation defined between entities of the same category:
  - from agents to agents part of represent groups of agents making the same action on a social object;
  - from social object to social objects part of represent the composition of social objects;
  - from concept to concept part of is an abstraction of the ontological relations between concepts.

Among the previous entities are defined the following relations:

Support given to social objects by agents through their actions.

Represent (representation) of concepts used in the social objects

The relations "support" and "representation" represent the following dynamic: agents create social objects from public acts (messages) about concepts. That scenario do not require concepts like models or roles of agents (called "Subjects" in [1]) so those where not considered. Furthermore, the presented ontology do not distinguish between "ideal objects" and "physical objects", both are considered linguistic "concepts" composing the message content. That solution was driven by the analysis of the dynamic of norm interpretation. In order to represent the norm/role relation we extend this ontology, the norm graph and the norm network.

3.2 Norms and Roles as Social Objects

Norms are social objects made of normative messages[17] in a juridical field[18]. The meaning of norms are the result of a dialectic process between juridical actors. In legal systems, roles are also social objects as part of the conten of norms. Furthermore, the assignment of roles of agents are also social objects because formal public acts like contracts.

A role assignes a position and a juridical function to an entity in a social structure: a) the role as position defines the scope of the function and the relations with other roles; b) the juridical function is the normative characterization of the role, assigning to situations in which the role is involved and to actions an effect.

Roles are not characterised by being assigned to agents but to the delegating prerogatives they hold and to assign the capabilities (powers) to archive an effect within a social structure. For instance also a norm can have a specific role within the normative system: relations with other norms and the delegation to have a specific effect on society. From now, we refer to agents as holders of roles but all the considerations we are going to present can be extended to any social objects.

A juridical function is one of the effect that a norm should archive, a role is what i put in action to do so. Thus as there is a connection between principles (the aim) and norms and between norms and roles, there is a connection from principles and roles. In particular, the juridical function of roles follows the principles behind the norms. Roles can be the result of several norms. Considering the hierarchy of the sources roles are the result of chain of social objects from constitutions to regulations. As roles assign new capabilities allowing agents to consider and make in action new strategies, those strategies are be related to the principles behind roles. Following, the meaning of a role (its juridical function) is the result of a chain of interpretations of the different sources by the holders of roles, and a role evaluation is the comparison between the principles and the effects of the holders' actions.

Example 2 (Contract 2). Considering a contract in wich a role r_j is assigned to an agent a_i, the contract uses the concept of role r_j relying on previous definitions and it socially describes the agent a_i assigning a role r_j within a context c_k. Searle's constitutive rule[19] describes the role assignment "X count as Y in C": agent a_i count as r_j in a context c_k. The context is the structure where the agent acts as r_j, the role r_j is the description of the agent a_i. The relation between r_j and a_i can be represented with the "represent" relation already defined. Nevertheless, it is necessary to reshape the ontology allowing the "represent" relation from social objects to agents.

3.3   The Social Characterization of Entities and Models

The current version of the ontology of social objects (as recalled in section 3.1) does not consent to represent the following two mechanisms: 1) the assignment of role to agents, 2) the connection between role as description in norms and role assigned. Those two are the basic mechanisms of the creation of social objects.

The mechanism of role assignment enriches an agent and bounds the agent to the social expectation and the other agents to role holder if they want to access to the role powers. The same mechanism of social characterization is involved when it is defined a social aspect of some entity. The meaning (interpretation) of social objects is always a message about other entities: social objects talk about something or someone. From this perspective a social object is a "social characterization" of other social objects, physical objects, agents or ideal objects.

The second mechanism we need is the one used to transfer the meaning between social objects or the construction of the meaning of social objects from other social objects. Social objects are instances of models that gives part of the meaning to the object, for instance we can another example about contracts:

Example 3. [Contracts 3] A selling contract between two agents a_1 and a_2 of an object o relies of the idea of "contract": it does not need to contain all details about the meaning of contracts, signatures or selling but only the information about the two parties a_1 and a_2, the object o and other contingent details.

To catch the mechanism in the previous example we need to extend the social ontology, in particular we need the relation between model and social object, e.g. the model of contracts and a contract between a_1, a_2 for o.

The mechanism of social characterization can be expressed using the "represent" relation if it is extended allowing the representation of agents. In order to represent roles, it is also required to catch the model/instance dynamic introducing a relation between models and social objects. The transferring of meaning using models is the base of the incremental growth of social structures. Norms involving a role can be considered incremental descriptions of the role. We can consider again an example about contracts.

Example 4 (Contract 4). Considering the example 3 we expect:

(1) several norms about contracts temporally and hierarchically ordered,
(2) examples of standard contracts made between different parties,
(3) examples of special contracts made, for instance, for real estates,
(4) examples of real estate selling contracts between different parties.

The contract as described in norms (1) is a model for the contracts (2) but also a model for a specialized contract for real estate (3). The real estate contract (3) is a model for the contracts (4) even if it is not considered yet a standard model of contracts (3) or in court it is found illegal. There is a connection between (1), (2), (3) and (4) created through an abstraction process from a specific social object to a concept used as model for a new social object. For each step of abstraction and use, models involve agents' interpretation about what is the model of contract (1) according with the current norms (an conflicts), what they consider in those contracts (2), how a contract should be extended for real estate (3) and what take in account from contract (1) and real estate contract (3) in order to make single contracts (4).

To represent the contract scenario it is requires a "model of" relation between concepts and social objects. For instance a social objects (a norm) "represent(s)" a role as abstract object (agent interpretation of the role) that is "model of" the assignment of the role to a specific agent.

4   A Social Ontology of Roles

Now we apply the discussion in section 3 introducing important changes to the recalled social ontology. First of all, we revise the assumptions behind the current version of the social ontology. In legal systems the type and number of inscriptions actually matters[1]. For instance the different copies of a document can have different normative status such as an original compared to a copy. Physical objects can give support to social objects, figure 1 summarizes the required social ontology changing the relations.

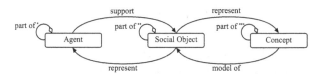

Fig. 1. Conceptual extension of the social ontology for models and roles

In this new version of the social ontology, agents and physical objects can both "support" social objects, agents' memory is a form of inscription. In this set up agents' actions and inscriptions are equivalently described with the relation "represent". Still there are differences between agents and physical objects in particular in term of "support", in this set up it is still possible to distinguish two types of support considering the two class types "physical object" and "agent". Agents' memory cannot be considered just an inscription for two reasons:

1. agents' memory embodies also their own evaluation of the objects, so it can be more or less important according with the type of reasoning footnoteFor instance during a trial a witness can considered more or less important than a signed document according to the context, the trial and the witness.;
2. physical object are not necessary to create social objects while agents are.

Summarizing we consider "agent" a specialization of "physical object" in order to make distinction among them defining interpretation theories.

---

[1] Considering digital media inscriptions are not important, they are always digital and in multiple copies across the web.

4.1 Creation of Models

The last issue we need to address is about the relation between models and entities. In society models change over time, that is possible because their current use: a model meaning is the result of its instances. On the other hand, legal models are defined through formal acts[2] and they can change also through new formal acts.

Physical objects and agents have their own life cycle independently from society. Differently, models are shared, they can survive their instances single objects and they can also be defined. Now the question is which kind of entity a model is? Social objects are social representations but they also have a life cycle involving their inscriptions (as physical objects) and their meaning (related to agent interpretations). Ideal objects represent shared concepts without time and a specific definition: they refer to a meaning that changes with the context. For instance consider the "rights", the meaning in court is different from its moral meaning but still "right" is a used and understandable concept. "Rights" as ideal object can represent different models of good behaviour in different domains because its meaning can be replace by agents' interpretation. This effect is what we want to represent, so we conclude that be a model is a relation between "model of" relation need to be defined from "ideal object" to "social object". Following the interpretation of a social object is a grounding problem. Now we describe how our model works considering a semiotic perspective and the Peirce's triadic signs:

a. an entity can be represented with a social object, for instance considering a physical object as the "object", the inscription of social object (i.e. the communication) is the "signifier" while the meaning of the social object is the "signified", figure 2 (a).
b. considering the abstraction of a concept used in social objects, a set of social object inscription are the "object", the ideal object referred by them the "signifier" and the meaning given to the ideal object (the result of the social object interpretation) the "signified", figure 2 (b).
c. an ideal object can be used a model of new social objects, the "object" is the meaning given to an ideal object (figure 2 (b)), the "signifier" is the social object inscription and the "signified" is the meaning of the social object, figure 2 (c).

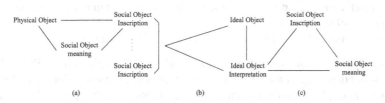

(a)   (b)   (c)

Fig. 2. The creation of meaning with social ontology, three semiotic cycles: (a) from physical object to social object, from a set of social object to concepts and (c) from concepts to new social objects

---

[2] It is easy to find example where social models are quite different from legal models, for instance contracts can offer quite anti-intuitive cases.

In this scenario, ideal objects represent agents' interpretation of social objects: they stand for the meanings used by agents. Any entity can be used to create social objects, for instance, an agent *a* playing a role can be the example for its successor: the some aspects of the predecessor's behaviour can be selected and become the meaning of the role as model. Summarizing, a model can be created from agents or physical objects in two steps: 1) interpretation step, from any "agent", "physical object" or "ideal object" to a "social object"; 2) abstraction step, from a "social object" to an "ideal object". Concluding, "agent" are not considered a specialization of "ideal object" but still they can be source of models through the creation of social object and them ideal objects.

4.2 Extended Social Ontology

In figure 3 is represented the extended social ontology. First of all, the class "concept" is split back in "ideal object" and "physical object". In [1] is used a class "Time" as specialization of "ideal object", we indicate "Time Interval" as specification of "ideal object". The class "agent" is a specialization of "physical object" represented with the relation "is-a".

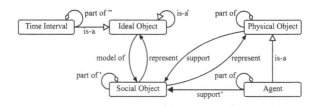

Fig. 3. The new social ontology extended with entities and relations

Relations changes as follows:

a. as consequence of splitting concept, the relation "represent" is defined from "social object" to "ideal object" and from "social object" to "physical object";
b. now the relation "support" is defined in general from "physical object" to "social object", with "support*" we indicates agents' support that involves actions and evaluation of social objects;
c. the new "model of" relation is from "ideal object" to "social object".

We do not discuss the details of the "part of" and "is-a*" relations, we will recall the general idea of this abstract model and give an intuition of the meaning of such relations.

The authors of [1,20] do not present a specific proposal for the "part of" and "is-a*" relations. The reason is that for each perspective in which norms can be looked at there are several specific formalisms. An abstract representation is required to represent the whole concept of norm while the specific formalism is required to reason about the specific aspect. Furthermore the semantic of the relations should chose considering the capability and level of details of possible tools for document analysis. We indicate the "part of" and "is-a" relations in order to provide some further considerations:

is-a relations is used to indicate "agent" as specialization of "physical object" and "Time Interval" as specialization of "ideal object", the consequence is only relation inheritance.

part of is used to describe agents groups (a group is an agent composed by agents), and physical objects like the wheels of a car.

is-a* abstracts the ontology relations, ideal objects are considered external entities from specialist ontologies with their own relations.

part of* abstracts the structure of social objects. Due to the big difference that can be found in non linguistic and linguistic documents (like paints and contracts), it is difficult to specify a set of possible semantic relations. Considering juridical entities, social objects have linguistic contents so a proposal can to use relations from semantic networks like kl-one or conceptual graphs[21]. Thus, there are norms aspects such as prescriptions that can be represented with more specific formalisms like deontic logic.

part of** represent all the possible relations between time interval such as Allen's interval algebra[22].

Now we define the network of social object and the social-object graph ignoring the class "Time Interval" and considering the "part of" relations as general relation among nodes of the same class.

Definition 1 (social object network). *Let be $G_S(V, E, \varphi)$ a directed graph with:*

V *set of nodes with* $V^1, V^2, V^3, V^4 \subseteq V \mid V^i \cap V^j = \emptyset \; \forall i, j \in [1, 4]$
E *set of directed edges* (v_k^i, v_q^j) *with* $v_k^i \in V^i, v_q^j V^j, ij$ *and* $i, j \in [1, 4]$
$\varphi : V \to [1, 4]$ *function assigning a label to each vertex* $v \in V$

We call $A = V^1$ *set of agents,* $S = V^2$ *set of social objects,* $I = V^3$ *ideal objects ("Time Interval" included) and* $P = V^4$ *physical objects. Considering the edges* E, *we call*

support relations is the set of all edges $(v_k^i, v_q^2) \in E$ *with* $i \in 1, 4$
represent relations is the set of all edges $(v_k^2, v_q^j) \in E$ *with* $j \in 1, 3, 4$
part of relations is the set of all edges $(v_k^i, v_q^i) \in E$ *with* $i \in 1, 2, 4$
is-a* relations is the set of all edges $(v_k^3, v_q^3) \in E$

Following we define the social object graph over a root $s \in S$. The social object graph contains the social objects connected to s, all directly connected group of agents and ontologies of physical or ideal objects.

Definition 2 (Social object graph). *Let be* $G_S(V, E, \varphi)$ *a social object network and* $s \in S$ *a social object* $G_s(V_s, E_s, \varphi)$ *is called social object graph of* s *and it is defined as follows:*

$G_s \subseteq G_S$
$V_s^1 = \{v_k^1 \in V \mid \exists \text{ path from } v_k^1 \text{ to } s \text{ or from } s \text{ to } v_k^1\}$
$V_s^2 = \{v_k^2 \in V \mid \exists \text{ path from } v_k^2 \text{ to } s\}$
$V_s^3 = \{v_k^3 \in V \mid \exists \text{ path from } s \text{ to } v_k^3 \text{ or from } s \text{ to } v_k^3\}$
$V_s^4 = \{v_k^4 \in V \mid \exists \text{ path from } s \text{ to } v_k^4 \text{ or from } s \text{ to } v_k^4\}$

$$V_s = \cap_{i=1}^{4} V_s^i$$
$$E_s = \{\forall (v_k^j, v_q^i) \in \exists \mid v_k^j, v_q^i V_s\}$$

Those definitions overrides the definitions in [1] consenting to build the same representations and use.

5   Reasoning about Norms and Roles

In order to reason about roles we need to represent both norms and principles. Principles are social objects connected to norms in the same way as roles are connected to norms. The relation between principles, norms and roles follows the schema presented in section 3. For the sake of compactness, in the next examples we focus only on the relations between agents, social objects, physical objects and ideal objects omitting their structure and the relations among their components. What we show can be applied to the substructure of entities to build complex structures.

Now we present two examples, the first one is about the implementation of norms, or more generally it is about how create new social objects using concepts as models.

Example 5 (Chain of norms). In Europe every state need to implement European norms in their national normative systems and in some case this is extended at local level involving regions, public institutions, municipalities, etc. Usually European norms comes with an introduction about the goal and sources of the norm explaining the principles behind defining scope, goals, limits, etc. The legal texts of the national norms can be quite off from the goal and the scope of European normative. This distance increase with each step down to the local level.

Figure 4 (a) represents a chain of norms: a principle p promotes a norm n_1 implemented at national level with norm n_2 as result of an interpretation p' of n_1. Figure 4 (b)

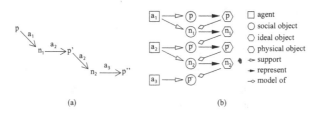

(a)   (b)

Fig. 4. (a) agent a_1 creates n_1 from p, agent a_2 interprets n_1 as p' and creates n_2 from from it. An agent a_3 interprets n_2 as p''. (b) agent a_1 is responsible for converting principle p into n_1, agent a_2 for using its interpretation p' of n_1 to create n_2. If an agent a_3 what to use n_3 it will be do it on the base of its own interpretation p''.

shows how to represent the chain as a network of social objects: when a norm in been implemented ($p \rightarrow n \rightarrow p'$) agents' interpretation occurs ("support" given to a social object) the ideal objects stands for agents' interpretations used in the social object

Evaluating the distance between principle and effects of norms involves several interpretation steps. Considering the previous example, the comparison require to check p and p': a) n is been created interpreting p, the "model of" indicates the passage from an ideal concept to a social object with a specific use of it; b) p' is also an interpretation of n effects. Every step from ideal objects to social objects is the result of an interpretation of the involved ideal objects. Roles use the same mechanism with more steps. The evaluation of agents "playing" roles is usually made considering norms as set of rules. In this set-up the whole process from principles to agent' actions is involved. This enables several different type of reasoning, for instance the miss-use of powers, handling the conflicts between roles, the role scope and much more than norm compliance.

Finally, the in this last example we discuss the relation between roles and principles:

Example 6 (The principles behind roles). The role of teacher involves the mission of "education" but schools regulation sets rules on measurable parameters like teaching hours and students rates. Good teachers end up doing extra work and being involved with students family to pursue the educational goal of their roles even if it there is not a within rules. Society evaluate more important the teachers' attitude than their compliance to school rules forgiving some rules breaking. On the other hand bad teachers will not be forgiven for even small rule brake that can also become a pretext to fire them.

Considering a, Figure 5 (a) represent a role r defined through a norms n and an agent a empowered with a role r creating a social object o like a norm or employment contract.

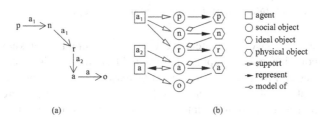

(a)   (b)

Fig. 5. (a) agent a_1 creates a norm n defining the role r, agent a_2 assigns r to agent a and a creates the document o. (b) agent a is connected with a double arrow indicating a "represent" relation to assign the role r to agent a and the "support" give by agent a to the assignment of role r.

. This scenario can be represented with the network in figure 5 (b). The principles p of role r are inherit from n, r is also the result of interpretation of n content. The result of agent's action o can be compared respective with p, n or r according to the scope of the evaluation

6   Concluding Remarks

In this contribution we discussed the creation of models and roles in legal systems. In particular we addressed the relations between roles, norms and principles. The result is an ontology to build representation of social entities related to normative systems. The presented ontology is not an upper ontology but it is inspired to social object theory and sociology of law.

The presented ontology follows a discussion about the role and models in normative systems. In particular we focused on how roles are created from legal norms and what is the influence of agents' interpretation occurs. Reasoning about norms and roles required the introduction of models and the exposure of the mechanism behind the chain connecting principles, norms and roles. The ontology of social objects allows the representation the mechanisms involved in roles dynamics:

1. About the social characterization of agents and other entities, we addressed the mechanism of "social characterization" of entities using the "represent" relation. In particular we discussed what can be represented with social objects and consequently extend the "represent" relation to agents.
2. The creation of models was described with a two-phases mechanism: 1) social characterization and 2) abstraction of social objects as ideal objects. Therefore we extended the social ontology with the "model of" relation enabling the construction of complex chains of social and ideal objects.
3. Using the "model of" and "represent" relations, we described the assignment of roles and how a specific assignment is related to the general meaning of a role.
4. The connections between principles and norms can be represented with the social ontology consenting also many pattern of legal reasoning. The ontology consent to represent the steps between principles and norms in a normative system. Also it shows with agents' interpretation is involved in each step.

The general approach we use is intent to avoid strong assumptions in knowledge representation, to combine together representations in different formalisms about different aspects of entities and to allow different kind of reasoning. We proposed an abstract representation that do not require to address the vagueness of norms. Moreover it distinguish the representation from the interpretation and use of norms. The goal of this approach in general is to focus on the creation of the social reality along with the legal systems instead of focusing on a specific aspect on law (obligations, arguments, rule revision, etc.)

This proposal is mainly based on the prospective of sociology of law. This systematization of model dynamics will be analysed and validated considering primary law practitioners. We plan to include the model in the norm management system Eunomos[23]. It can be extended according to our proposal in order to enable a wider use, recognising different prospective according to the user role. For instance a lawyer can use a norm database to find norm exceptions based on the interpretation of principles, or based on cases of use in courts[3]. In future development we will focus on extending the idea of ontology dynamics showed (the creation of concepts), on defining a semantic for graph operators a methodology for a quantitative comparison between social objects.

---

[3] Two cases involving the concepts of norm principles, model and roles discussed in this contribution.

References

1. Antonini, A., Boella, G., van der Torre, L.: Beyond the rules representation of norms: norms as social objects. In: Proceedings of Rules 2013 Conference (2013)
2. Pacheco, O., Carmo, J.: A role based model for the normative specification of organized collective agency and agents interaction. Autonomous Agents and Multi-Agent Systems 6(2), 145–184 (2003)
3. Esteva, M., Padget, J., Sierra, C.: Formalizing a language for institutions and norms. In: Meyer, J.-J.C., Tambe, M. (eds.) ATAL 2001. LNCS (LNAI), vol. 2333, pp. 348–366. Springer, Heidelberg (2002)
4. Cavedon, L., Sonenberg, L.: On social commitment, roles and preferred goals. In: Proceedings of the International Conference on Multi Agent Systems, pp. 80–87. IEEE (1998)
5. Fasli, M.: On commitments, roles, and obligations. In: Dunin-Keplicz, B., Nawarecki, E. (eds.) CEEMAS 2001. LNCS (LNAI), vol. 2296, pp. 93–102. Springer, Heidelberg (2002)
6. Dastani, M., Dignum, V., Dignum, F.: Role-assignment in open agent Societies. In: Proceedings of the Second International Joint Conference on Autonomous Agents and Multiagent Systems, pp. 489–496 (2003)
7. Masolo, C., Vieu, L., Bottazzi, E., Catenacci, C., Ferrario, R., Gangemi, A., Guarino, N.: Social roles and their descriptions. In: KR, pp. 267–277 (2004)
8. Boella, G., Van Der Torre, L.: The ontological properties of social roles in multi-agent systems: Definitional dependence, powers and roles playing roles. Artificial Intelligence and Law 15(3), 201–221 (2007)
9. Boella, G., Damiano, R., Hulstijn, J., van der Torre, L.: Role-based semantics for agent communication: embedding of the'mental attitudes' and'social commitments' semantics. In: Proceedings of the Fifth International Joint Conference on Autonomous Agents and Multiagent Systems, pp. 688–690. ACM (2006)
10. Liebwald, D.: Law's capacity for vagueness. International Journal for the Semiotics of Law 26(2), 391–423 (2013)
11. Dworkin, R.M.: The model of rules. The University of Chicago Law Review (1967)
12. Alexy, R.: On the structure of legal principles. Ratio Juris 13(3) (2000)
13. Raz, J.: Legal principles and the limits of law. Yale Law Journal, 823–854 (1972)
14. Ferraris, M.: Documentality: Why It Is Necessary to Leave Traces. Oxford University Press (2012)
15. Marcin, M.: Why legal rules are not speech acts and what follows from that. In: Rules (2013)
16. Searle, J.R.: Speech Acts. Cambridge University Press (1969)
17. Ferrari, V.: Lineamenti di sociologia del diritto. Laterza (1997)
18. Bourdieu, P.: La force du droit. eléments pour une sociologie du champ juridique. Acted de la recherche en sciences sociales 64 (1986)
19. Searle, J.R.: The construction of social reality. Free Press, New York (1995)
20. Antonini, A., Vignaroli, L., Schifanella, C., Pensa, R.G., Sapino, M.L.: Mesoontv: a media and social-driven ontology-based tv knowledge management system. In: Proceedings of the 24th ACM Conference on Hypertext and Social Media, HT 2013, pp. 208–213. ACM, New York (2013)
21. Sowa, J.F.: Conceptual graphs. Knowl.-Based Syst. 5(3), 171–172 (1992)
22. Allen, J.F.: Maintaining knowledge about temporal intervals. Commun. ACM 26(11), 832–843 (1983)
23. Boella, G., Humphreys, L., Martin, M., Rossi, P., van der Torre, L.: Eunomos, a legal document and knowledge management system to build legal services. In: Palmirani, M., Pagallo, U., Casanovas, P., Sartor, G. (eds.) AICOL-III 2011. LNCS, vol. 7639, pp. 131–146. Springer, Heidelberg (2012)

Integrating Legal-URN and Eunomos: Towards a Comprehensive Compliance Management Solution

Guido Boella[5], Silvano Colombo Tosatto[4,5], Sepideh Ghanavati[1],
Joris Hulstijn[2], Llio Humphreys[4,5], Robert Muthuri[3,5], André Rifaut[1],
and Leendert van der Torre[4]

[1] CRP Henri Tudor, Luxembourg City, Luxembourg
{sepideh.ghanavati,andre.rifaut}@tudor.lu
[2] Delft University of Technology, Delft, The Netherlands
j.hulstijn@tudelft.nl
[3] University of Bologna, Bologna, Italy
robertkevin.kiriiny2@unibo.it
[4] University of Luxembourg, Luxembourg City, Luxembourg
{silvano.colombotosatto,llio.humphreys,leon.vandertorre}@uni.lu
[5] University of Torino, Torino, Italy
guido@di.unito.it

Abstract. Business process compliance with regulations has been a topic of many research areas in Computer Science such as Requirements Engineering (RE), Artificial Intelligence (AI), Logic and Natural Language Processing (NLP). This work aims to provide a systematic way of establishing and managing compliance to assist decision-making and reporting. Despite many notable advances, few systems deal adequately with legal interpretation and modeling norms in an expressive way that is well-integrated with business modeling practices. In this paper, we bring together two leading systems, LEGAL-URN and Eunomos, for a comprehensive compliance management solution.

Keywords: Compliance, Legal Interpretation, Requirements Engineering.

1 Introduction

Organizations are motivated to comply with legislation since failure to do so leads to undesirable consequences such as lawsuits, loss of reputation and financial penalties. With the rapid increase and evolution of regulations and policies relevant to business processes, it becomes difficult for organizations to constantly keep their goals, policies and business processes compliant with applicable legislation.

The legal documents that dictate how a corporation must behave are usually complex. This complexity originates from the cross-referential nature of legal documents; the inherent (some say intentional [28]) vagueness of legal documents to cover different scenarios; and the ever-changing nature of the law due

P. Casanovas et al. (Eds.): AICOL IV/V 2013, LNAI 8929, pp. 130–144, 2014.

to legislative amendments and interpretation by other legal authorities. Applying generic laws to the business processes of different organizations is fraught with difficulties, creating the need to rely on expert advice from lawyers and regulators.

The dynamic nature of laws and business creates problems in large organizations, where different stakeholders may introduce goals that conflict with existing ones or with each other. These goals may even unknowingly conflict with the law. Consider also that being fully compliant with legislation may not be feasible or in the organization's best interests. In such cases, organizations may wish to consider alternative solutions based on top-level goals or strict literal reading and aim for minimum compliance with the law while accepting the penalty.

Much effort has been invested in Computer Science to solve some of the issues mentioned, specifically in Requirements Engineering (RE), Artificial Intelligence (AI), Logic and Natural Language Processing (NLP). Two leading systems, from AI & Law and RE respectively, are Eunomos [9] and LEGAL-URN [16]. Both are suitable for compliance monitoring, but each looks at the problem from a different perspective. Eunomos is a legal knowledge and document management system focused on identifying 1) norms 2) related norms 3) legislative modifications 4) different interpretations of the same norms. Menslegis[1], a commercial version of this system, is targeted towards the banking sector in Italy. LEGAL-URN enables business analysts or software engineers to factor in legal requirements as part of their strategic planning by modeling legal norms in the same way as goal and business process management notations, albeit with deontic extensions. LEGAL-URN has been tested in the healthcare domain in Ontario, Canada by modeling four Ontario regulations for healthcare and analyzing the compliance of the business processes of a research hospital in Ontario to these regulations. The result of this case study has been published in [16] in detail.

This paper aims to analyze how to integrate the state of the art from AI & Law and RE for complete traceability from legal sources to business process models, representing regulatory conversations at each level, thereby allowing informed analysis and design of compliant business processes. The rest of the paper is as follows: Section 2 provides a background in contemporary issues in regulatory compliance. Section 3 introduces the case study, Section 4 describes Eunomos, Section 5 the URN-Framework, Section 6 how to integrate the two systems, Section 7 Related Work, and Conclusion ends the paper.

2   Regulatory Compliance

The law evolves with the involvement of different authorities. Organizations may take into account interpretations of legislation from many different sources - case law, subsidiary laws, ministerial decrees, government authority, legal scholars, self-regulatory bodies, industry bodies, internal regulator and external regulators. Stakeholders may use a variety of legal reasoning techniques, as identified by Bobbit [8] and described by Bartrum [5]: *Historical*, relying on the purpose

---

[1] http://www.nomotika.it/EN/MensLegis/Flyer

behind the written law; *Textual,* relying on the actual text; *Structural,* taking into account relations between bodies issuing the law; *Doctrinal,* applying rules generated by precedent; *Ethical,* tied to the ethos of the community; and *Prudential,* aiming to avoid absurd outcomes.

The most important conversation [7] about regulatory compliance is between companies and external auditors. Interpretation of how norms apply takes place at two stages: first, by the auditee in designing business processes, and secondly, by the auditor when assessing the compliance. During the first stage, auditees sometimes do not initially have a clear view of what constitutes compliance to a particular norm, as the legal community works through the issues on a case-by-case basis. There are certain areas, such as IT security, where there is high mutual trust and transparency between auditors and auditees, so that an honest dialogue can take place about proper interpretation [11]. Case studies in the financial sector have shown the importance of regulatory conversations to provide valid models agreed by all. These models can be used later to support decisions taken on operational aspects of compliant business systems [34].

Letterman [45] highlights the challenges of interpretation from a different angle, the trend towards laws that prescribe the achievement of goals while leaving it to the organization to concretize these goals into finer and more concrete goals and targets. Such concretization occurs in two stages, firstly, analysing abstract goals and subdividing them into their component parts, and secondly, development of criteria to indicate to what extent this goals should be realized. At all levels, the emphasis remains on what should be achieved rather than how it should be done.

Cunningham [13] argues that all laws contain a mixture of principle-based and rule-based legal provisions. Moreover, rule-like rules can be treated like principles or vice versa depending on their application and interaction with other provisions. Often, there are rules and principles about the same issues - this can, but does not necessarily, address the problem of legal loopholes.

3   Case Study

Our case study comes from the European Union's Markets in Financial Instruments Directive (MiFID). Among many articles subject to different interpretations is Article 13(6) of Directive 2004/39/EC which states: *"An investment firm shall arrange for records to be kept of all services and transactions undertaken by it which shall be sufficient to enable the competent authority to monitor compliance with the requirements under this Directive, and in particular to ascertain that the investment firm has complied with all obligations with respect to clients or potential clients."*

In a consultation paper [30], The Committee of European Securities Regulators (CESR) proposed that for the purpose of Implementing Directive 2006/73/EC, Article 51(3) (which concretised Article 13(6) of Directive 2004/39/EC), investment advice should be regarded as a type of financial service. The proposal received approval from consumer groups but was rejected by some banking organizations.

The European Savings and Retail Banking Group (ESBG) [23] complained that "It will be extremely difficult for entities to organize and keep records of this type, as the information to be included in such records may be provided through different channels which will be difficult to compile". It added that "If the relationship is ruled by an agreement, this agreement will be recorded in the clients agreements record, and therefore it will not be necessary to keep this additional record." In other words, no additional records should be taken. Regulators may refer to records on investment advice for trading "by agreement" (i.e. for a negotiated deal) as the advice will already be in the agreement. However, where advice is given merely regarding "what is on offer", the advice need not be recorded (presumably due to the extra workload and associated costs).

From a legal point of view, it is possible to argue either way. A teleological or principle-based interpretation would regard investment advice as services, even if financial organizations would not define it as such, to ensure effective compliance monitoring. A literal or rule-based interpretation would avoid its inclusion. Where legal uncertainty exists, organizations need a mechanism to analyze different interpretations in the context of their own business processes.

4   Eunomos

The Eunomos Legal Management System is a web-based interface for managing knowledge about laws and legal concepts in different sectors and different jurisdictions. Legislation from official web portals can be downloaded via web spiders or uploaded by a web interface to the Eunomos database, where they are then stored in legislative XML [2], making it easy to reference individual articles or paragraphs from the text of the law. References are extracted to build a network of internal and external citations. When viewing legislation, the Cosine Similarity technique is used to provide a sorted list of the most similar legislation in the database. This can be useful for finding legislations implicitly modified by later ones. Eunomos has an interface to make comments about legislation and all its paragraphs and articles. This feature is especially useful for annotating elements that have been implicitly modified. The system also includes an alert messaging system to identify knowledge engineers of new legislation, so that they can begin to analyze the impact of the legislative changes.

An important feature of Eunomos is its lightweight legal ontology. Specialist terms within legislation are hyperlinked to jurisdiction-specific multilingual ontologies based on European Legal Taxonomy Syllabus [35]. Legislation-specific and generic definitions can co-exist, with generic definitions grouping legislative definitions together with doctrinal interpretation. Given the ever-evolving nature of legal concepts in an increasingly multi-jurisdictional legislative environment, different definitions are linked by relations such as substituted_by,

---

[2] Currently in accordance with the Norme in Rete standard using the ITTIG CNR parser (http://www.ittig.cnr.it), with a view to developing an Akoma Ntoso parser (the emerging international standard) as part of the EU Cases project.

or `transposed_into`, or `group_by` for generic definitions created by gathering different definitions.

While constitutive norms are used for definitions of legal concepts, prescriptive norms are represented in Eunomos as special composite concepts in the ontology called 'prescriptions' ([9]) with the following relations:

- Deontic clause: obligation, prohibition, permission, exception.
- Active role: the addressee of the norm (e.g., director, employee).
- Passive role: the beneficiary of the norm (e.g., customer).
- Description: the prescription reworded as necessary to aid comprehension
- Norm Identifier: hyperlink to relevant provision in the source document
- Violation: the crime or tort resulting from violation (often defined in other legislation such as a Penal Code).
- Sanction: the sanction resulting from violation (e.g., a fine of 1 *quote*, where emphquote is defined in other legislation).

A similar mechanism to that described in Ajani et al. [1] for ontological terms is used to model change over time in Eunomos for prescriptions. Legislation is amended continually, and, thus, prescriptions need to be changed to align with the new text. The modification link is maintained in the Eunomos knowledge-base based on the identifiers of the NormaInRete standard.

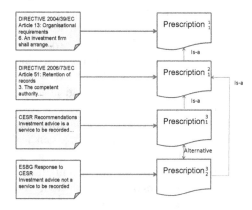

Fig. 1. Conceptual Model of Eunomos to Support Interpretation

The Eunomos system is a very rich legal knowledge management system that allows users to access laws and understand their meaning with user-friendly, well-structured ontologies. A key distinguishing aspect of Eunomos is that the premise that laws and legal terminology are inherently context-sensitive and replaceable is built into the system. However, the system addresses only the normative side. To better support compliance management, the natural next step is to map norms to business processes in an integrated environment.

Figure 1 illustrates how Eunomos relates regulations and prescriptions. Article 51(3) of Directive 2006/73/EC is interpreted in more than one way by CESR and ESBG: *Prescription*$_3^1$ and *Prescription*$_3^2$ respectively are alternative interpretations - as represented by the relation in the ontology. Although Eunomos is a lightweight ontology to be used by lawyers, and lacks formal semantics, the *alternative* relation is inspired from specifications hierarchies where specifications of a concept can be labeled as disjoint. Interpretations should be specified as being candidate or non-candidate, where in the latter case the company has determined that the interpretation is unlikely, undesirable or irrelevant. Graphically, this is represented by the dotted line. Where there are more than one candidate interpretations, this represents an area of possible conflict, which requires careful analysis and consultation with domain and legal experts to resolve.

5   The Legal-URN Framework

LEGAL-URN supports business process compliance by extending the model-based compliance framework ([17,18]) based on User Requirements Notation (URN) Language [25]. The LEGAL-URN framework has four layers for legal and organizational models, which are shown in the left-hand side of Figure 2:

1. **Official Source Documents** that define the legislation on one side and organizational structures, policies and processes on the other side.
2. **A Hohfeldian Model** which consists of a set of Hohfeldian statements [44] together with structured elements of legal statements.
3. **Goal Models** based on URN's Goal-oriented Requirement Language (GRL), which capture the objectives and requirements of both organization and legislation.
4. **Business Process Models** based on URN's Use Case Maps (UCM), which define the business processes that implement organizational policies on the one hand and represents steps mandated by legislation on the other hand.

Different pieces of the framework are connected with five types of links introduced by LEGAL-URN. To build this framework, first, the relevant regulations, organizational policies and procedures are identified manually. This step is usually done by the legal expert in the organization. Next, the Hohfeldian model for the legal documents is created. For this, first, each legal statement in each legal document is annotated with one of the Hohfeldian correlative classes of rights: *duty-claim, privilege-no-claim, power-liability,* or *immunity-disability* and next, the legal statement breaks into the elements as followed: Subject, Modal Verb, Clause, Precondition, Exception and Cross-references.

Although the Hohfeldian ontology is not without its critics, essentially based on redundancy ([24]) or lack of elegance for formal modeling ([39]), it is used in LEGAL-URN to help identify the type of modal verb, the type of the legal statement and the priorities between legal statements (through power or disability). The Hohfeldian model is broader than the Hohfeldian ontology with its

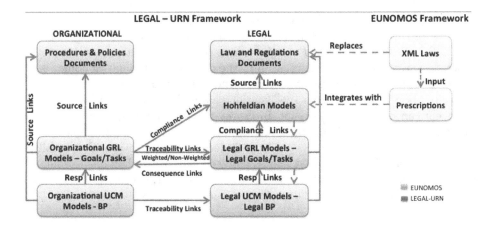

Fig. 2. LEGAL-URN Framework Overview and Eunomos Integration

additional elements introduced above. This layer provides the formalism to goal modeling in GRL and creating the legal extension of GRL called Legal - GRL. That is, subjects in the Hohfeldian model are mapped to Actors in Legal-GRL, clauses are mapped to softgoals, goals and tasks, modalities to permission and obligation stereotypes, and precondition, exception and cross-references are mapped to precondition, exception and cross-reference stereotypes in Legal-GRL. To build the Hohfeldian model, the following rules need to be considered:

- **Rule 1 -** Each legal statement shall be atomic. This means that each legal statement contains one <actor> (the subject), one <modal verb>, one to * <Clause> (<verb> & <actions>), 0 to * optional <crossreference>, 0 to * optional <precondition> and 0 to * optional <exception>.
- **Rule 2 -** If a legal statement contains more than one modal verb, it must be broken down into atomic statements.

In the next step, the Hofeldian classifications (i.e. duty-claim, privilege-noclaim, etc.) are transformed into Permission and Obligations and the Legal-GRL model of regulation and the GRL model of organization are developed. GRL's main concepts come from management and from socio-technical systems and include actors, which have intentional elements (goals, softgoals, tasks, and resources) and indicators, linked through various relationships (AND/OR decompositions, dependencies, and weighted contributions) [12]. The compliance analysis is done in this step.

Figure 3 shows the Hohfeldian models structure and its mapping to Legal-GRL. Modalities in the Hohfeldian model are transformed to *Permission* and *Obligation* softgoals or goals in the Legal-GRL. Power-liability and immunity-disability statements are also of type Permission and Obligation with additional conditions and priorities. More detail of the mapping is explained in [36]. Figure 4 illustrates a Hohfeldian model and Legal-GRL model of the case study.

HOHFELDIAN MODEL	HOHFELDIAN MODEL	LEGAL-GRL MODEL
SECTION	SECTION	-
ARTICLE #	ARTICLE #	-
SUBJECT	SUBJECT	ACTOR, EXPECTIONACTOR
MODAL VERB	MODAL VERB	OBLIGATION, PERMISSION STEREOTYPE
CLAUSE	CLAUSE	INTENTIONAL ELEMENT
PRECONDITON	PRECONDITON	PRECONDITION INTENTIONAL ELEMENT
POSTCONDITION	POSTCONDITION	POSTCONDITION INENTIONAL ELEMENT
EXCEPTION	EXCEPTION	EXCEPTION INTENTIONAL ELEMENT
XREF	XREF	CROSSREFERENCE IE

a) Hohfeldian Structure   b) Mapping between Hohfeldian Model and Legal-GRL Model

Fig. 3. Hohfeldian Model Structure and Mapping with Legal-GRL

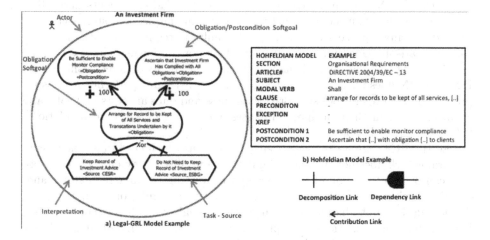

Fig. 4. Hohfeldian Model Structure and Mapping with Legal-GRL

At the last layer, the business processes of organization and regulations are built in Use Case Maps (UCMs). The benefit of using UCM over other business process modeling notations is that it has the ability to link its elements to GRL elements (as both views are part of URN). In other words, tasks and actors in GRL can be linked to responsibilities and components in UCM maps. Having such business processes for legal clauses helps to capture the sequential aspects of laws and, as a result, this helps to identify violations of the procedural laws. The detail on how to build the UCM models are documented in the literature [43].

LEGAL-URN contains three types of compliance analysis algorithms which are: 1) Quantitative Analysis 2) Qualitative Analysis 3) Hybrid Analysis. These compliance analysis algorithms extend the GRL analysis algorithms of Amyot et al. [2]. These algorithms are all bottom-up which means that the satisfaction value of each of the intentional elements in the model are propagated from the lowest level to the highest level in the model. In the LEGAL-URN compliance analysis, the satisfaction values are propagated from lowest-level of organization to the highest level of intentional elements in both organizational and legal models. Actor satisfaction values are calculated from the satisfaction and importance of intentional elements embedded in each actor. After analyzing the GRL models quantitatively or qualitatively, these values are propagated to the UCM models of organization through "responsibility" links. As a result, it is possible to identify the non-compliant business processes and modify them.

LEGAL-URN has several unique characteristics to help organizations achieve compliance. One the major benefits of LEGAL-URN is the use of the same modeling notation for both organizations and regulations which helps achieve a shared understanding of the regulations and enable better comparisons. It promotes reuse across organizations in the same sector through annotating non-relevant parts of the legal models with «No» tags, contributing to this reusability. LEGAL-URN adds precision and formalism to legal statements and URN models via Hohfeldian model structures, deontic modalities and stereotypes. It supports business process compliance with multiple regulations with the pair-wise comparison algorithm and it has a tool support [3] for modeling, verifying, and analysing compliance, and change management. It is worth mentioning that LEGAL-URN does not aim to replace the lawyer or legal experts but it aims to provide guidelines and tool support for business and data analysts and software engineers to align their business processes and softwares with the regulations before the audits happening.

LEGAL-URN framework has yet some limitations which need to be addressed. The framework does not include a regulations repository. Having such a repository can help automation of the process of identifying relevant regulations and ensuring business processes compliance with relevant regulations. Furthermore, developing the Hohfeldian model is currently manual. With the help of an XML-based version of regulations, this process can be semi-automated. It also lacks legal interpretations [8]. Legal interpretations help identify sets of business process patterns which can be legally compliant. These patterns can be used by organizations to build business processes which satisfy the legal goals and the high-level goals of the organization simultaneously.

6   System Integration

LEGAL-URN and Eunomos are complementary systems for compliance monitoring. Our preliminary investigation suggests that integration is perfectly feasible but would require modifications to both systems. Figure 2 shows the integration of the two systems.

The Eunomos repository of laws - with legislative XML for clickable cross-references, definitions of terms and their inter-relationships in specialist ontologies - would replace the LEGAL-URN "Law and Regulation Documents" level. At the legal provisions level, there is a new representation that integrates Eunomos prescriptions and LEGAL-URN Hohfeldian models. Table 1 shows the mapping of fields and relations between the two representations. Many fields can be mapped directly, some require adaptation, and others are taken from one representation.

The integrated solution classifies provisions according to Hohfeldian modalities rather than Deontic Logic because they allow a more refined characterization

Table 1. Integration of Prescriptions and Hohfeldian Models

Prescriptions	Hohfeldian Model	Integrated Representation
Deontic Clause	Hohfeldian Modality	Hohfeldian Modality
Active role	Subject	Responsible Actor
Passive role	-	Beneficiary
Violation	-	Violation
Sanction	-	Sanction
-	Modal Verb	-
Description	Clause	Clause
-	Precondition	Precondition
-	-	Postcondition
IsA Relation	-	IsA Relation
PartOf Relation	-	PartOf Relation
Exception Relation	Exception	Exception Relation
Norm Identifier	Section + Article	Norm Identifier
-	Cross-reference	Cross-reference
-	-	Stakeholder

of legal provisions with an explicit way to represent the hierarchy of norms. The Active Role, or Subject, are essentially the same, and can be more clearly expressed as the Responsible Actor - who has the responsibility for ensuring the provision is fulfilled. This field is essential at the GRL or UCM level. The Passive Role here is renamed as Beneficiary for clarification. Beneficiaries do not need to be represented at the next levels, unless they also have legal responsibilities that need to be modeled. The question of what is violated and what are the possible sanctions are important considerations in compliance decisions, and are represented at the legal provisions level. In LEGAL-URN, sanctions are modeled as "Consequence" goals which have links from Legal-GRL to organizational models.

The modal verb can provide useful clues for the knowledge engineer to classify legal provisions, but is not required as information about the provision in the final analysis. The Description in Eunomos corresponds to the Clause in the Hohfeldian model - simplifying the syntax and adding information from citations. The Precondition from the Hohfeldian model is maintained as it is useful for describing applicability and sequential information. Postcondition is the correlative. The ontological relations from Eunomos - IsA, PartOf and Exception - are used to show the interaction between legal provisions. Clickable hyperlink norm identifiers are used instead of textual citations (Section and Article fields) to enable easy referencing to legal sources.

The major innovation in the integrated solution is the addition of a stakeholder field which classifies the source of the legal provision as constitutional law, legislation, case law, subsidiary laws, ministerial decrees, legal scholars, self-regulatory bodies, industry bodies, internal regulator or external regulator. Different stakeholders have different levels of authority and/or persuasiveness in different jurisdictions and different domains, which is important to take into account in compliance decisions.

At the Legal-GRL level, different interpretations are associated with relevant stakeholders (derived from the legal provisions) and are modeled as alternative realizations of softgoals, just as there are alternative business processes to realize organizational goals. Different interpretation modalities can be integrated in the Legal-GRL model to simulate the possible modalities of interpreting the regulations that an auditor can adopt. Following from how a Legal-GRL is constructed, explained by Ghanavati et al. [36], the softgoals contained in the model can be associated with the purpose of the law intended by its creators. The different interpretation modalities can then be applied (via capturing them as "Source" tasks), while determining whether an execution is compliant with the Legal-GRL model, to identify whether the executions being analysed are compliant with an interpretation of the law.

7   Related Work

The most comprehensive research project in this area is COMPAS [38], which aims to support the entire compliance life-cycle. The project is inspired by the work of Ghanavati et al. [17] on conformance checking.

Contributions in the AI & Law field more often focus on sub-problems rather than the comprehensive system that the integration of Eunomos and LEGAL-URN provides. Bianchi et al. [6] designed a system to help the readers of legal documents to classify terms and view laws, however this approach lacks Eunomos's legislative XML conversion feature. On the other hand, Lu et al. [29] and Kharbili et al. [14], have sought to develop a sophisticated notation for norms and business process models, with the unfortunate drawback that the models are too general for use in legal settings and the notation difficult for legally trained people. Other relevant work in the area are that of Weiss et al. [42], who sought to separate the domain knowledge from the sequence of activities, and Gong [20], who proposes to use agent technology for mapping legal rules onto business processes. While the structure is simple and elegant in theory, in practice the repositories can get unmanagable as organizational procedural rules are mixed with legal rules. Our solution allows a clearer separation between organizational and legal goals.

Combining ontologies with other techniques to study legal documents is not new. For instance *Carneades*, combining ontologies and rules, studies open source compatibility issues [22]. The LKIF ontology set out to model basic concepts of law identified by citizens, legal professionals and legal scholarsis with a reasoning mechanism. However, the system finds its limits on EU Directive 2006/126 on driving licences, a relatively straightforward regulation. One of the biggest challenge for creating ontologies for machine reasoning, as opposed to merely for human understanding, is the sheer amount of basic knowledge and interconnections a machine needs to be provided with.

Francesconi [15] presents an RDF/OWL implementation of Hohfeldian representations of legal provisions to aid information retrieval and automated reasoning. The representation is similar to the integration of Eunomos prescription and

LEGAL-URN hohfeldian models. The main difference is that our work is aimed at legal compliance and use lightweight ontologies rather than semantic web formalism, because they are easier to understand for legal and business practitioners [35].

The use of Requirements Engineering (RE) techniques for compliance monitoring is well-established - a recent systematic literature review [19] shows that Requirements Engineering (RE) techniques, especially, Goal-Oriented Requirements Engineering (GORE) methods have been used to extract and model legal requirements or build business process compliance frameworks. Among these, Rifaut et al. [33] integrate *i** with the ISO/IEC 15540 standard to measure business process compliance with regulations, Siena et al. [41] introduce a new *i**-based language called *Nòmos* modeling normative statements in terms of 8 classes of rights (Hohfeldian ontology), Breaux et al. [10] develop a process to map natural language domain descriptions to deontic logic descriptions.

Beside GORE approaches, some work in RE aim to integrate regulatory compliance with business processes: Karagiannis [26] uses a meta-modeling platform to integrate Business Process Management (BPM) and Enterprise Risk Management (ERM), Kharbili et al. [27] propose a framework for semantic policy-based compliance management for business processes and Schleicher et al. [37] define a refinement process based on *compliance templates*, consists of abstract business processes iteratively refined according to the requirements.

None of the current systems caters adequately for the ever-changing nature of the law, which can result in an unwieldy rules model. Norms and the interpretation of norms need to have a status, active or inactive, and to be linked to explanations and sources for clarification as needed. As the systematic literature review [19] mentioned, while the work mentioned above apply RE and GORE techniques to extract and model legal requirements and establish the compliance, they focus on only few aspects of compliance. The LEGAL-URN framework [16], however, covers all these aspects by providing a formal structure for legal statements and guidelines for mapping procedure for creating Legal-GRL and Legal-UCM models, and by developing semi-automatic compliance analysis.

8 Conclusions and Future Work

This paper proposed integrating two complementary compliance management and monitoring approaches (i.e. LEGAL-URN and Eunomos) to achieve a comprehensive business process compliance solution. Eunomos, a legal knowledge and document management system based on AI techniques, focuses on identifying norms, cross-references and semantic similarities, with a clear structure for representing multiple interpretations and normative change.

LEGAL-URN, on the other hand, applies Requirements Engineering techniques to model regulations in the same notation as business process modeling notations as a business-focused means to analyze business process compliance. We demonstrated that an integration at the level of legal provisions allows for complete traceability from legal sources to business process models, representing regulatory conversations at each level, thereby allowing informed analysis and design of compliant business processes.

LEGAL-URN includes GRL analysis algorithms which can help analyze the impact of different interpretation modalities on organizational business processes and high-level goals. We aim to extend these analysis algorithms to help organizations select a set of business process templates that satisfy concrete rules regulations as well as their high-level goals.

We aim to extend the use of interpretations and comparisons between different regulations in the context of economic globalization. To increase the effectiveness of international regulations, regulatory bodies and researchers are studying different international regulatory contexts such as harmonization, standardization, normalization, reconciliation and mutual recognition for regulations in the financial sector [4]. Laws and regulations are ever-changing. Thus, being more proactive in compliance management and monitoring would better address the complexity of change management. Our system could also integrate techniques that can identify changes in advance for new versions of regulatory text [21].

For a number of years, it has been recognized that the creation and uses of laws and regulations occurs in complex network of stakeholders having different objectives or intentions for regulating complex socio-technical systems (see e.g. [32] for the maritime, aeronautics or nuclear sectors). A main shift that has been made recently and that should be better addressed by our proposal is the focus on evidence-based methods in the legal process [31]. Key Performance Indicators (KPI) are extensively used to link regulations and evidence. KPI should be added to our integration between LEGAL-URN and Eunomos. In the context of GRL and URN, a proposal has been made in the work of Shamsaei et al. [40] on measuring compliance with goal-based legal provisions with key performance indicators.

Acknowledgements. Silvano Colombo Tosatto, Llio Humphreys and Sepideh Ghanavati are supported by the National Research Fund, Luxembourg. Guido Boella is supported by the ITxLaw and European Legal Culture projects, both funded by Compagnia di San Paolo.

References

1. Dworkin, R.M.: The model of rules. The University of Chicago Law Review (1967)
2. Amyot, D., Ghanavati, S., Horkoff, J., Mussbacher, G., Peyton, L., Yu, E.S.K.: Evaluating goal models within the goal-oriented requirement language. Int. J. Intell. Syst. 25, 841–877 (2010)
3. Amyot, D., Mussbacher, G., Ghanavati, S., Kealey, J.: GRL Modeling and Analysis with jUCMNav. In: 5th International i* Workshop (iStar 2011), vol. 766, pp. 160–162. CEUR-WS, Trento (2011)
4. Baker, C.R., Barbu, E.M.: Evolution of research on international accounting harmonization: a historical and institutional perspective. Socio-Economic Review 5(4), 603–632 (2007)
5. Bartrum, I.C.: The modalities of constitutional argument: A primer. In: Edwards, L., Kluwer, W. (eds.) Readings in Persuasion: Briefs that Changed the World (2011)

6. Bianchi, M., Draoli, M., Gambosi, G., Pazienza, M., Scarpato, N., Stellato, A.: ICT tools for the discovery of semantic relations in legal documents. In: Proc. of the 2nd Intl. Conf. on ICT Solutions for Justice, ICT4Justice (2009)
7. Black, J.: Regulatory conversations. Journal of Law and Society 29, 163–196 (2002)
8. Bobbitt, P.: Constitutional Interpretation. Oxford University Press (2006)
9. Boella, G., Humphreys, L., Martin, M., Rossi, P., Violato, A., Torre, L.v.d.: Eunomos, a legal document and knowledge management system for regulatory compliance. In: ITAIS 2011, pp. 571–578. Springer (2012)
10. Breaux, T.D., Antón, A.I.: Analyzing goal semantics for rights, permissions, and obligations. In: RE 2005, pp. 177–188. IEEE Computer Society (2005)
11. Burgemeestre, B., Hulstijn, J., Tan, Y.-H.: Value-based argumentation for justifying compliance. Artificial Intelligence and Law 19(2-3), 149–186 (2011)
12. Cai, Z., Yu, E.: Addressing performance requirements using a goal and scenario-oriented approach. In: Pidduck, A.B., Mylopoulos, J., Woo, C.C., Ozsu, M.T. (eds.) CAiSE 2002. LNCS, vol. 2348, pp. 706–710. Springer, Heidelberg (2002)
13. Cunningham, L.A.: Prescription to retire the rhetoric of principles-based systems in corporate law, securities regulation, and accounting, a. Vand. L. Rev. 60, 1409 (2007)
14. El Kharbili, M., Ma, Q., Kelsen, P., Pulvermueller, E.: Corel: Policy-based and model-driven regulatory compliance management. In: EDOC 2011, pp. 247–256. IEEE (2011)
15. Francesconi, E.: Axioms on a semantic model for legislation for accessing and reasoning over normative provisions. In: Palmirani, M., Pagallo, U., Casanovas, P., Sartor, G. (eds.) AICOL-III 2011. LNCS, vol. 7639, pp. 147–161. Springer, Heidelberg (2012)
16. Ghanavati, S.: Legal-URN framework for legal compliance of business processes, Ph.D. thesis, UOttawa, Canada (2013), http://hdl.handle.net/10393/24028
17. Ghanavati, S., Amyot, D., Peyton, L.: Towards a framework for tracking legal compliance in healthcare. In: Krogstie, J., Opdahl, A.L., Sindre, G. (eds.) CAiSE 2007 and WES 2007. LNCS, vol. 4495, pp. 218–232. Springer, Heidelberg (2007)
18. Ghanavati, S., Amyot, D., Peyton, L.: Compliance analysis based on a goal-oriented requirement language evaluation methodology. In: RE 2009, USA, pp. 133–142 (2009)
19. Ghanavati, S., Amyot, D., Peyton, L.: A systematic review of goal-oriented requirements management frameworks for business process compliance. In: Fourth Int. Workshop on Requirements Engineering and Law, pp. 25–34 (2011)
20. Gong, Y., Janssen, M.: From policy implementation to business process management: Principles for creating flexibility and agility. Government Information Quarterly 29(1), 61–71 (2012)
21. Gordon, D.G., Breaux, T.D.: Assessing regulatory change through legal requirements coverage modeling. In: 2013 21st IEEE International Requirements Engineering Conference (RE), pp. 145–154. IEEE (2013)
22. Gordon, T.F.: Combining rules and ontologies with carneades. In: Proceedings of the 5th International RuleML2011@BRF Challenge (2011)
23. The European Savings Banks Group. Esbg response to cesr on the record keeping requirements arising from the mifid (2006)
24. Husik, I.: Hohfeld's jurisprudence. University of Pennsylvania Law Review and American Law Register, 263–277 (1924)
25. ITU-T. Recommendation Z.151 (10/12), User Requirements Notation (URN) - Language definition, http://www.itu.int/rec/T-REC-Z.151/en (2012)

26. Karagiannis, D.: A business process-based modelling extension for regulatory compliance. In: Multikonferenz Wirtschaftsinformatik, pp. 1159–1173 (2008)
27. Kharbili, M.E., Pulvermüller, E.: A semantic framework for compliance management in business process management. In: Business Process, Services Computing and Intelligent Service Management. LNI, vol. 147, pp. 60–80. GI (2009)
28. Liebwald, D.: Law's capacity for vagueness. International Journal for the Semiotics of Law-Revue Internationale de Sémiotique Juridique 26(2), 391–423 (2013)
29. Lu, R., Sadiq, S., Governatori, G.: Measurement of compliance distance in business work practice. Information Systems Management 25(4), 344–355 (2009)
30. The Committee of European Securities Regulations. The list of minimum records in article 51(3) of the mifid implementing directive: Public consultation (2006)
31. Rachlinski, J.J.: Evidence-based law. Cornell L. Rev. 96, 901 (2010)
32. Rasmussen, J.: Risk management in a dynamic society: a modelling problem. Safety Science 27(2), 183–213 (1997)
33. Rifaut, A., Dubois, E.: Using goal-oriented requirements engineering for improving the quality of ISO/IEC 15504 based compliance assessment frameworks. In: 16th Int. Conf. on Requirements Engineering (RE 2008), Spain, pp. 33–42 (2008)
34. Rifaut, A., Ghanavati, S.: Measurement-oriented comparison of multiple regulations with GRL. In: 5th Int. RELAW Workshop, pp. 7–16. IEEE (2012)
35. Rossi, P., Vogel, C.: Terms and concepts; towards a syllabus for european private law. European Review of Private Law (ERPL) 12(2), 293–300 (2004)
36. Ghanavati, A.R.S., Amyot, D.: Legal goal-oriented requirement language (Legal-GRL) for modeling regulations. In: MiSE @ICSE, India (to be appeared, 2014)
37. Schleicher, D., Anstett, T., Leymann, F., Schumm, D.: Compliant business process design using refinement layers. In: OTM Conferences (1), pp. 114–131 (2010)
38. Schumm, D. (ed.): Compliance-driven Models, Languages, and Architectures for Services (COMPAS): Achievements and Lessons Learned (2011)
39. Sergot, M.: A computational theory of normative positions. ACM Transactions on Computational Logic (TOCL) 2(4), 581–622 (2001)
40. Shamsaei, A., Pourshahid, A., Amyot, D.: Business process compliance tracking using key performance indicators. In: Muehlen, M.z., Su, J. (eds.) BPM 2010 Workshops. LNBIP, vol. 66, pp. 73–84. Springer, Heidelberg (2011)
41. Siena, A., Mylopoulos, J., Perini, A., Susi, A.: Designing law-compliant software requirements. In: Laender, A.H.F., Castano, S., Dayal, U., Casati, F., de Oliveira, J.P.M. (eds.) ER 2009. LNCS, vol. 5829, pp. 472–486. Springer, Heidelberg (2009)
42. van der Pol, J.: Rules-driven business services: Flexibility within the boundaries of the law. In: Ashley, K. (ed.) Proceedings of the Thirteenth International Conference on Artificial Intelligence and Law, ICAIL 2011 (2011)
43. Weiss, M., Amyot, D.: Designing and evolving business models with URN. In: Proc. of Montreal Conf. on eTechnologies (MCETECH 2005), pp. 149–162 (2005)
44. Wenar, L.: Rights (2011), http://plato.stanford.edu/entries/rights/ (Online; accessed July 15, 2011)
45. Westerman, P.: Legal or non-legal reasoning: the problems of arguing about goals. Argumentation and the Application of Legal Rules, page 87 (2009)

Criminal Liability of Autonomous Agents:
From the Unthinkable to the Plausible

Pedro Miguel Freitas[1], Francisco Andrade[1], and Paulo Novais[2]

[1] Law School, Universidade do Minho, Braga, Portugal
{pfreitas,fandrade}@direito.uminho.pt
[2] Informatics Department, Universidade do Minho, Braga, Portugal
pjon@di.uminho.pt

Abstract. The evolution of information technologies have brought us to a point where we are confronted with the existence of agents - computational entities - which are able to act autonomously with little or no human intervention. And their behavior can damage individual or collective interests that are protected by criminal law. Based on the analysis of different models of criminal responsibility of legal persons - which constituted an interesting advance in the criminal law in relation to what was hitherto traditionally accepted -, we will appraise whether the necessary legal elements to have direct criminal liability of artificial entities are present.

Keywords: Criminal liability, Software Agents, Autonomous agents, Objects, Legal Persons.

1   Software Agents and Objects – An Introduction

An "agent" is a computational entity (software/hardware) that "being located in a defined environment acts upon it by autonomous actions, having a defined goal to accomplish" [1]. Thus being, the "agent" performs tasks on behalf of a user in a pre-defined computational environment, with little or no human intervention at all [2]. The agent is capable of analysing its environment and the problem data and of deciding accordingly, in an independent way [3-4].

An important distinction must be considered between "agents" and "objects" [1]. Obviously, the degree of autonomy is much bigger in "agents" [1]. But also the definition of communication mechanisms and used language must be considered [1]. An object is an entity capable of storing an inner state, of using a set of methods acting upon that inner state and of communicating through messages [3-4]; the object has autonomy in the sense that it controls its own state but, contrarily to agents, is not capable of controlling its own behaviour [3-4]. Decision control centres are different in objects and in agents. And it can be said that objects have a static behaviour while "agents" have a dynamic behaviour [3-5]. Another difference arises out of the definition of the dialog mechanism, more complex in "agents" than in objects [4], [6]. And it may be said that while both "agents" and objects have an identity, a state and a behaviour of its own, actually

P. Casanovas et al. (Eds.): AICOL IV/V 2013, LNAI 8929, pp. 145–156, 2014.
© Springer-Verlag Berlin Heidelberg 2014

"agents" may be described in terms of a set of characteristics integrating knowledge, beliefs, desires, intentions, aims and even obligations [7-8].

An "agent" is thus a program capable of acting in a flexible way, on behalf of its owner, user or client[1], in order to reach defined goals. So, it must present a set of properties or characteristics such as autonomy (capacity of taking decisions on which actions to undertake without having to be constantly inquiring the user), reactivity [9]. (capacity of properly responding to prevailing circumstances in dynamic and unpredictable environments), proactivity [9] (capacity of acting in anticipation of future goals), communication, cooperation and sociability [9] and adaptive behaviour. This said, it must be stated that "agents" are not limited to data interchange (such as EDI – Electronic Data Interchange) but are capable of communicating in complex conversational environments and of assuming different roles, as well as adapting to diverse situations [7].

Autonomy is one of the most relevant features of "software agents", implying the possibility of acting and performing tasks without any human intervention. A "software agent" is independent and acts autonomously, having control both of its inner state and of its behaviour, being capable of clearly understanding the goals of its mission and of defining a strategy in order to reach the defined goals. Of course, the levels of autonomy may greatly vary [10] and although the "software agent" may decide autonomously, without any human intervention, the user may have more or less capacity of controlling the parameters influencing the behaviour of the agent [11]. But the greater autonomy of new generations of "software agents", capable not only of acting within pre-established parameters but also of having initiative and deciding, by themselves, what, when and how to do, upon favourable conditions (in the perspective of the "software agent"!) may force us to distinguish the situations in which the user will still have some control upon the strategy to be followed by the "software agent" or at least upon the main parameters of decision [11] from the cases when this control will be totally lost and only the trust (or lack of trust) of the user in the "software agent" capabilities remains. And we may even have to face the possibility of the "software agent", reasoning upon the available data, overcoming what the user may reasonably have foreseen [12].

Software agents are not considered as persons. Yet, they have this capacity of autonomous acting and their acting may well modify the legal position of legal persons. Furthermore, it may be considered that software agents have something more or less equivalent to a "will" or at least what may be called "intentional states" [13-14].

The intentionality of software agents brings along the issue of the legal consideration of the acts of software agents [15]. For the moment being, software agents are not considered as legal persons and the most plausible solution for the consideration of their legal acts is the one suggested by Giovanni Sartor [16] of having commercial corporations specially created for the use of software agents. Thus being, liability for the acts of software agents would impend on commercial corporations [17-18]. This will force us, when concerning criminal responsibilities, to analyse the issue of the consideration of criminal liability of the corporations.

---

[1] Obviously, we are not considering a legal framework (which does not exist) for the actuation of software agents.

2   Criminal Liability of Legal Persons

In fact, one of the most troublesome questions that criminal law is currently facing is the criminal liability of corporations or legal persons. Should criminal penalties be solely imposed upon an individual or should also a legal person be subjected to those penalties and, being so, in what way would that occur?

It is a rather old question but still widely disputed in the context of criminal policy and criminal law, for which, to use an example that is closer to us, only in 2007, with several amendments to the Criminal Code, the Portuguese legislator gave a pragmatic and definitive answer.

Despite that, the replacement of the old principle of Roman law *societas delinquere non potest* has been gradually accepted in countries of Anglo-Saxon legal tradition, such as the United States, which is believed to be one of the first countries to do it, or the UK and in many countries of different legal traditions.

Such a solution – the criminal liability of legal persons – appears to be not only essential for a timely and adequate response of criminal law to an increasingly complex human society, but it also meets the requirements imposed by various international bodies such as the European Union, the Council of Europe and the United Nations, which require States to adopt the necessary legislative measures in order to sanction legal persons for acts that constitute certain offenses.

However, despite this demand for the accountability of legal persons, there is no consensus on the actual manner this should be done. Many different models of responsibility of legal persons exist, and they range from mere tort liability to criminal liability.

The specific reason why there are doubts whether criminal law, a body of law which, as we all know, should only be applied as last resort, when all legal remedies are insufficient, should apply, has to do with two fundamental concepts in criminal law theory: agency/conduct (the common law *actus reus* or the german *Handlung*) and blameworthiness (*mens rea* or Schuld[2]). These are the core challenges necessary to overcome in order to legitimate criminal liability of legal entities.

On the one hand, some authors, including most German authors, defend that the notion of action in the criminal law framework demonstrates that legal entities are not able to act for themselves. Only the natural or physical persons may carry out behaviours that are criminally relevant. And as such, criminal responsibility cannot fall on a legal entity, but rather on the individual [19-20].

On the other hand, many say that it is impossible to morally and ethically judge legal entities for not acting lawfully (despite having the opportunity to abide by the law), due to the fact that blameworthiness for an unlawful action demands the existence of an agent that has free and conscious will and chooses to break the law in an

---

[2] Despite not being totally equivalent, the English legal term mens rea and the german legal term Schuld are, for simplicity reasons, treated as functional equivalents for the purpose of this paper. It should be noted however that depending on the perspective that one assumes on the concept of crime, more especifically, whether it is a classic approach (of such authors as Liszt, Beling and Berner), a neoclassical approach (Mezger), a teleological theory (Welzel) or a functional-teleological and rational system (Schünemann and Roxin), the translation of mens rea to german can be subjektiver Tatbestand and/or Schuld.

hypothesis where he/she could and should have acted differently [21] - in this sense only individuals possess "personal qualities necessary to be censured for not acting differently" [21]. Therefore, the conclusion should be obvious: regarding the lack of ontological unfitness to be blamed, legal entities cannot be held criminally accountable [22]. Accordingly, only the individuals that have committed the relevant criminal acts on behalf of legal entities or in their interest can suffer criminal sanctions, and not the legal entities themselves.

However, one must not neglect that we live in a rapidly evolving society characterized by the discourse of the global risk society [23], which entails a profound paradigm shift in our cultural, economic, sociological and technological dimensions as a community, and brings paramount changes to the way criminality materializes. There is an increasing criminality that involves a greater complexity and organization, frequently having corporations, societies and associations as key actors. Thus, it seems accurate the idea expressed by Figueiredo Dias [24] if we chose to only prosecute and punish physical or biological persons acting on behalf of legal entities, completely waiving criminal accountability of the latter, that would mean that (given the degree of complexity, not only of the committed crimes, especially those against the economy, but also of these legal entities' organizational structures) it would be impossible to specifically determine the individuals that should be held responsible. And so there would be absolute impunity.

Thus, assuming that there is a need for a real and autonomous criminal responsibility of legal persons, how can we overcome these dogmatic obstacles upheld by the traditional thinking of criminal law?

The answers vary. In the Portuguese legal order, Figueiredo Dias rejects the arguments of inability of agency and blameworthiness of legal entities and considers adequate the implementation of a so-called *analogic model* [22]. According to this Author, "the individuals can be replaced, as criminally responsible, both objectively and subjectively, ethical and social hubs, by their collective work and materialization, such as legal entities, associations, groupings or corporations, in which free beings express themselves" [22].

Faria Costa, on the other hand, although not recognizing in his first writings the criminal liability of legal entities [25], later admits the plausibility of their punishment in light of the theory he coined as material rationality of opposite places [26]. The legitimacy of this type of criminal liability is based on a material analogy between the behaviour of natural persons and legal persons: if under 16, natural persons, although having capability to act, are exempt from criminal responsibility, as stated by article 19 of the Portuguese Criminal Code, then it would not be totally unreasonable to accept the punishment of legal persons despite not being physically and anthropologically capable of acting. According to Faria Costa, the criminal justice system, through "axioms developed by criminal dogmatics" [26], constructs "a space of normativity whose essential feature is represented by the absence of a particular characteristic" [26]. This space of normativity can "enlighten and justify, in terms of material rationality, its opposite place" [26]. And so, if with the infancy defence "we have the curtailment of ontological segments of action, here, inversely, there is an extension of a communicational act, criminally relevant; if with infants we limit and remove blameworthiness, here, inversely, the notion blame is reconstructed and the legal person becomes a true centre of imputation" [26].

These theoretical solutions, of greater expressiveness in Portugal, are obviously not exclusive. Examining comparative law, particularly civil law (continental law) countries, we observe that this topic has been debated to exhaustion and to the same exhaustion answers have been offered which aim to support and implement the notion of blameworthiness of legal entities [22]: imputation model – *Zurechnungsmodell* – according to which guilt and action of the responsible corporation boards are imputed to the legal person; model of the culpability of the organization – *Modell des Organisationsverschuldens* [27] – that recognizes the existence of a specific and autonomous blameworthiness of the legal entity, which derives from the idea that the legal entity provides a favourable environment for the practice of certain crimes; model of prevention – *Präventionsmodel* [28] – which acknowledges the possibility of sanctioning legal entities with *security measures*; and, finally, the model of analogue blameworthiness [29], where an analogue imputation of blameworthiness is shed on the legal entity, having as a criteria of criminal imputation an appraisal of the way business was carried out (*Betriebsführungsschuld*).

From the point of view of Common Law [30], we reach the conclusion that opinions are mainly divided between the doctrine of identification and the vicarious liability (or agency doctrine)[3]. The first theory sets up an overlap between the conduct and blameworthiness of individuals in positions of leadership and the conduct and blameworthiness of the legal entity. In other words, those individuals represent the "body" and "mind" of the legal entity and therefore the acts carried out by them must be regarded as being done by the legal entity itself. Meanwhile, the advocates of the second theory stress the liability of the legal entity for the conduct of its agents, meaning that the legal entity is charged with criminal responsibility for the actions of agents such as directors, supervisors, etc.

3   Models of Criminal Liability of Autonomous Agents: An Appraisal

Having outlined the main points of interest on criminal liability of legal entities, a few questions arise: Are there such substantial differences between legal entities and autonomous agents that justify the exemption from criminal responsibility of the latter? Is it plausible to conceive criminal responsibility of autonomous agents or AI entities?

Traditionally AI entities are considered not to have legal personhood. They are said to be mere objects[4]. And this is may be the *punctum crucis* of this question. Throughout the different branches of law (civil law, administrative law, etc.), and in particular criminal law, there is one key distinction that is commonly made between subject (or agent) and object. According to George P. Fletcher, "[a] subject is someone who acts, and an object is someone or something that is acted upon" [31]. Although simple in its

---

[3] There are other approaches to corporate liability, such as the aggregation theory (also termed as collective knowledge doctrine), the culpable corporate culture, the reactive corporate fault.

[4] Here we use the common notion of objects and not "objects" in the sense we referred in the beginning of this paper.

wording, it encompasses complex issues, which we face namely when addressing "software agents" or "artificial entities".

The criminal liability of artificial entities has been a rather unknown territory for legal scholars[5]. There are a few exceptions however.

Gabriel Hallevy has proposed three models of the criminal liability of artificial intelligence entities: Perpetration-via-Another Liability Model; Natural-Probable-Consequence Liability Model; Direct Liability Model [32-35]. They present a sound foundation on which this topic could be further developed and, as such, we should therefore understand the main characteristics of each model.

The first model considers that the AI does not possess any human attribute and so denies the possibility of having the AI as a perpetrator of an offense. It is seen as akin to mentally limited persons, such as a child, a person who is mentally incompetent or one who lacks a criminal state of mind. The AI entity is an innocent agent that is a mere instrument used by the real perpetrator, who architects the offense and constitutes the real mastermind behind it. As such, the person behind the AI is to be held accountable for the conduct (*actus reus*) of the AI, albeit the subjective or internal element (*mens rea*) is determined by the perpetrator-via-another's mental state.

The perpetrator-via-another can either be the programmer or the user: the programmer, when he designs an AI entity with the purpose of committing criminal offenses, or the user (end-user), that, albeit not designing the AI entity, is in control of it, and uses it to commit offenses.

It should be noted that this model assumes that the AI is completely dependent on either the programmer or the user. It is not self-ruling or self-determining, but solely an instrument (equivalent to a hammer or even a dog used for illicit purposes) for which no specific mental state is required, *e.g.* a programmer creates an AI entity to destroy computer data.

Accordingly, this model would not be implemented in hypotheses where the AI entity decides to commit an offense based on its own accumulated experience or knowledge; commits an offense despite not being programmed to do so; acts as a semi-innocent agent[6] [36].

---

[5] We can find an interesting account on this topic on LEGAL-IST Consortium's Report on Legal Issues of Software Agents, Doc. No. D14, Rev. No. 2, 29 March 2006, which for liability purposes drafts a fruitful analogy between software agents (owner of certain cognitive capabilities and mental states) and trained dogs, coined by the authors as the dog model. This model starts by assuming that both software agent and trained dog are programmed to autonomously pursue assigned tasks and goals. Depending on the direction and level of training/programming, the dog's and agent's cannot be completely foreseen in advance, which in turn can lead to unwanted results or even illicit. Disregarding the possibility of holding the AI (Artificial Intelligence) entity directly liable for its actions, there could be criminal responsibility of the developer or user (or in dog's case, the trainer or the owner) for their negligence, imprudence or unskillfulness – this is in essence what is described in the Natural-Probable-Consequence Liability Model. In the hypothesis of wilful misconduct by the trainer/owner/developer/user, criminal liability would rest upon the subject to whom the fact can be led back.

[6] Gabriel Hallevy describes a semi-innocent agent as "a negligent party that is not fully aware of the factual situation while any other reasonable person could have been aware of it under the same circumstances".

The second model – coined Natural-Probable-Consequence Liability Model – presupposes that the programmer or user of the AI entity, despite not programming or using it for the purpose of committing a certain crime, might be held accountable for the crime committed by the AI entity, if the offense is a natural and probable consequence of the AI's conduct. Even though the programmer or user was not aware that the offense was committed until it had already been committed, did not plan to commit any offense and did not take part in the commission of the offense, if there is evidence that they could and should foresee the potential commission of offenses, then they might be prosecuted for the offense.

So, this model does not require the criminal intention of the programmers or the users, as the first model does, but only their negligence, which is criminally relevant due to the fact that a diligent and reasonable programmer and user should be able to foresee the offense and prevent it from happening[7], e.g. a programmer sets up an AI entity to protect a computer system and the latter decides, as part of its mission, to seriously hinder a computer system which it considers a potential threat.

Finally, the Direct Liability Model – the third and last model – aims at providing a theoretical framework for a functional equivalence between AI entities and humans for criminal liability purposes. For this reason, this model deserves greater attention in our analysis, as it constitutes the main focus of our paper. Gabriel Hallevy's reasoning stems from the idea that criminal liability implicates solely the fulfilment of two different requirements: *actus reus* (external element) and *mens rea* (internal element) and if AI entities were able to fulfil them both then criminal accountability would follow.

We have no doubt that if such liability of AI entities were to exist, it should not replace the programmer or user's liability. Both could co-exist, if all the legal requirements were fulfilled, meaning that the criminal liability of AI entities would not exclude the individual responsibility of programmer or users nor would it depend on the criminal accountability of those – similar to what is commonly done when punishing legal entities, where criminal punishment of the individuals behind the legal entity does not constitute a requirement to have the criminal punishment of legal entities themselves. But the problem remains: do AI entities fulfil all necessary requirements to trigger criminal liability?

On one side, regarding the *actus reus* requirement, it is insufficient to propose its fulfilment only when AI entities control a mechanical or other mechanism to move its moving parts (*e.g.* robots). In our view, this argument should clearly be regarded as unbearably limited. If we were to establish the criminal liability of AI entities, why

---

[7] Under this model, Gabriel Hallevy devises two situations that bring different outcomes. The first situation is when programmers or users did not want to commit any offense but negligently programmed or used the AI entity and an offense occurred. In this hypothesis, programmer and user should be held accountable for an offence, as long as there is a negligent offense stated by criminal law for that type of cases. The second situation deals with accomplice liability cases, namely when programmers or users programmed or used the AI to commit one offense, but the latter committed another, in addition or instead of the planned one. The author proposes the punishment of the programmer or the user as if they acted with knowledge and intent. Alongside the criminal liability of the programmer or the user, the AI entity, provided that did not act as an innocent agent, could be directly held liability for its actions.

should those be solely responsible when it is proved that they controlled mechanical instruments or others of the same sort? It seems to be nothing more than an unjustifiable overlap between AI entities and robots. The former, as we know, is not the same as the latter. One example that clearly shows that this confusion between terms can lead to unjust results has to do with computer offenses. Let us imagine that the AI entity, merely software, intentionally decides to target a computer system with a denial-of-service attack (DoS attack). Shouldn't the AI entity be held criminally responsible here as well?

To perceive the fulfilment of the *actus reus* requirement as having willed muscular movement (in this case, mechanical) or bodily movement is to ignore that there are crimes without *actus reus* or *acts* in a traditional sense – *e.g.* computer crimes. Unless we consider that the physical act in computer crimes resides in electronic impulses – which seems to be a far-fetched and unnecessary argument –, to suggest that *actus reus* equals the traditional definition of act is inadmissible. As Figueiredo Dias [22] and David Ormerod [30]remind us, it is misleading or even strange to say that, for example, in the crime of defamation the relevant act corresponds to the movement of one's tongue, mouth and vocal chords. For these reasons, the traditional view of acts as willed voluntary movements is seen, in recent years, as outdated[8] [37].

More importantly we should emphasize the fact that in order to occur the criminal liability of an agent, the conduct proscribed by a certain crime must be done voluntarily. What this actually means it is something yet to achieve consensus, as concepts as consciousness, will, voluntariness and control are often bungled and lost between arguments of philosophy, psychology and neurology, leading the judiciary and legal scholars alike to prefer stating the cases where there is not a voluntary act [38-39]. In these cases, as Jonathan Herring affirms, "an involuntary action is one for which not only is the defendant not responsible, it is not even properly described as *his* act" [37]. So, the voluntariness requirement serves the purpose of excluding from criminal liability those acts that are mere automatisms [22], [30] or done unconsciously. This fact makes clear that AI entities should only be made criminally accountable if they voluntarily acted, which means that must be an act done with will, volition or control. Accordingly, we cannot say that an AI entity voluntarily acted if the presence of one of these internal elements, depending on what particular theory one follows on the characterization of the "voluntarily" concept, is not found in a certain situation. While these elements describe a certain internal state of the agent, they should not however be confused with *mens rea* [39]. There can be volition without *mens rea*, but the contrary is not true[9]. Thus, before turning to a closer insight on *mens rea*, it becomes necessary to call volition (or will or control) into question. While we may find easy to note that volition and human acts generally appear hand in hand, and so in the acts of legal entities, to plunge into the same conclusion as to AI entities' acts would arguably be precipitated.

---

[8] When it comes to punish an absence of behaviour (omission) it must be proved that there was a duty to act and the agent failed to perform such a duty.

[9] Saunders [39] gives the example of the athlete who, during an athletic competition, throws a javelin, after being sure that no person was in his path, but a bystander is hit by the javelin and dies. Despite not having *mens rea* in causing the death of the bystander, there is a voluntary act which consists in throwing the javelin.

Additionally, criminal courts and legal scholars demand the existence of a human action, which means that this voluntary act, whatever it may be, must be carried out by humans and not inanimate objects or animals. This, for us, shows that voluntariness being expressed as a requirement is deeply tangled with demanding human agency. But, as we stated previously, human agency is no longer an absolute and unsurpassable criteria: legal entities are now criminally liable for certain offenses – which could open the path for having criminal responsibility of AI entities.

Finally, recognizing *mens rea* of AI entities can pose a difficult challenge to overcome. There is first a matter of determining the specific level of development of a particular AI entity. Not all AI entities bear the same capabilities, *e.g.*, cognitive skills and abilities, and this should be reflected on whether *mens rea* can be attributed to an AI entity. Secondly, a certain state of mind, which differs from one crime to another, must be attributed to the accused. Some Authors remind us that the only mental requirements needed to impose criminal liability are knowledge, intent, negligence, among others, and peremptorily affirm that knowledge and specific intent can be attributed to AI entities when these have sensory receptors of factual data, which in turn is analysed by the AI entity [33]. Even if AI have sensors which provide them with data that could be processed internally, can we say that the AI entity understands or comprehends what is being processed? This would lead us to the highly controversial "Chinese Room Argument" of John Searle, which is the subject of a never-ending debate with inconclusive results.

Additionally there is the problem that predicates on determining blameworthiness of AI entities. *Mens rea* can be referred to in its general sense or in its special sense [37], [40-41]. To demand the presence of a certain mental state in the agent, which is described by the offense, is to demand *mens rea* in its special sense. But this is not sufficient. Criminal law must ensure that there is only punishment when the agent is at fault [42]. So we must pose the question: can there be any blameworthiness in AI entities' actions that enables their legal punishment?

Criminal conviction encompasses a censure [37] of the agent for acting in a certain fashion. And this relates to the element of guilt/blame/*Schuld* that has to be present. Guilt or *Schuld* is seen, by some Authors (*e.g.* Kaufmann), as censuring someone for acting unlawfully when he could have acted differently; or for acting unlawfully as a result of not promoting a law abiding character or personality (e.g. Mezger). But blameworthiness supposes a free being – with conscious and free will [22] – that has a choice in determining his essence. Although criminal law was used, until late eighteenth century, to punish animals for crimes such as homicide and theft [41], it seems now that invoking criminal law for these cases is, in light of the reasons behind criminal punishment – either retribution, deterrence, rehabilitation or restoration, rather useless and unjust[10]. But as far as science goes, animals lack this ability to become cognizant and influence the "self", at least at the same level humans do [43]. On the other hand, remembering what was stated above on criminal liability of legal entities, there is a theory that could well be called into action: the analogic model [22]. Indi-

---

[10] There are however recent studies that challenge the traditional deterministic view of animal behaviour [44]. And those who proclaim the idea that animals share with humans the possession of neurological substrates that generate consciousness, see The Cambridge Declaration on Consciousness, July 7, 2012.

viduals, or biological people, are free beings that, for criminal purposes, can and should be replaced by their work - as ethical and social cores that too are "products of freedom" or "materialization of free beings" [22]. Provided that AI entities have self-awareness, self-consciousness, free and conscious will, ability to apprehend the (un)lawfulness of their behaviour and means to guide themselves by law, the minimum requirements to call forth their blameworthiness and, hence their criminal responsibility, are present, since they too - AI entities - could embody social and ethical cores, as they are human creations, either directly or indirectly. As a result, in this hypothesis, we reach the dogmatic, juridical and technological apparatus to enable AI entities as active legal actors in criminal justice.

4   Conclusion

The criminal liability of legal person persons has constituted an innovative breakthrough in criminal law and the models used to support such an advance can provide us with invaluable clues to unveil a plausible dogmatic framework for the criminal responsibility of artificial entities. But more importantly, it demonstrates a certain degree of flexibility shown by criminal law when criminal policy demands so.

A flexibility that can be used provided that certain dogmatic premises are met, to justify the punishment of AI entities. The question then will not be anymore whether "can we do it?" but "should we?", "why?" and "how"?

Relying on previous studies put forwarded by Reynolds and Ishikawa, Ugo Pagallo considers three examples of criminal robots [45]: Picciotto Roboto[11]; Robot Kleptomaniac[12] and Robot Falsifier[13], and then points out that today's state-of-the-art in technology is not capable of producing a "Robot Kleptomaniac". It may be so. Legal personality and criminal accountability of AI entities may be nowhere soon. But, living in an ever-evolving world as we do, means that the notion of fully autonomous AI entities or robots is not totally unthinkable, either in battlefields or in our civil life. This argument surely gives grounds to further legal and technical investigation on this topic.

Acknowledgments. This work is part-funded by CROWDSOURCING project (Reference: DER2012-39492-C02-01).

---

[11] The Picciotto Robot hypothesis deals with a robot security guard, deprived of free will or moral sense, which is used by a gang to carry out criminal enterprises. Reynolds and Ishikawa conclude: "As such, it seems that the robot is just an instrument just as factory which produces illegal products might be. The robot in this case should not be arrested, but perhaps impounded and auctioned" [46].

[12] The Robot Kleptomaniac has free will and self-chosen goals and, when confronted with a fixed supply of energy that is running low, chooses to rob batteries from a local convenience store.

[13] The Robot Falsifier example creates awareness for the fact that the Legal Tender project claimed that viewers could remotely operate a robotic system to physically alter purportedly authentic money.

References

1. Wooldridge, M., Jennings, N.R.: Agent Theories, Architectures, and Languages: A Survey. In: ECAI Workshop on Agent Theories, Architectures, and Languages, pp. 1–39 (1994)
2. Wong, H.C., Sycara, K.: Adding Security and Trust to Multi-Agent Systems. In: Proceedings of Autonomous Agents 1999 - Workshop on Deception, Fraud and Trust in Agent Societies, pp. 1–13 (1999)
3. Durfee, E.H., Rosenschein, J.: Distributed Problem Solving and Multi-Agent Systems: Comparisons and Examples. In: Proceedings of the International on Distributed Artificial Intelligence, Seattle, USA, pp. 1–10 (1994)
4. Wooldridge, M.: An Introduction to MultiAgent Systems, 2nd edn. Wiley, UK (2009)
5. Brito, L., Neves, J.: A execução paralela em sistemas multiagente: comunicação, distribuição, coordenação e coligação, vol. I, pp. 4–5. Universidade do Minho, Braga (2000)
6. Jennings, N.R.: Agent-oriented software engineering. In: Imam, I., Kodratoff, Y., El-Dessouki, A., Ali, M. (eds.) IEA/AIE 1999. LNCS (LNAI), vol. 1611, pp. 4–10. Springer, Heidelberg (1999)
7. Fasli, M.: Agent Technology for e-commerce. John Wiley and Sons Ltd, England (2007)
8. Georgeff, M., Pell, B., Pollack, M.E., Tambe, M., Wooldridge, M.J.: The Belief-Desire-Intention Model of Agency. In: Papadimitriou, C., Singh, M.P., Müller, J.P. (eds.) ATAL 1998. LNCS (LNAI), vol. 1555, pp. 1–10. Springer, Heidelberg (1999)
9. Weitzenboeck, E.: Electronic agents and the formation of contracts. International Journal of Law and Information Technology 9(3), 204–234 (2001)
10. Russel, S.J., Norvig, P.: Artificial Intelligence: a modern approach. Prentice Hall, USA (1995)
11. Chavez, A., Maes, P.: Kasbah: an agent marketplace for buying and selling good. AAAI Technical Report. 8-12 (1996)
12. Dowling, C.: Intelligent agents: some ethical issues and dilemmas. In: Proceedings of 2nd Australian Institute of Computer Ethics Conference (AICE2000), pp. 1–5 (2001), http://crpit.com/confpapers/CRPITV1Dowling.pdf
13. Sartor, G.: L'intenzionalitá dei sistemi informatici e il diritto. Rivista Trimestrale di Diritto e Procedura Civile, Anno LVII. 23-51 (2003)
14. Sartor, G.: Cognitive automata and the law: electronic contracting and the intentionality of software agents. Artificial Intelligence and Law 17, 253–290 (2009)
15. Andrade, F., Novais, P., Machado, J., Neves, J.: Contracting Agents: legal personality and representation. Artificial Intelligence and Law 15, 357–373 (2007)
16. Sartor, G.: Agents in Cyberlaw. Proceedings of the workshop on the law of electronic agents – LEA 2002 (2002)
17. Wettig, S., Zehendner, E.: The electronic agent: a legal personality under German law? In: Proceedings of 2nd Workshop on Law an Electronic Agents – LEA 2003, pp. 57–112 (2003)
18. Wettig, S., Zehendner, E.: A legal analysis of human and electronic agents. Artificial Intelligence and Law 12(1-2), 111–135 (2004)
19. Correia, E.: Direito Criminal, I. Almedina, Portugal (2010)
20. Martín, G.: La responsabilidad penal de la propias personas jurídicas. In: Puig, M., Peña, L. (eds.) Responsabilidad penal de las empresas y sus órganos y responsabilidad por el producto. Bosch, Spain, pp. 35–74 (1996)
21. Castro e Sousa, J.: As pessoas colectivas em face do direito criminal e do chamado direito de mera ordenação social. Coimbra Editora, Portugal (1985)

22. Dias, F.: Direito Penal, Parte Geral, vol. I, 2nd edn. Coimbra Editora, Portugal (2012)
23. Beck, U.: Risk society: towards a new modernity. Sage Publications, London (1992)
24. Dias, F.: Pressupostos da Punição e Causas que Excluem a Ilicitude e a Culpa. In: Centro de Estudos Judiciários (org.): Jornadas de Direito Criminal I, 41-83 (1983)
25. Costa, F.: Aspectos fundamentais da problemática da responsabilidade objectiva no direito penal português. Coimbra Editora, Portugal (1981)
26. Costa, F.: Responsabilidade jurídico-penal da empresa e dos seus órgãos (ou uma reflexão sobre a alteridade nas pessoas colectivas, à luz do direito penal). Revista Portuguesa de Ciência Criminal 2(4), 537-559 (1992)
27. Tiedemann, K.: Die "Bebußung" von Unternehmen nach dem 2. Gesetz zur Bekämp-fung der Wirtschaftskriminalität. Neue Juristische Wochenschrift, 1169-1232 (1988)
28. Schünemann, B.: Die Strafbarkeit der juristischen Person aus deutscher und europäischer Sicht. In: Schünemann, B., Gonzalez, C.S. (eds.) Bausteine des europäischen Wirt-schaftsstrafrechts. Madrid-Symposium für Klaus Tiedemann, pp. 265–295. Heymanns, Berlin (1994)
29. Heine, G.: Die strafrechtliche Verantwortung von Unternehmen. Nomos, Germany (1995)
30. Ormerod, D.: Smith and Hogan, Criminal Law, 12th edn. Oxford University Press, Oxford (2008)
31. Fletcher, G.P.: Basic Concepts of Criminal Law. Oxford University Press, Oxford (1998)
32. Hallevy, G.: I, Robot - I, Criminal – When Science Fiction Becomes Reality: Legal Liabil-ity of AI Robots Committing Criminal Offenses. Syracuse Sci. & Tech. L. Rep. 22, article 1, 1-37 (2010)
33. Hallevy, G.: The Criminal Liability of Artificial Intelligence Entities – from Science Fic-tion to Legal Social Control. Akron Intellectual Property Journal 4(2), 171–201 (2010)
34. Hallevy, G.: Virtual Criminal Responsibility. Original Law Review 6(1), 6–27 (2010)
35. Hallevy, G.: Unmanned Vehicles – Subordination to Criminal Law under the Modern Concept of Criminal Liability. J. of Law, Info. & Sci. 21(2), 200–211 (2012)
36. Hallevy, G.: The Matrix of Derivative Criminal Liability. Springer, Heidelberg (2012)
37. Herring, J.: Criminal Law: Text, Cases, and Materials, 5th edn. Oxford University Press, Oxford (2012)
38. Hamilton, M.: Reinvigorating Actus Reus: The Case for Involuntary Actions by Vet-erans with Post-Traumatic Stress Disorder. Journal of Criminal Law 16(2), art. 2, 346–390 (2011)
39. Saunders, K.W.: Voluntary Acts and the Criminal Law: Justifying Culpability Based on the Existence of Volition. U. Pitt. L. Rev. 49, 443–476 (1988)
40. Kadish, S.H., Schulhofer, S.J.: Criminal Law and Its Processes: Cases and Materials, 7th edn. Aspen Publishers, Inc., New York (2001)
41. Wilson, W.: Criminal Law, 3rd edn. Longman, England (2008)
42. Jefferson, M.: Criminal Law, 8th edn. Longman, England (2008)
43. Morin, A.: Levels of consciousness and self-awareness: A comparison and integration of various neurocognitive views. Consciousness and Cognition 15(2), 358–371 (2006)
44. Brembs, B.: Towards a scientific concept of free will as a biological trait: spontaneous ac-tions and decision-making in invertebrates. Proc. R. Soc. B. 278(1707), 930–939 (2011)
45. Pagallo, U.: Robots of Just War: A Legal Perspective. Philos. Technol. 24, 307–323 (2011)
46. Reynolds, C., Ishikawa, M.: Robotic thugs. Ethicomp Proceedings, Global e-SCM Re-search Center & Meiji University, 487–492 (2007)

Extraction of Legal Definitions and Their Explanations with Accessible Citations

Makoto Nakamura[1], Yasuhiro Ogawa[2,3], and Katsuhiko Toyama[2,3]

[1] Japan Legal Information Institute, Graduate School of Law, Nagoya University,
Nagoya, Japan
mnakamur@law.nagoya-u.ac.jp
[2] Information Technology Center, Nagoya University, Nagoya, Japan
[3] Graduate School of Information Science, Nagoya University, Nagoya, Japan
{toyama,ogawa}@is.nagoya-u.ac.jp

Abstract. The aim of this paper is to produce a Japanese legal terminology consisting of legal terms and their explanations that includes accessible citations. Although we have succeeded in finding over 14,000 terms with high precision, 23.1 percent of the correct explanations included citations that were inaccessible due to context-dependent format. We propose a method for revising explanatory sentences that takes into account XML-tag annotation for context-independent format for all citations. The effectiveness of this method is confirmed by our experimental results.

Keywords: Japanese statutes, Definitions, XML, Citations.

1 Introduction

The goal of this research is to construct a legal terminology for translators in which each entry consists of a legal term and its explanation. This terminology is expected to improve not only systematic translations, but also appropriate word selection depending on context. This study is related to the Ministry of Justice's Japanese Law Translation Database System project[1], which was released in 2009 [1]. The number of laws translated into English for publication has increased slowly and, as of August 7, 2013, only 339 of the over 7,800 (< 5%) acts and regulations have been translated. One of the most crucial issues remaining for translation is disunity in word selection. Since a number of human translators are involved, many Japanese legal terms have a variety of English translations. We have even found that act titles are often translated differently in citations [2]. Obviously, each expression should have a specific translation for consistency of meaning. Although the government has compiled a standard translation dictionary for legal terms, the number of entries (3,594 in the latest version) is not sufficient for unified translation.

---

[1] http://www.japaneselawtranslation.go.jp

P. Casanovas et al. (Eds.): AICOL IV/V 2013, LNAI 8929, pp. 157–171, 2014.
© Springer-Verlag Berlin Heidelberg 2014

We first focused on collecting legal terms as defined by statute. We started by compiling a Japanese act corpus consisting of all acts enacted between 1947 and 2012, from which all the legal definitions can be extracted. This completeness of processing for all the acts is significant in our study. We developed an automatic method for extracting the tuples of a legal definition and its explanation from the Japanese act corpus. As a result, we succeeded in finding over 14,000 terms with high precision [3]. However, we found that 23.1 percent of the correct explanations included inaccessible citations. The notation of citations in statutes varies for provisions or acts, some of which are written with an abbreviation or a relative address from the explanation location. In other words, citations found in the explanation written in particular notations can no longer access the specified provision or act. If written with an accessible notation, the explanations would be readable and refer to the specified provision or act.

Our aim is to provide a legal terminology containing the tuples of a legal term and its explanation that also includes an accessible citation. This is achieved by replacing the citation with particular expressions to an independent notation located apart from the original location of the explanation. Our main idea is to XMLize Japanese acts and annotate citation tags in an accessible format. This task is an application of natural language processing to Japanese legal texts. Some studies on reference resolution in legal texts have been written in several European languages [4–7]. In this paper, we deal with Japanese acts by considering the characteristics of the Japanese language and Japanese statutes.

Our paper is organized as follows: In Section 2, we provide the background of this study. Section 3 discusses the types of citations included in the explanatory sentences. In Section 4, we propose a method for annotating in a context-independent format. We describe how this method works with experiments in Section 5, and we conclude and discuss our future work in Section 6.

2 Japanese Legal Text Processing

In this section, we briefly explain Japanese legal text processing. We introduce the basic structure of Japanese laws and the Japanese legislative system in the first two subsections, and conclude with a discussion of our previous study using our Japanese act corpus.

2.1 Basic Explanation of Japanese Laws

In general, laws are roughly divided into written and unwritten categories. Although unwritten laws include local customs and judicial precedents, we do not deal with these in this paper. Written laws are also called statutes, which are further divided into acts and bylaws. While acts are enacted by the National Diet (Parliament), bylaws consist of orders enacted by the cabinet and ordinances and regulations enacted by various ministries.

In this paper, we focus on statutory texts. A statute consists of a number of articles, each of which may be further subdivided into a number of paragraphs or items. Articles, paragraphs, and items have sequential numbers with different typefaces. A provision denotes an independent article or a paragraph.

2.2 Basic Explanation of the Japanese Legislative System

The rational nature of the legislative system of Japanese law maintains the notation of expressions of statutes.[2] Although the Cabinet and Diet members can submit a bill to the National Diet, most bills are introduced by the Cabinet. In this case, the proper authority for that law basically makes a draft of the bill. Once this is accomplished, the authority negotiates with other authorities. The Cabinet Legislation Bureau then closely examines the draft in terms of inconsistency with other statutes, expressions, formats, and so on. As a result, even the usage of commas and periods is maintained. When a Diet member submits a bill, it is reviewed by the Legislation Bureau of the House of Representatives or Councilors.

Not every country's legislative system is similar to that of Japan. In the United Kingdom, the legislature's description check is not as strict, as in most cases the bill is drafted outside of the ministry. In the United States, there is no organization or system for the legislature's description check. In Asian countries other than Japan and Korea, often each ministry independently prepares a draft of a bill without coordinating with other ministries. As a result, the notation of bills differs among ministries. Moreover, in some countries bills are often modified during deliberation in the national assembly, while bills mostly pass the National Diet in Japan as drafted.

Since this political process results in consistencies in notation, this strict wording style may be an idiosyncratic feature of Japanese statutes. This suggests that simple text processing is sufficient to locate important terms or phrases with conventional expressions.

2.3 Definition of Legal Terms and Their Explanations

Although several methods for the extraction of legal terms have been proposed [8–11], the organization of terminology differs depending on their purposes. In this paper, a legal term is a term explicitly defined in an act prior to use and includes both the legal term and its explanation. These terms typically take the form of

- An independent provision or
- A statement in parentheses.

Figure 1[3] shows examples of definitions found in both provisions and parentheses, where Article 2, paragraphs (1), (2), and (3) are independent provisions that define the terms *"administrative organs," "incorporated administrative agencies, etc.,"* and *"official statistics,"* respectively. A defined term is placed in quotations

---

[2] This section is based on our discussion with Prof. Matsuura of the Graduate School of Law, Nagoya University.

[3] Hereinafter, Japanese sentences are immediately followed by their English translations. We referred to the Japanese Law Translation Database System for English translations [1].

統計法（平成十九年法律第五十三号）/ Statistics Act (Act No. 53 of 2007)

第二条　この法律において「行政機関」とは、法律の規定に基づき内閣に置かれる機関若しくは内閣の所轄の下に置かれる機関、宮内庁、内閣府設置法（平成十一年法律第八十九号）第四十九条第一項若しくは第二項に規定する機関又は国家行政組織法（昭和二十三年法律第百二十号）第三条第二項に規定する機関をいう。/

Article 2　(1) The term *"administrative organs"* as used in this Act means organs established within the Cabinet or organs established under the jurisdiction of the Cabinet pursuant to the provisions of laws, the Imperial Household Agency, organs provided in Article 49, paragraph (1) or paragraph (2) of the Act for Establishment of the Cabinet Office (Act No. 89 of 1999) or organs provided in Article 3, paragraph (2) of the National Government Organization Act (Act No. 120 of 1948).

2　この法律において「独立行政法人等」とは、次に掲げる法人をいう。/

(2)　The term *"incorporated administrative agencies, etc."* as used in this Act means juridical persons listed as follows:

　一　独立行政法人（独立行政法人通則法（平成十一年法律第百三号）第二条第一項に規定する独立行政法人をいう。次号において同じ。）/

　　(i)　*Incorporated administrative agencies*　(meaning incorporated administrative agencies provided in Article 2, paragraph (1) of the Act on General Rules for Incorporated Administrative Agencies　(Act No. 103 of 1999; the same shall apply in the following items); (snip)

3　この法律において「公的統計」とは、行政機関、地方公共団体又は独立行政法人等（以下「行政機関等」という。）が作成する統計をいう。/

(3)　The term *"official statistics"* as used in this Act means statistics produced by administrative organs, local public entities, or incorporated administrative agencies, etc.　(hereinafter referred to as *"administrative organs, etc."*).

Fig. 1. Example of definitions

(*"term"*) and its explanation is underlined. Defined terms are properly extracted using pattern match [12]. In addition, definitions appear in parentheses in Article 2, paragraph (2), item (i) and paragraph (3) for the terms *"incorporated administrative agencies"* and *"administrative organs, etc.,"* respectively.

The second item is further divided into two types:

- A defined term appears in parentheses following its explanation in the main text, as shown in Article 2, paragraph (3). The term *"administrative organs, etc."* is the defined term in this example. Abbreviations are often defined in this way.
- A sentence in parentheses explains the legal term just before the parentheses, as shown in Article 2, paragraph (2), item (i) *"Incorporated administrative agencies."*

簡易生命保険法の一部を改正する法律（平成三年法律第三十号）／

Act for Partial Amendment of the Postal Life Insurance Act (Act No. 30 of 1991)

簡易生命保険法（昭和二十四年法律第六十八号）の一部を次のように改正する。／

The Postal Life Insurance Act (Act No. 68 of 1946) shall be partially amended as follows:

第二十四条第二項中「七十二万円」を「九十万円」に改める。／

The term "720,000 yen" in Article 24, Paragraph (2) shall be replaced with the term "900,000 yen."

Fig. 2. Example of an amendment act

Legal terms or explanations in parentheses are easily extracted by analysis of the character string, but analysis of the content outside of the parentheses is not so simple [13]. The difficulty comes from wording that is peculiar to statutory sentences. Despite the presence of high-quality dependency parsers for Japanese, we cannot count on their performance with legal texts. Since legal sentences are designed to avoid ambiguity of expression, they are likely to be long and syntactically complicated, which often leads to a parsing failure. Therefore, we employ a simple method based on pattern match and thus do not rely on a syntactic parser.

2.4 Japanese Act Corpus

We compiled a corpus of all of the Japanese acts, consisting of 9,915 acts enacted up to 2012 since enforcement of the new constitution of Japan in 1947. The size of the corpus is 252 MByte. This Japanese act corpus is based on articles of legislation in official gazettes. Since most of these acts, especially the older ones, are digitally scanned, there are many typographical errors that are not included in the published versions. We developed a preprocessor to address these typographical errors.

Since amendment acts describe how to revise pre-existing acts using amendment language [14], it is difficult to properly extract legal terms unless accurately consolidated. Figure 2 shows an example of an amendment act. Despite the terms in quotes, there is no term definition in this provision. Since the terms in quotes are supposed to be consolidated into the main clause of the original act, a tuple of a legal term and its explanation is unlikely to be extracted from the amendment act. Therefore, we eliminate in advance all acts concerning the amendment or repeal of pre-existing acts, which can be inferred from their titles, as well as supplementary provisions in other acts, which may include amendment.

2.5 Evaluation of Our Previous Study

Table 1 shows the total number of definitions and their explanations collected in our previous experiment [3]. Since some terms are defined in multiple acts, the

Table 1. Analysis of collected definitions and their explanations

Definitions	#Tokens	#Types	Correct	Incorrect	Precision	Recall
in provisions	5,250	3,799	98 (15)	2	0.980	0.980
in parentheses	9,624	6,030	84 (27)	16	0.840	0.392
Total	14,874	9,368	182 (37)	18	0.910	0.511

number of types differs from that of tokens. The precision scores were calculated from 100 samples chosen at random. The figures in parentheses under "Correct" denote the number of correct explanations, including citations in which their specified provisions or acts are inaccessible, as will be described in detail in Section 3. The recall scores were calculated based on the assumption that all legal terms in quotations are perfectly obtained with our method. Since 11,004 terms in parentheses are still unprocessed by the current method despite detection of only the terms, the recall score for the terms in parentheses was estimated. According to our experimental results, 14,874 tokens and 9,368 types of terms were extracted with high precision.

3   Citations Included in Explanatory Sentences

A number of the correctly extracted explanations include citations to other items, paragraphs, articles, or acts. Our analysis revealed that 23.1 $(= (15 + 27)/(98 + 84) \times 100)$ percent of the correct explanations include citations for which the specified provisions or acts are inaccessible. Further investigation revealed that 23.9 percent of the whole includes the inaccessible citations.

The citation format is categorized as follows:

1. The absolute addressing method (**TYPE1**), which is expressed as the full notation of the location for the reference consisting of the title of the act, the article number, and so on, as shown in Article 2, paragraph (1) in Fig. 1.
2. The absolute addressing method with abbreviated expression (**TYPE2**), which shows only the notational difference between the reference address and the current address or the address previously referred to, as shown in Article 2, paragraph (4) in Fig. 3.
3. The relative addressing method (**TYPE3**), which shows the relative distance from the current address, such as a previous article, following items, and so on, as shown in Article 5, paragraph (2) in Fig. 3.

TYPE2 and TYPE3 must coincide with the current address.

From the provisions shown in Fig. 1 and Fig. 3, the above-mentioned tuples are enumerated with their citations underlined as follows:

1. *Administrative organs*: Organs established within the Cabinet or organs established under the jurisdiction of the Cabinet pursuant to the provisions of laws, the Imperial Household Agency, organs provided in Article 49, paragraph (1) or paragraph (2) of the Act for Establishment of the Cabinet Office (Act No. 89 of 1999), or organs provided in Article 3, paragraph (2) of the National Government Organization Act (Act No. 120 of 1948).

統計法（平成十九年法律第五十三号）/ Statistics Act (Act No. 53 of 2007)

第二条 / **Article 2**

4　この法律において「**基幹統計**」とは、次の各号のいずれかに該当する統計をいう。/

(4)　The term "*fundamental statistics*" as used in this Act means statistics falling under any of the following items:

一　第五条第一項に規定する国勢統計/

(i)　Population census statistics provided in Article 5, paragraph (1);

二　第六条第一項に規定する国民経済計算/

(ii)　National accounts provided in Article 6, paragraph (1); (snip)

第五条　総務大臣は、本邦に居住している者として政令で定める者について、人及び世帯に関する全数調査を行い、これに基づく統計（以下この条において「国勢統計」という。）を作成しなければならない。 /

Article 5　(1) With regard to persons specified by a Cabinet Order as those residing in Japan, the Minister of Internal Affairs and Communications shall conduct a complete census concerning individuals and households and produce statistics based on such a census (hereinafter referred to as "*population census statistics*" in this Article).

2　総務大臣は、前項に規定する全数調査（以下「国勢調査」という。）を十年ごとに行い、国勢統計を作成しなければならない。（以下略）/

(2)　The Minister of Internal Affairs and Communications shall conduct a complete census as specified in the preceding paragraph　(hereinafter referred to as the "*population census*") every ten years and produce population census statistics. (snip)

Fig. 3. Example of explanations including specified citations

2-1. *Fundamental statistics*: Population census statistics provided in Article 5, paragraph (1).

2-2. *Fundamental statistics*: National accounts provided in Article 6, paragraph (1).

3. *Population census*: A complete census as specified in the preceding paragraph.

We intend to construct a terminology from these tuples in which each explanation is written in an independent phrase apart from the original act. In other words, citations should not be written in the notation of TYPE2 or TYPE3. From this standpoint, we can review explanations as follows:

- Item 1. requires citing to specified provisions for which the full notation of TYPE1 is accessible.
- Both items 2-1. and 2-2. refer to provisions in other articles belonging to the same act as these provisions. In this case, addresses are expressed in TYPE2, that is, only the notational difference between the reference address and that of the explanation location or the address previously referred to.

– Item 3. employs TYPE3, referring to the preceding paragraph. It is impossible to reach the designated provision unless the location of this explanation is clear.

In this paper, we propose a method to replace the TYPE2 and TYPE3 notations to that of TYPE1.

4 Approach to Revision of Explanatory Sentences

We propose a method for revising the explanatory sentences that takes into account annotation of the absolute address for all the citations, as follows: (1) XMLizing Japanese acts; (2) extending the XML format to absolute addressing; and (3) running the extraction method with revision of the addresses. We address these issues in the following subsections.

4.1 XML Tagging to the Japanese Act Corpus

The Japanese Law Translation Database System project provides law data in XML format, as well as document type definition (DTD), for Japanese statutory laws including definitions for 103 elements and 75 attributes. According to the *One Source Multi Use* policy, users can easily reuse and reformat this law data for their own purposes [1].[4]

We applied the XML format to Japanese acts in our corpus introduced in Section 2.4. Figure 4 shows an example of an act in the XML format. Note that this act is translated into English for readability.

The project also developed a tool for automatic annotation using pattern match based on the strict wording style mentioned in Section 2.2. Although actual statutory data in the XML format are released after manual modification, only five percent of all the statutes has been completed so far due to the progress of translation. We divert this tool to our Japanese act corpus.

4.2 Method for Annotating Absolute Addresses

We extend the XML format to the annotation of absolute addressing. Although we dealt with all the strings in a plain text file in the previous version, the new method is restricted to strings in the 'Sentence' tag or other content tags. The string is annotated with an 'a' tag if it matches an act title in the list of act titles or the notation of TYPE2 or TYPE3. If the newly annotated 'a' tag is nested, the inside annotation is eliminated. Finally, all the 'a' tags are reviewed and an attribute of 'href' is added. Obviously, both 'a' and 'href' are loanwords from HTML.

Figure 5 shows XML tags for citations where the provisions correspond to the ones shown in Fig. 3 and part of Fig. 1. The new attribute 'id' is added

---

[4] The DTD is downloadable at http://www.japaneselawtranslation.go.jp/

Statistics Act (Act No. 53 of 2007)

```
<?xml version="1.0" encoding="UTF-8"?>
<!DOCTYPE Law PUBLIC "-//JaLII//DTD J-STATUTE 1.0//EN" "jstatute.dtd">
<Law OriginalPromulgateDate="May 23, 2007" LawType="Act" Lang="en" Year="19"
                                                  Era="Heisei" Num="053">
  <LawNum>Act No. 53 of May 23, 2007</LawNum>
  <LawBody>
    <LawTitle>Statistics Act</LawTitle>
    <EnactStatement>
      All provisions of the Statistics Act (Act No. 18 of 1947) shall be revised.
    </EnactStatement>
    <TOC>
      <TOCLabel>Table of Contents</TOCLabel>
                        *** snip ***
    </TOC>
    <MainProvision>
      <Chapter Num="1" >
        <ChapterTitle>Chapter I General Provisions</ChapterTitle>
                        *** snip ***
        <Article Num="2" >
          <ArticleCaption>(Definitions)</ArticleCaption>
          <ArticleTitle>Article 2</ArticleTitle>
          <Paragraph Num="1" >
            <ParagraphNum>(1)</ParagraphNum>
            <ParagraphSentence>
              <Sentence>The term "administrative organs" as used in this Act means
                        *** snip ***
                Act (Act No. 120 of 1948).</Sentence>
            </ParagraphSentence>
          </Paragraph>
          <Paragraph Num="2" >
            <ParagraphNum>(2)</ParagraphNum>
            <ParagraphSentence>
              <Sentence>The term "incorporated administrative agencies, etc." as used in
                this Act means juridical persons listed as follows:</Sentence>
            </ParagraphSentence>
            <Item Num="1">
              <ItemTitle>(i)</ItemTitle>
              <ItemSentence>
                <Sentence>Incorporated administrative agencies (meaning incorporated
                        *** snip ***
                  1999; the same shall apply in the following items);</Sentence>
              </ItemSentence>
            </Item>
            <Item Num="2">
              <ItemTitle>(ii)</ItemTitle>
              <ItemSentence>
                <Sentence>Juridical persons specified by a Cabinet Order among those
                        *** snip ***
                  administrative agencies is required for their incorporation.</Sentence>
              </ItemSentence>
            </Item>
          </Paragraph>
        </Article>
                        *** snip ***
      </Chapter>
    </MainProvision>
    <SupplProvision >
      <SupplProvisionLabel>Supplementary Provisions</SupplProvisionLabel>
                        *** snip ***
    </SupplProvision>
  </LawBody>
</Law>
```

Fig. 4. Example of an act in XML format (English version)

```
Statistics Act (Act No. 53 of 2007)
        <Article Num="2" id="at2">
         <ArticleCaption>(Definitions)</ArticleCaption>
         <ArticleTitle>Article 2</ArticleTitle>
         <Paragraph Num="1" id="at2pr1">
          <ParagraphNum>(1)</ParagraphNum>
          <ParagraphSentence>
           <Sentence>The term "administrative organs" as used in this Act means
                     organs established within the Cabinet or organs established
                     under the jurisdiction of the Cabinet pursuant to the provisions
                     of laws, the Imperial Household Agency, organs provided in
                     <a href="H11HO089.html#at49pr1">Article 49, paragraph (1)</a> or
                     <a href="H11HO089.html#at49pr2">paragraph (2) of the Act for
                     Establishment of the Cabinet Office (Act No. 89 of 1999)</a> or
                     organs provided in <a href="S23HO120.html#at3pr2">Article 3,
                     paragraph (2) of the National Government Organization Act
                     (Act No. 120 of 1948)</a>.</Sentence>
          </ParagraphSentence>
         </Paragraph>
                            *** snip ***
         <Paragraph Num="4" id="at2pr4">
          <ParagraphNum>(4)</ParagraphNum>
          <ParagraphSentence>
           <Sentence>The term "fundamental statistics" as used in this Act means
                     statistics falling under any of the following items:</Sentence>
          </ParagraphSentence>
          <Item Num="1" id="at2pr4it1">
           <ItemTitle>(i)</ItemTitle>
           <ItemSentence>
             <Sentence>Population census statistics provided in
                       <a href="#at5pr1">Article 5, paragraph (1)</a>;</Sentence>
           </ItemSentence>
          </Item>
          <Item Num="2" id="at2pr4it2">
           <ItemTitle>(ii)</ItemTitle>
           <ItemSentence>
             <Sentence>National accounts provided in
                       <a href="#at6pr1">Article 6, paragraph (1)</a>;</Sentence>
           </ItemSentence>
          </Item>
                            *** snip ***
         </Paragraph>
        </Article>
         <Article Num="5" id="at5">
          <ArticleCaption>(Population Census Statistics)</ArticleCaption>
          <ArticleTitle>Article 5</ArticleTitle>
          <Paragraph Num="2" id="at5pr2">
           <ParagraphNum>(2)</ParagraphNum>
           <ParagraphSentence>
            <Sentence Num="1" Function="Main">The Minister of Internal Affairs and
                      Communications shall conduct a complete census as specified
                      in <a href="#at5pr1">the preceding paragraph</a> (hereinafter
                      referred to as the "population census") every ten years and
                      produce population census statistics.</Sentence>
            <Sentence Num="2" Function="Proviso"> *** snip *** </Sentence>
           </ParagraphSentence>
          </Paragraph>
         </Article>
```

Fig. 5. Example of XML tags for citation (English version)

to the tags, denoting articles, paragraphs, items, and so on for the absolute address. For example, the attribute 'id="at2pr4it1"' means the item is located at 'Article 2, paragraph (4), item (1).' Likewise, a citation is tagged with its absolute address denoted by the attribute 'href' in the 'a' tag. The attribute takes on the notation of *"(Statute)#(Provision),"* where the former part denotes the cited statute expressed by a part of the uniform resource identifier (URI), which enables access to the statutory database run by the Ministry of Internal Affairs and Communications[5]. It is left blank if the source statute is cited. The latter uses the same format as the 'a' tag. For example, the attribute 'href="H11HO089.html#at49pr1"' shows Article 49, paragraph (1) of the Act for Establishment of the Cabinet Office (Act No. 89 of 1999). This act can be seen on the online database[6].

4.3 Extraction Method with Revision of Citation Addresses

In the process of extraction, a text may include a citation expression with an 'a' tag, which needs to be replaced with its absolute address. Given an 'a' tag with an absolute address, the citation is correctly decoded to a Japanese expression.
   Examples are shown as follows:

1.′ `<a href="H11HO089.html#at49pr1">Article 49, paragraph (1)</a>`
   ⇒ Article 49, paragraph (1) of the Act for Establishment of the Cabinet Office (Act No. 89 of 1999)

2-1.′ `<a href="#at5pr1">Article 5, paragraph (1)</a>`
   ⇒ Article 5, paragraph (1) of Statistics Act (Act No. 53 of 2007)

2-2.′ `<a href="#at6pr1">Article 6, paragraph (1)</a>`

   ⇒ Article 6, paragraph (1) of Statistics Act (Act No. 53 of 2007)

3.′ `<a href="#at5pr1">the preceding paragraph</a>`
   ⇒ Article 5, paragraph (1) of Statistics Act (Act No. 53 of 2007)

Item 1.′ shows a part of a citation that is tagged because the rest is shared with another connected with a coordinate conjunction[7]. According to the attribute href, the phrase "Article 49, paragraph (1)" is decoded and the act title is added by referring to the list of act titles. The remainder in the other items can also be replaced with absolute addresses corresponding to their href tags. Since the statute title is left blank, the title of the source statute, Statistics Act (Act No. 53 of 2007), is added. This can also be replaced with Japanese expressions according to the same procedure.

---

[5] http://www.e-gov.go.jp/ (in Japanese)

[6] http://law.e-gov.go.jp/htmldata/H11/H11HO089.html (in Japanese)

[7] The regions of the citations tagged for Article 49, paragraph (1) and paragraph (2) do not correspond to that of the Japanese version due to a difference in word order. They are actually separated into "Article 49, paragraph (1) of the Act for Establishment of the Cabinet Office (Act No. 89 of 1999)" and "paragraph (2)," respectively.

5   Experimental Results

We examined the collected explanations. In the XML-tagging process, 148 out of 9,915 acts failed due to the presence of rare styles. These are still under investigation. Moreover, 15 percent of the citations were not annotated correctly according to calculation using a random sampling of 100. One reason is the failure of annotation for statutory titles.

In the previous study [3], the precision scores were calculated from 100 samples chosen at random for each of the definitions in provisions and in parentheses, of which 42 out of 182 correct explanations (23.1%) included citations with a context-dependent format. Table 2 shows an experimental result for the inaccessible citations included in the previous study [3]. A new trial with the same samples succeeded in replacing 27 out of these 42 citations with those in context-independent format except for 13 for failure of annotation and 2 for need of anaphora resolution.

Examples of revised explanations are:

2-1." *Fundamental statistics*: Population census statistics provided in <u>Article 5, paragraph (1) of Statistics Act</u> (Act No. 53 of 2007).

2-2." *Fundamental statistics*: National accounts provided in <u>Article 6, paragraph (1) of Statistics Act</u> (Act No. 53 of 2007).

3." *Population census*: A complete census as specified in <u>Article 5, paragraph (1) of Statistics Act</u> (Act No. 53 of 2007).

Figure 6 shows an act that caused a failure of annotation. The citations with underlines are expected to be replaced as follows:

- *The level of consumption*: The calculated level, prescribed in <u>Article 1, paragraph 7 of the Protocol</u>, of consumption prescribed in <u>Article 1, paragraph 6 of the Protocol</u>.

However, the tuple was extracted as follows:

- *The level of consumption*: The calculated level, prescribed in <u>Article 1 of Act on the Protection of the Ozone Layer Through the Control of Specified Sub Substances and Other Measures</u>, paragraph 7 of the Protocol, of consumption prescribed in paragraph 6 of <u>Article 1 of Act on the Protection of the Ozone Layer Through the Control of Specified Substances and Other Measures</u>.

Since the Protocol is written not conforming to the Japanese legislative rules, the notation of paragraphs differs from that of Japanese statutes[8]. In addition, the Protocol is not registered in the list of act titles. As a result, the underlined parts were replaced in the wrong way, as only articles were detected and were recognized as the ones of the current act. Updating the list of act titles and accepting an additional notation can solve this error.

---

[8] Paragraph numbers are denoted in Arabic numerals. While paragraphs in Japanese acts are cited in a text using Chinese numerals with the suffix for paragraph, ones in the Protocol àre cited in Arabic numerals as they are.

Table 2. Experimental result for inaccessible citations included in the previous study

Evaluation	#Citations
Successfully replaced with an absolute address	27
Failed in annotation	13
Need anaphora resolution	2
Total	42

特定物質の規制等によるオゾン層の保護に関する法律（昭和六十三年法律第五十三号）
/ Act on the Protection of the Ozone Layer Through the Control of Specified
Substances and Other Measures (Act No. 53 of 1988)

第二条
5　この法律において「消費量」とは、議定書第一条6に規定する消費量の同条7に
規定する算定値をいう。/

Article 2
(5) The term "the level of consumption" as used in this Act means the calculated level, prescribed in Article 1, paragraph 7 of the Protocol, of consumption
prescribed in paragraph 6 of the said Article.

Fig. 6. Example of failure in annotation

6 Conclusion and Future Work

In this paper, we focused on Japanese statutory sentences. As long as boilerplate
expressions are commonly used, a simple method for surface pattern recognition
is sufficient for legal text processing. Based on these characteristics, we proposed
the following methods:

– Extraction of legal terms and their definitions, the number of which exceeds
  14,000 tokens;
– XML tagging in terms of the document structure; and
– Annotation of absolute addresses to citations.

We were faced with the problem that the extracted explanations include citations, some of which are not accessible apart from the original statutes. This
is because they are expressed in context-dependent format. We replaced these
ambiguous expressions with the absolute addressing method. The effectiveness
of our method was demonstrated in our experimental results.

Our goal is to provide a terminology for translation. Although explanations
including citations in context-independent format are accessible for a specified
provision or act, their readability is insufficient. Explanatory sentences should
be independently readable. For example, the term *population census* shown in
Fig. 3 should be explained as "a complete census concerning individuals and

households," taking into consideration the expression in the preceding paragraph. One method of replacing citations with referential expressions has already been proposed [15] using a machine learning method, although it dealt with only the National Pension Act. In our future work, we will integrate this method for reference resolution with our XML corpus.

Acknowledgements. The authors would like to give special thanks to Prof. Matsuura of Nagoya University for his discussion on the Japanese legislation system. This research was partly supported by the Japan Society for the Promotion of Science KAKENHI Grant-in-Aid for Scientific Researches (S) No.23220005, (A) No.26240050 and (B) No.23300094.

References

1. Toyama, K., Saito, D., Sekine, Y., Ogawa, Y., Kakuta, T., Kimura, T., Matsuura, Y.: Design and Development of Japanese Law Translation Database System. In: Proceedings of Law via the Internet, 12 p. (2011)
2. Sekine, Y., Toyama, K., Ogawa, Y., Matsuura, Y.: The Development of Translation Memory Database System for Law Translation. In: Proceedings of 2012 Law via the Internet Conference, 21 p. (2012)
3. Nakamura, M., Ogawa, Y., Toyama, K.: Extraction of Legal Definitions from a Japanese Statutory Corpus – Toward Construction of a Legal Term Ontology. In: Proceedings of 2013 Law via the Internet Conference, 11 p. (2013)
4. Palmirani, M., Brighi, R., Massini, M.: Automated Extraction of Normative References in Legal Texts. In: Proceedings of International Conference on Artificial Intelligence and Law (ICAIL), pp. 105–106 (2003)
5. Bolioli, A., Dini, L., Mercatali, P., Romano, F.: For the Automated Mark-Up of Italian Legislative Texts in XML. In: Legal Knowledge and Information Systems - JURIX 2002: The Fifteenth Annual Conference, pp. 21–30 (2002)
6. de Maat, E., Winkels, R., van Engers, T.: Automated Detection of Reference Structures in Law. In: Legal Knowledge and Information Systems - JURIX 2006: The Nineteenth Annual Conference, pp. 41–50 (2006)
7. Martínez-González, M., de la Fuente, P., Vicente, D.-J.: Reference extraction and resolution for legal texts. In: Pal, S.K., Bandyopadhyay, S., Biswas, S. (eds.) PReMI 2005. LNCS, vol. 3776, pp. 218–221. Springer, Heidelberg (2005)
8. Lame, G.: Using NLP techniques to identify legal ontology components: Concepts and relations. In: Benjamins, V.R., Casanovas, P., Breuker, J., Gangemi, A. (eds.) Law and the Semantic Web. LNCS (LNAI), vol. 3369, pp. 169–184. Springer, Heidelberg (2005)
9. Höfler, S., Bünzli, A., Sugisaki, K.: Detecting Legal Definitions for Automated Style Checking in Draft Laws. Technical Report CL-2011.01, University of Zurich, Institute of Computational Linguistics (2011)
10. Winkels, R., Hoekstra, R.: Automatic Extraction of Legal Concepts and Definitions. In: Legal Knowledge and Information Systems - JURIX 2012: The Twenty-Fifth Annual Conference, pp. 157–166 (2012)
11. Le, T.T.N., Nguyen, M.L., Shimazu, A.: Unsupervised Keyword Extraction for Japanese Legal Documents. In: Legal Knowledge and Information Systems - JURIX 2013: The Twenty-Sixth Annual Conference on Legal Knowledge and Information Systems, pp. 97–106 (2013)

12. Nakamura, M., Kobayashi, R., Ogawa, Y., Toyama, K.: A Pattern-based Approach to Hyponymy Relation Acquisition for the Agricultural Thesaurus. In: Proceedings of International Symposium on Agricultural Ontology Service 2012 (AOS 2012), pp. 2–9 (2012)
13. Nakamura, M., Ogawa, Y., Toyama, K.: Extraction of Legal Terms and Their Explanations Defined in Parenthesis in Statutes. In: Proceedings of 19th Annual Meeting on Natural Language Processing (NLP 2013), pp. 670–673 (2013) (in Japanese)
14. Ogawa, Y., Inagaki, S., Toyama, K.: Automatic consolidation of japanese statutes based on formalization of amendment sentences. In: Satoh, K., Inokuchi, A., Nagao, K., Kawamura, T. (eds.) JSAI 2007. LNCS (LNAI), vol. 4914, pp. 363–376. Springer, Heidelberg (2008)
15. Tran, O.T., Ngo, B.X., Nguyen, M.L., Shimazu, A.: Automated Reference Resolution in Legal Texts. Artificial Intelligence and Law, 1–32 (2013)

Representing Judicial Argumentation in the Semantic Web

Marcello Ceci

GRCTC, University College Cork, Ireland
marcello.ceci@ucc.ie

Abstract. This paper presents part of a Semantic Web framework for precedent modelling. The research applies theoretical models of legal knowledge representation and rule interchange for applications in the legal domain to a set of real legal documents. The aim is to represent the legal concepts and the argumentation patterns contained in a judgement, as expressed by the judicial text. The bases of the framework are a set of metadata associated with judicial concepts and an ontology library, providing a solid ground for an argumentation system based on defeasible rules. The present paper shortly presents the metadata and ontology layers, focusing on the rules and argumentation layers. In the example provided (an application of the Carneades Argumentation System) the framework reconstructs the legal interpretations performed by the judge in a specific judicial decision, presenting its reasoning path, and suggesting possible different interpretations in the light of relevant code- and case-law.

Keywords: Legal Argumentation, Case-law, Carneades, Semantic Web, OWL.

1 Introduction

Judicial decisions represent a paramount of legal argumentation. Trials are, in fact, formalized discussions, and the whole procedural law is devoted solely to laying down clear and equilibrate rules for the development of the judicial process and for the adjudication of the competing claims. Moreover, case-law is a main element of legal knowledge worldwide: by settling conflicts and sanctioning illegal behaviours, judicial activity enforces law provisions within the national borders, supporting the validity of laws as well as the sovereignty of the government that issued them.

The work of Wigmore [1] and Toulmin [2] can be considered as a first attempt to visually represent some aspects of case-law, such as the validity of evidential arguments and the application of statutory rules to facts. Formalizing this complex environment has later represented a primary task for the research on Information Technologies (ITs), and the AI & LAW community has presented very significant outcomes in this topic since the '80, with different approaches: argumentation as in [3-4]; legal case-based reasoning as in [5-6]; legal concepts representation through logics as in [7]; rule interchange for applications in the legal domain as in [8].

Goal of this paper is to present part of the author's research, aimed at representing judicial argumentation in the Semantic Web. The next section introduces the research

P. Casanovas et al. (Eds.): AICOL IV/V 2013, LNAI 8929, pp. 172–187, 2014.
© Springer-Verlag Berlin Heidelberg 2014

and its constituting parts. Section 3 shows, through a modelling example, the part of research related to the use of argumentation graphs for rule instantiation and argument evaluation, with particular attention paid to the representation of the role played by the precedent in the judicial reasoning. The paper is concluded by highlighting the issues related to the research.

2   The Judicial Framework Project

The research presented in this paper stems from the author's work on a framework for case-law semantics whose goal is to exploit Semantic Web technologies to achieve isomorphism between a text fragment (the only legally binding expression of a norm) and a legal rule, thus *filling the gap* between the semantics of legal documents and the syntax of legal norms, as explained in [9]. More precisely, the framework models the content of judicial documents, i.e. the decisions issued by the courts of judgement.

The aim of the framework is to formalize the legal concepts and the argumentation patterns contained in a judgement in order to check, validate and reuse the legal concepts as expressed by the judicial decision's text. It relays on four layered models along the Semantic Web stack of technologies (Fig. 1):

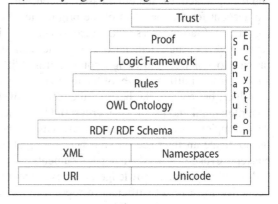

Fig. 1. The Semantic Web stack of technologies

- a **document metadata structure**, capturing the main parts of the judgement to create a bridge between text and semantic annotation of legal concepts;
- an **legal ontology framework** further divided into a *legal core ontology*, describing the legal domain's main elements in terms of general concepts through an LKIF-Core extension, and a *legal domain ontology*, an extension of the legal core ontology representing the legal concepts of a specific legal domain concerned by the case-law, including a set of sample precedents;
- a **set of defeasible rules** representing legal norms involved in case-law;
- **logics for normative argumentation**, representing the structure and dynamics of argumentation.

Cornerstone of the framework is the ontology, intended in its computer science meaning: shared vocabulary, taxonomy and axioms representing a domain of knowledge by defining objects and concepts together with their properties, relations and semantics.

The research is based on a middle-out methodology: top-down for modeling the core ontology, bottom-up for modeling the domain ontology and the ruleset. It has been practically realized through the JudO ontology set (section 2.1) and an application of the Carneades Argumentation Sytem (section 2.2), both built upon a sample set consisting in 27 decisions of Italian case-law, from different courts (Tribunal, Court of Appeal, and Cassation Court), concerning the legal subject of consumer law[1]. The research relies on the previous efforts of the community in the field of legal knowledge representation as in [10] and rule interchange for applications in the legal domain as in [8].

Many projects tried to represent case-law during the nineties, most of which are related to the work of Prof. Kevin Ashley such as [5-6]. Their main focus is similar to the one of the present research: capturing the elements that contribute to the decision of the judge. They were meant to support legal argumentation teaching in law classes, and the approach was therefore based on concepts rather than on the legal documents. No account for the metadata of the original text is given, and no ontology underlies the argumentation trees that reconstruct the judge's reasoning. Rather than representing a single judicial decision, the approach presented in this paper allows instead to connect knowledge coming from different decisions, and to highlight similarities and differences between them, not only on the basis of factors, dimensions or values, but also on the basis of the efficacy of the legal documents involved (e.g. under temporal and hierarchical criteria). Of course, *templatizing* legal documents is a very complex task: the intention, in any case, is not to provide a complete NLP tool but to create an interface through which a legal expert can easily identify the legal concepts evoked by words in legal documents.

Modelling judicial knowledge involves the representation of situations where strict deductive logic is not sufficient to reproduce the legal reasoning as performed by a judge. In particular, defeasible logics [13] seem needed to represent the legal rules underlying judicial reasoning. For example, many norms concerning contracts could be overruled by a different legal discipline through specific agreements between the parties. The problem of representing defeasible rules, in fact, is a core problem in legal knowledge representation. The present approach tries to tackle such problem through specific design of the OWL/DL Knowledge Base, to make it compatible with syntactically powerful defeasible reasoning. It is thus possible to build argumentation automatically upon the knowledge base, suggesting incomplete arguments to successfully deal with entymemes.

Deontic defeasible logic systems, such as those presented in [12-13], constitute indeed a powerful tool for reasoning on legal concepts. Most of them are explicitly built to import RDF triples, which means that they can perform reasoning on knowledge bases contained in ontologies such as the one presented section 2.1. The ontology, in the perspective of the present research, represents basic document semantics and relations, performing shallow reasoning mostly oriented to data completion, enhanced by the open world assumption. Over such knowledge base, rule systems based on advanced logic dialects perform complex reasoning with engines such as Carneades (see below) or SPINdle (see [14]) by importing only the set of

---

[1] The matter is specifically disciplined in Italy through the "Codice del Consumo" (Consumer Law) and articles 1341-1342 of the Italian Civil Code.

triples that best suits their syntactic needs. This may be preferable to approaches that try to extend OWL/DL to perform defeasible reasoning such as [15]: JudO shows that it is possible to perform shallow reasoning while staying within OWL2, and in order to perform an efficient reasoning on legal concepts it is not sufficient to implement defeasible reasoning, being also necessary to rely on argumentation schemes [16].

The same considerations apply to the approach in [17], which interestingly provides a simple and intuitive way to encode default knowledge on top of terminological KBs: such a reasoning system does not, however, reach the complexity needed to manage legal concepts (for which deontic defeasible logics are required, with an account for argumentation schemes), suggesting the need for a distinct layer.

The argumentation system described in the present paper (first presented in [18]) combines the features of the DL-based ontology with non-monotonic logics such as Defeasible Logics. In particular, T. F. Gordon's Carneades [19] is based on Walton's theory [20] and gives account for most of Prakken's considerations on the subject [21], including argumentation schemes and burden of proof, tools that are fundamental to perform effective reasoning on legal issues (see section 2.2). The present approach adopts a procedural view on argumentation, which is necessary in order to properly represent those processes in an argument graph (see [22]).

2.1   The Metadata and Ontology Layers: The JudO Ontology Library

The "JudO ontology library"[2] is composed of two OWL/RDF ontologies conceived to model the semantics of judicial interpretations [23]. Here, judicial interpretation is intended in the meaning of judicial subsumption, the act through which the judge takes into consideration a fact (material circumstance) and subsumes it under an abstract legal category (i.e., applies an abstract legal status to it).

The aim of the JudO ontology library is to apply a model for judicial interpretations to a set of real legal documents, stressing the definitions of OWL axioms as much as possible in order to provide a semantically powerful representation of the legal document for an argumentation system which relies on a defeasible subset of predicate logics.

The ontology set is composed by a core ontology and a domain ontology. The core ontology (an extension of the LKIF-Core Ontology) introduces the main concepts of the legal domain, defining the classes which will be later filled with the metadata of judicial decisions. Following the structure outlined in the Core Ontology, the metadata taken from judicial documents are represented in the Domain Ontology. The modelling was carried out manually by an expert in the legal subject, actually the only viable choice in the legal domain: automatic information retrieval and machine learning techniques, in fact, do not yet ensure a sufficient level of accuracy. Building a domain ontology is similar to writing a piece of legal doctrine, thus it should be manually achieved in such a way as to maintain a reference to the author of the model, while at the same time keeping an open approach.

The ontology library, tested on its sample taken from real judicial decisions, meets the following requirements:

---

[2] Available at https://code.google.com/p/judo/

- **Text-to-knowledge morphism**: the ontology can correctly classify all instances representing fragments of text. The connection to the Akoma Ntoso markup language ensures the identification and management of those fragments of text and of the legal concepts they contain.
- **Distinction between document layers**: The Qualifying Legal Expression class constitutes the main expressive element, introducing an n-ary relation that ignites the reasoning engine. Its instances can refer to the same text fragment, yet represent different (and potentially inconsistent) interpretations of that text.
- **Shallow reasoning on judgement's semantics**: the Domain Ontology can perform reasoning on the relevance of a material circumstance under a certain law. The OWL2 property chain *judged_as* and the axioms for law relevance and legal consequence application allow the reasoner to complete the framework, easening the effort needed to model all knowledge contained in the judicial decisions through semi-automatical legal knowledge elicitation.
- **Querying**: the *considers/applies* properties allow complex querying on the knowledge base, and the *judged_as* shortcut provides semantic sugar in this perspective. Querying on temporal parameters is not yet possible due to limits in LKIF-Core language: solutions for this are being achieved through emerging standards for rules such as LegalRuleML.
- **Modularity:** the layered (core/domain) structure of the ontology library renders domain ontologies independent between each other - and yet consistent, through their compliance to the core ontology template.
- **Supporting text summarization**: the ontology library supports the identification of dispositions and decision's groundings inside a judicial decision.

These design choices make OWL/DL fit for legal reasoning, relying on redundancy of information to enrich the Knowledge Base as much as possible. Moreover, the JudO Ontology library creates an environment where the knowledge extracted from the decision's text can be used to perform deeper reasoning and argumentation on the interpretation instances grounding the decision itself, using statements (either manually inserted or automatically imported from the ontology) in combination with a defeasible rulebase (see next section). For a complete presentation of the ontology see [24].

2.2 The Rules and Logic Layers: The Application of the Carneades Argumentation System

Built upon the ontology, the ruleset introduces defeasible logics, necessary to mediate conflicting dispositions contained in case-law and to support argumentation towards a desired solution for the case. Section 3 describes an application of the Carneades Argumentation System, used to extend the framework to the rules and logic layers of the Semantic Web.

Carneades[3] is a set of open source software tools for mapping and evaluating arguments, under development since 2006. Carneades contains a logical model of argumentation based on Doug Walton's theory of argumentation [20], and developed in

---

[3] Presented in [19], available at http://carneades.berlios.de.

collaboration with him. In particular, as shown in [25], it implements the argumentation schemes of Walton not only to reconstruct and evaluate past arguments in natural language texts, but also as templates guiding the user as she generates her own arguments graphs to represent ongoing dialogues. It can therefore be used for studying argumentation from a computational perspective, but also to develop tools supporting practical argumentation processes. The main application scenario of Carneades is that of dialogues where claims are made and competing arguments are put forward to support or attack these claims, but it also takes into account the relational conception of argument[4] of [26].

In the application described in section 3, Carneades' potentialities are exploited to conduct reasoning on a fragment of case-law, whose semantics have been previously modeled in an OWL/RDF ontology and in a set of rules in LKIF-Rule language[5]. The aim of this application is to create a reasoning environment with a high level of human-machine interaction: the user can start from some basic concept (a legal concept, a fact, an exception, a law prescription) and query OWL KB to get pilot cases in return: those are presented in a graph which shows not only the logical process followed by the judge and the laws which she applied but also those who could be, and the precedents which – if accepted - could lead to a different judgement.

This is possible thanks to the mix of OWL/DL reasoning, semantically managing static information on the elements of the case, and rule-based defeasible reasoning, representing the dynamics of norms and judicial interpretations. The application focuses on the argument from ontology feature of Carneades, presented in [27]: the program is in fact capable of accepting (or rejecting) the premises of arguments on the basis of the knowledge contained in the imported ontology, and of the defeasible rules. This allows building complex argumentation graphs, where the argument nodes represent legal rules and the statements are accepted or rejected on the basis of knowledge coming from the ontology and/or data inserted by the user.

In this perspective, the Carneades argument graph represents both:

- a reconstruction of a judicial decision's contents in terms of laws applied, factors taken into consideration, interpretations performed by the judge. The conclusion of the argumentation represents the final adjudication of the claim, and the Carneades reasoner is expected to accept or reject the claim by applying the semantics of judicial interpretations contained in the decision's groundings (this is the kind of representation which will be shown in Section 3);
- a collection of argumentations paths leading to a given legal statement (such as "contract x is inefficacious"). On the basis of manually-inserted statements

---

[4] The main difference between the two conceptions is that a proposition which has not been attacked is acceptable in the relational model of argument, while in most dialogues it would be not acceptable, since in most schemes making a claim involves having the burden of proof on it. See Chapter 5 of [29] for a presentation of the full theoretical model of reasoning with cases adopted by the present research.

[5] See [28]. A newer version of Carneades (the "Policy Modelling Tool") supports rules written in a Lisp dialect similar to Clojure.

concerning the object of the case (statuses or factors concerning the material circumstance, i.e. contract x) the Carneades reasoner suggests possible argumentation paths leading to (the acceptation or rejection of) the desired legal statement.

In both cases, the framework presents to the user not only argumentation paths which have been proven valid (i.e. rules whose conditions have all been met), but also incomplete argumentation paths where one or more of the premises is still undecided, or whose status is unknown (enthymemes): under this perspective, the framework creates semantic environment where different laws, legal statuses and precedents are semantically related to each other, thus highlighting critical aspects of the case which were not been taken into consideration by the judge (in the precedent case) or by the user (in the query). Given a set of judicial decisions encoded in the OWL ontology, the framework is capable not only to represent the argumentation path followed by the judge, but also possible alternative paths which lead to different outcomes. This is where DL and defeasibility are combined together, and monotonicity of OWL/DL is mediated by the instantiation of statements into PRO and CON arguments, which can be overridden in the argument graph.

From that point, the user can go further by investigating precedents where similar (or different) interpretations are made: in this way, she can realize which differences – if any – exist between two or more precedents. It is like browsing case-law in a law journal in order to compare different decisions, but in the Carneades environment this can be done directly with legal concepts, not only to verify a combination of circumstances and laws under a logical point of view, but also to receive suggestions from the system on which law, precedent or circumstance could lead to a different outcome.

3   Representing the Judicial Argumentation Process

The Judiciary Framework, in its components of the JudO ontology library and the Carneades application, has been tested on a set of 27 judgements concerning the interpretation of a fragment of Italian Consumer Law[6]. The objectives are the following:

- retrieve the relevant case-law, i.e. case-law concerning legal concepts involved in the law fragment;
- analyze the interpretations made by the judges in the retrieved case-law;
- search possible alternative outcomes.

The present section presents one of the 27 judgments and then rapidly shows the construction of the rule set, with related ontology concepts, focusing on the reasoning taking place in the Carneades argumentation environment.

---

[6] Actually regulated through legislative decree n. 206 of September 6th, 2005 - even though some cases still fall under artt. 1341 and 1342 of Italian Civil Code, which are still in force.

3.1   The Case

In the decision given by the 1st section of the Court of Como on May 20th, 2004[7], concerning contractual obligations between Mr. M. E. and La Sorgente sas (from now on α and β), the judge had to decide whether clause 8 of α/β contract, concerning the competent judge, could be applied. The judge cites art. 1341 subsection 2 of Italian Civil Code:

> Are inefficacious, unless specifically signed by writing, clauses establishing, in favor of the proponent, limitations to responsibility, right to withdraw from contract or to suspend the execution, or clauses establishing on the counterpart sunsets, limitations to exceptions, restrictions to contracting freedom towards third parties, tacit renewal, arbitration or competence derogation.[8]

In the contract signed by the parties there is a distinct box for a "specific signing" where several clauses of the contract are recalled by their object and number, including clauses with no oppressive content. The judge, with the support of precedents (he cites two Cassation Court sentences: 6976/2005 and 5860/1998) interprets the "specific signing" as not being fulfilled through a generic recall of a group of clauses including with mixed content, and therefore declares clause 8 of α/β contract invalid and inefficacious. The claim of inefficacy of clause 8, brought forward by α, is thus accepted, undercutting the claim of a lack of competence by the judge of Como, brought forward by β, which is rejected.

3.2   Modelling of the Law

The modelling of article 1341 subsection 2 of Italian Civil Code is based on both the ontology and the rules. In particular, the ontology contains "static" information on the law (such as the enacting authority, the subject, the legal concepts contained in the text, the URI of the legal expression), while defeasible rules are used to classify the material circumstances (in this case, the contract clauses) which share certain legal statuses as being relevant under that law, and to apply the legal consequences to those clauses. Following is an example of a rule stating the relevancy for comma 2 of Article 1341:

> **Rule**: LAW_Art1341co2
> If
> (C1 applies S1) and (S1 is Oppressive_Status) and (C1 applies General) and (C1 applies Unilateral) and (not C1 applies SpecificallySigned)
> Then
> (C1 is Relevant_ExArt1341co2) and (C1 is considered_by Art1341co2)

---

[7] Sent. N. 304, Tribunale di Como, giudice dott. Mancini.

[8] "non hanno effetto, se non sono specificamente approvate per iscritto, le condizioni che stabiliscono, a favore di colui che le ha predisposte, limitazioni di responsabilità, facoltà di recedere dal contratto o di sospenderne l'esecuzione, ovvero sanciscono a carico dell'altro contraente decadenze, limitazioni alla facoltà di opporre eccezioni, restrizioni alla libertà contrattuale nei rapporti coi terzi, tacita proroga o rinnovazione del contratto, clausole compromissorie o deroghe alla competenza dell'autorità giudiziaria."

Please notice that the rule does not contain the list of statuses considered as oppressive by the law, rather refering to a class of "Oppressive statuses" (a naming acknowledged by the legal doctrine), whose modelling is left to the ontology (Fig. 2). This distribution in the representation of the law allows an open organization of legal knowledge, keeping rules simple and general, and devolving classifications to the ontology.

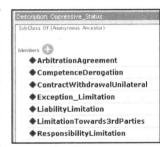

Fig. 2. Members of the Oppressive_Status class

The rule presented above only states which circumstances are subsumed under that legal rule; to represent the application of legal consequence(s) another rule comes into play, verifying if any exceptions apply: if not, the consequence of the legal rule (in this case, inefficacy) is applied to the circumstance (the contract clause):

Rule: LAWCONS_InefficacyRule

If

(C1 is Relevant_ExArt1341co1) or (C1 is Relevant_ExArt1341co2) or (C1 is Relevant_ExArt1342co2) and ((not C1 applies ReproducingLawDisposition) or (not C1 applies International))

Then

C1 is Inefficacious

3.3   Modelling of the Contract

The material circumstances taken into account by a precedent are modelled only in the domain ontology (not in the rules), and semantically classified depending on its characteristics (in the case of a contract clause: containing contract, contract parties, object of the contract, object of the clause). Two different OWL properties represent the relation between a circumstance (the clause) and a legal status: "applies", which means that the status has been recognised by both parties as applicable to the circumstance, and "judged_as" (an OWL2 property chain, see 2.1), which means that status has been interpreted as being applicable to the circumstance by a judge.

Fig. 3. Properties of a contract clause instance

So, for example, the contract clause ME/LaSorgente_Clause8 (Clause 8 of the contract between M.E. and "La Sorgente") has the characteristics indicated in Fig. 3.

We can see that "Unilateral", "CompetenceDerogation", "General" are three characteristics which are acknowledged by both parties; "recalled_by ME/LaSorgente_box" represents a relation of this clause with another part of the contract; the "considered_by" property links to the precedent (and therefore the authority) which produced the subsumptions marked as judged_as: a legal status ("NotSpecificallySigned") and two precedents of the Cassation Court (Cass. 6976/1995 and Cass. 5860/1998). In Section 3.4 we will see how this knowledge is managed in Carneades. Finally, please notice that the clause also applies the status of "CompetenceDerogation", an oppressive status.

3.4   Modelling of the Precedent

The judicial interpretation instances are modelled both into the domain ontology and into the rules. The domain ontology contains static knowledge (enacting authority, object of the case, classification, a URI) as well as basic information such as those presented above: the circumstance taken into consideration, the legal status under which the circumstance is subsumed, the precedents cited.

The mechanics underlying the judicial interpretation are contained in the rules. This is an example of a rule representing a judicial subsumption:

Rule: JINT_RecallNonOppressiveClauses
If
(C1 recalled_by B1) and (B1 hasfactor RecallsNonOppressiveClauses) and
(ASSUMPTION_C1 judged_as Cass.5860/1998)
Then
(not C1 applies SpecificallySigned) and (C1 RecallException)

Please notice the particular role given to the precedent "Cass.5860/1998": it is an assumption[9], so it does not prevent the system from suggesting this particular interpretation (and the precedent) as a result of the reasoning process on cases which share the other conditions, even if that precedent is not explicitly recalled. At the same time, if the precedent is directly cited in the case, the system is capable of putting a stronger accent on that interpretation, not only by assuming its applicability but by directly stating it. This is a central mechanism for representing precedent in the present application, as it will be made clear in the following explanation.

3.5   Reconstructing Precedents and their Reasoning

The above presented system is capable of automatically creating argument graphs on the basis of the application of the rulebase to the knowledge contained in the ontology set. The reasoning is performed automatically by the reasoner of the Carneades System.

---

[9] See section 4.3 of [29] for an explaination on the role of assumptions in Carneades' argument evaluation.

Fig. 4 (second column) shows two applicable laws to argument the inefficacy of the clause. If the conditions of one of these two laws are met, and no exception exists (in this case, possible exceptions - broken lines - are the contract being an international contract, and the contract reproducing law dispositions), the clause is inefficacious. The requirements for a clause to be relevant under one of these two laws are presented in the central part of Fig. 4:

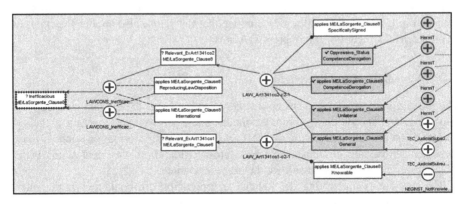

Fig. 4. Argumentation PRO the contract clause being inefficacious visualized in Carneades following [30]. The graph shows the argumentation towards the inefficacy of a specific clause (first column on the left), in the light of applicable laws (second column) and judicial interpretations made by the judge in the precedent (third column).

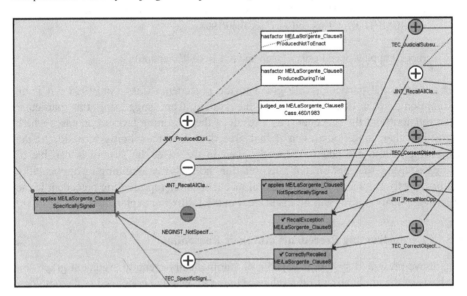

Fig. 5. Argumentation PRO and CON the clause being specifically signed

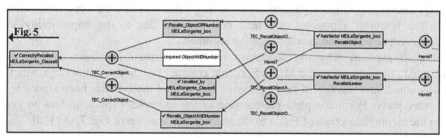

Fig. 6. Argumentation PRO the clause being correctly recalled

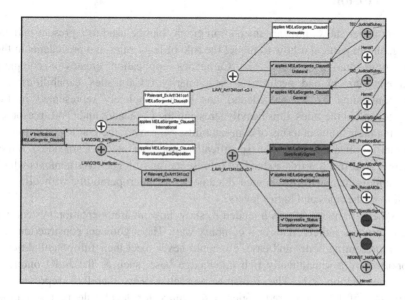

Fig. 7. The complete argumentation graph

- In order for the clause to be relevant under Article 1341co1, it must be general, unilateral and not knowable by the other party by using ordinary diligence (we will not further investigate this norm).
- In order for the clause to be relevant under Article 1341co2, it must be general, unilateral, oppressive and not specifically signed.

Some of these premises appear as "accepted" (dark grey boxes). The arguments towards their acceptance come directly from the ontology set's knowledge base, which means that the relative information has been manually inserted in (or inferred by) the OWL ontology and it is not possible to further explain those positions (the "HermiT" argument followed by dotted lines pointing to the premise "valid [ontologyURI]"). Some of the premises' statuses, however, have not been decided (white boxes). It is possible to ask the system to search for argumentation paths for them: Fig. 5 shows the argumentation produced by Carneades towards the rejection of the clause being specifically signed. Besides two accepted negative argument, the system shows a positive argument which is not valid due to an exception (second from the bottom), a positive argument not accepted due to the impossibility to evaluate its premises, and another negative argument.

Fig. 6 explains why the "correctly recalled" premise has been marked as "accepted" by Carneades: the ME/LaSorgente contract contains a distinct box which recalls object and number of the oppressive clause, and the box has been signed by the other party. For a complete presentation of the example, explaining how to get from the incomplete graph of Fig. 4 to the argument evaluation of Fig. 7, see [18].

4   Conclusions

The present paper shows how argumentation graphs can be used to represent judicial reasoning, and in particular how to model the role of legal rules and precedents in the judicial argumentation process. The Carneades application creates a complete juridical environment and performs a benchmark of Carneades' capabilities: the sample (constituted by 27 precedents) has been completely represented in the ontology set and in the rules, thus heavily stressing the Carneades and OWL reasoners and showing their limits in terms of computability. Moreover, the ontology set used in the present application was not specifically modelled upon Carneades, rather representing an effort towards a standard representation of legal text's contents which ensures isomorphism with the source document and interoperability with different applications in the rules and logics layers.

The Judicial Framework was intended to show how an argumentation system can be used to process judicial data in a complex way. The arguments construction and the rules representing code- and case-law could never meet their full potentialities if not supported by a semantically rich knowledge base, such as the JudO ontology presented in section 2.1. The research presented here therefore represents a demonstration of how a shared logics and syntax for legal rule representation, combined with a standard core ontology for legal concepts, would constitute powerful tools for case-law classification, browsing and management. It also includes an innovative way of representing the role of precedents in judicial argumentation (see Fig. 7). Computability was not an issue in the JudO ontology library (<5 seconds reasoning time on a Intel i5@3.30 Ghz), while the Carneades reasoner was moderately encumbered by the application of the rules to the ontology (8-15 seconds in the example described in section 3). This could be improved by optimizing the reasoner and/or with a further refinement of the ontology/rules structure.

The main issues which emerged in the application can be divided into KR-related and legal reasoning-related issues, and are analysed in details in Chapters 3 and 4 of [29]. In general, building the ontology set, adding factors and writing rules in the LKIF-Rule and Carneades languages highlighted some critical aspects in the modelling process and (as already explained in section 2.1) in the knowledge acquisition phase. The main issue is about the correct design and management of information between the ontologies and the rules: some of the axioms already modeled in the ontologies, in fact, could better meet their potentialities if modeled as an LKIF rules instead. This issue should be solved with general criteria, since its implications are many and important, the logic used (OWL/DL vs. defeasible logics) being very different to start with. This suggests the distinction between static information (thesauri, taxonomies, administrative and procedural data), to be included in the ontology, and legally relevant information (legal statuses, subsumptions, inclusion of a material circumstance into the scope of a norm), to be modeled as rules for the purposes of the argument evaluation. It is also possible to mod Carneades making him capable of translating OWL axioms into defeasible rules, but this solution would alter the logics underlying OWL inferences, therefore creating a significant risk of semantic shift. To summarize, OWL is used to provide a consistent, complete, redundant Knowledge Base, while defeasible rules are used to sort out this KB in a procedural environment.

References

1. Wigmore, J.H.: The Principles of Judicial Proof, 2nd edn. Little, Brown and Company, Boston (1931)
2. Toulmin, S.: The Uses of the Argument. Cambridge University Press, Cambridge (1958)
3. Loui, R.P., Norman, J., Olson, J., Merrill, A.: A Design for Reasoning with Policies, Precedents and Rationales. In: Proceedings of the Fourth International Conference on Artificial Intelligence and Law, pp. 202–211. ACM (1993)
4. Prakken, H., Sartor, G.: Argument-based extended logic programming with defeasible priorities. Journal of Applied Non-Classical Logics 7, 25–75 (1997)
5. Ashley, K.D.: Modeling Legal Argument: Reasoning with Cases and Hypotheticals. MIT Press, Cambridge (1990)
6. Rissland, E.L., Ashley, K.D., Branting, L.K.: Case-based Reasoning and Law. Knowledge Engineering Review 20(3), 293–298 (2003)
7. Boella, G., Governatori, G., Rotolo, A., van der Torre, L.: Lex minus dixit quam voluit, lex magis dixit quam voluit: A formal study on legal interpretation. In: Casanovas, P., Pagallo, U., Sartor, G., Ajani, G. (eds.) AI Approaches to the Complexity of Legal Systems. Complex Systems, the Semantic Web, Ontologies, Argumentation, and Dialogue, pp. 162–183. Springer, Berlin (2010)
8. Gordon, T.F., Governatori, G., Rotolo, A.: Rules and Norms: Requirements for Rule Interchange Languages in the Legal Domain. In: Governatori, G., Hall, J., Paschke, A. (eds.) RuleML 2009. LNCS, vol. 5858, pp. 282–296. Springer, Heidelberg (2009)
9. Palmirani, M., Contissa, G., Rubino, R.: Fill the Gap in the Legal Knowledge Modelling. In: Governatori, G., Hall, J., Paschke, A. (eds.) RuleML 2009. LNCS, vol. 5858, pp. 305–314. Springer, Heidelberg (2009)
10. Hoekstra, R., Breuker, J., Di Bello, M., Boer, A.: The LKIF Core Ontology of Basic Legal Concepts. In: Casanovas, P., Biasiotti, M.A., Francesconi, E., Sagri, M.T. (eds.) Proceedings of LOAIT (2007)
11. Governatori, G., Rotolo, A.: Defeasible Logic: Agency, Intention and Obligation. In: Lomuscio, A., Nute, D. (eds.) DEON 2004. LNCS (LNAI), vol. 3065, pp. 114–128. Springer, Heidelberg (2004)

12. Kontopoulos, E., Bassiliades, N., Governatori, G., Antoniou, G.: A Modal defeasible Reasoner of Deontic Logic for the Semantic Web. In: International Journal on Semantc Web and Information Systems, pp. 18–43 (2011)
13. Nute, D.: Norms, Priorities, and Defeasible Logic. In: Norms, Logics and Information Systems, pp. 201–218. IOS Press (1998)
14. Lam, H.-P., Governatori, G.: The Making of SPINdle. In: Governatori, G., Hall, J., Paschke, A. (eds.) RuleML 2009. LNCS, vol. 5858, pp. 315–322. Springer, Heidelberg (2009)
15. Antoniou, G., Dimaresis, N., Governatori, G.: A Modal and Deontic Defeasible Reasoning System for Modelling Policies and Multi-Agent Systems. Expert Systems With Applications 36(2), 4125–4134 (2009)
16. Walton, D., Reed, C., Macagno, F.: Argumentation Schemes. Cambridge University Press, Cambridge (2008)
17. Dao-Tran, M., Eiter, T., Krennwallner, T.: Realizing Default Logic over Description Logic Knowledge Bases. In: Sossai, C., Chemello, G. (eds.) ECSQARU 2009. LNCS, vol. 5590, pp. 602–613. Springer, Heidelberg (2009)
18. Ceci, M., Gordon, T.F.: Browsing case-law: An application of the Carneades Argumentation System. In: Aït-Kaci, H., Hu, Y., Nalepa, G., Palmirani, M., Roman, D. (eds.) Proceedings of the RuleML2012 Challenge, at the 6th International Symposium on Rules within the European Conference of Artificial Intelligence, vol. 874, pp. 79–95 (2012)
19. Gordon, T.F., Walton, D.: The Carneades Argumentation Framework: using presumptions and exceptions to model critical questions. In: Dunne, P.E. (ed.) Proceedings of COMMA 2006: Proceedings of COMMA 2006: 1st International Conference on Computational Models of Argument, The University of Liverpool, UK, September 11-12, pp. 195–207. IOS Press, Amsterdam (2006)
20. Walton, D.: The New Dialectic: Conversational Contexts of Argument. University of Toronto Press, Toronto/Buffalo (1998)
21. Prakken, H.: Formalizing Ordinary legal Disputes: a Case Study. In: Artificial Intelligence and Law, pp. 333–359 (2008)
22. Ceci, M.: The Role of Argumentation Theory in the Logics of Judgments. In: RULES 2013, Springer (2013) (publication pending)
23. Palmirani, M., Ceci, M.: Ontology framework for judgement modelling. In: Proceedings of AICOL, Frankfurt am Main, August 15-17 (2011)
24. Ceci, M., Gangemi, A.: An OWL Ontology Library Representing Judicial Interpretations. In: Semantic Web Journal. IOS Press (publication pending)
25. Gordon, T.F., Walton, D.: Legal reasoning with argumentation schemes. In: Proceedings of the Twelfth International Conference on Artificial Intelligence and Law, pp. 137–146. ACM Press, New York (2009)
26. Dung, P.M.: On the acceptability of arguments and its fundamental role in nonmonotonic reasoning, logic programming and n-person games. Artificial Intelligence 77(2), 321–357 (1995)
27. Gordon, T.F.: Combining Rules and Ontologies with Carneades. In: Proceedings of the 5th International RuleML2011@BRF Challenge, CEUR Workshop Proceedings, pp. 103–110 (2011)
28. Gordon, T.F.: Constructing Legal Arguments with Rules in the Legal Knowledge Interchange Format (LKIF). In: Casanovas, P., Sartor, G., Casellas, N., Rubino, R. (eds.) Computable Models of the Law. LNCS (LNAI), vol. 4884, pp. 162–184. Springer, Heidelberg (2008)

29. Ceci, M.: Interpreting Judgements Using Knowledge Representation Methods and Computational Models of Argument. Ph. D dissertatioN (2013), http://amsdottorato.cib.unibo.it/6106/1/Marcello_Ceci_tesi.pdf
30. Gordon, T.F.: Visualizing Carneades argument graphs. Law, Probability and Risk 6(1-4), 109–117 (2007)

On the Road to Regulatory Ontologies

Interpreting Regulations with SBVR

Elie Abi-Lahoud, Leona O'Brien, and Tom Butler

Governance, Risk and Compliance Technology Centre - University College Cork, Cork, Ireland
{e.abilahoud,leona.obrien}@ucc.ie, tbutler@afis.ucc.ie

Abstract. Regulatory compliance has proved to be difficult and time consuming across business domains. In Financial Services, the wide and complex spectrum of regulations calls for machine assistance in making sense of, and in consuming, the regulatory text. Semantic technologies, and Ontologies in particular, bring new solutions to the challenges in consuming financial services regulations that traditional technologies fell short in addressing. Current state-of-the-art related work is silent on the role of Legal/ Regulatory Subject-Matter-Experts in building these ontologies. This paper presents an on-going study on creating regulatory ontologies. It describes a Subject-Matter-Expert-centric approach to collaborative development of regulatory ontologies using structured natural language, Semantics of Business Vocabulary and business Rules (SBVR) in particular.

Keywords: Regulatory Ontology, Structured Natural Language, SBVR, Financial Industry, Regulations, Semantic Technologies, Subject Matter Experts, Common Vocabulary, Complex Regulations.

1   Introduction

The global financial regulatory environment is growing in complexity and scope in response to the financial crisis in 2008 [1]. The growth and complexity of national and international 'hard' and 'soft' regulation [2] is causing problems for organisations in the financial industry [3], with the "deep distributional implications of rule making in a world of competitive and globally integrated financial markets" little understood or appreciated [4]. Take, for example, that the "Dodd-Frank Wall Street Reform and Consumer Protection Act of 2010 increases the power of financial regulatory agencies, reduces regulatory gaps, develops better crisis management tools, and consolidates the regulation of systemically important institutions" [5]. It will do this using an estimated 1,500 provisions and 398 rules, which will be drafted by relevant regulatory agencies—approximately 40% of these rules are in force in 2013 at the time of writing.[1] The resultant rules can be extremely complex; take, for example, the Volker Rule, which was originally 10 pages, had "swelled to 298 pages and was accompanied

---

[1] http://www.usatoday.com/story/money/business/2013/06/03/dodd-frank-financial-reform-progress/2377603/

P. Casanovas et al. (Eds.): AICOL IV/V 2013, LNAI 8929, pp. 188–201, 2014.

by more than 1,300 questions about 400 topics" and was claimed by financial organisations as being "too complex to understand and too costly to adopt"[2]. The international reach of the Dodd-Frank Act is also significant, as non-U.S. banks will only be exempt from the Volker Rule's prohibitions if their activities have no link with the U.S. market—a truly rare scenario [6]. The problems are created by 'hard' regulations, such as Dodd-Frank, with 'soft' regulations based typically on standards and which focus on particular regulatory domains, such as capital adequacy or disclosure obligations. Nevertheless, these 'soft' regulations have been 'hardened' through their adoption by the EU and governments globally. All this presents significant problems for the regulators drafting the regulations and rules, legal practitioners who interpret them, and financial services practitioners who apply them.

There is significant interest in the concept of semantic technologies and legal ontologies to capture procedural legal knowledge [7-9], to deal with the flood of legal information [10] and to provide legal knowledge management services [27]. Section 2 focuses on this body of literature. While it is generally agreed that semantic technologies can help stem this flood, there is a paucity of research on semantic technologies to address the regulatory flood that faces the financial industry. Research on the use of semantic technologies in financial services is emergent [11], and focuses on the business domain [12]. At the Demystifying Financial Services Semantics Conference in New York, 2012, Wall Street executives and U.S. regulators call for the development of a 'common vocabulary' for the industry that would be human and machine readable. The need for such a vocabulary is indicated by [13] who calls for a taxonomy of global securities and for common definitions.   However, [14] echoes each of the points made above by arguing that the "looming train wreck" for regulatory compliance in the financial industry requires regulatory ontologies such as that described herein.

The remainder of this paper is structured as follows: section 2 describes the state of the art in Legal Ontologies. Section 3 explains the challenges faced when consuming a regulatory document for the purpose of knowledge representation. Section 4 describes the suggested approach to use SBVR to express regulations in structured natural language as means to bridge the gap between Subject Matter Experts and Ontology Engineers. Section 5 concludes and draws next steps.

2   Related Work

The typology of legal ontologies developed to date is diverse as a result of the purpose and focus of the ontology, the degree of formality, the various methodologies used, and the application of the ontology. Some of the relevant legal ontologies are briefly discussed here in terms of their completeness, reusability and availability, subject-matter and purpose.

Many of the early ontologies can be described as core ontologies. These were concerned with modelling knowledge that is common across various legal domains with the focus on jurisprudence and legal doctrine that didn't reflect the true nature of the

---

[2] http://www.nytimes.com/2011/10/22/business/volcker-rule-grows-from-simple-to-complex.html?_r=0

law in practice. The focus was on legal norms, legal actors and legal concepts. Some of these early core ontologies such as FBO [15], DOLCE, and FOLaw [16] are either legal ontologies or contain legal terms. The high-level nature of these ontologies dealing with legal theory has meant they have been reused in very limited circumstances due to the small number of legal concepts contained therein. This was highlighted in the Estrella project (European project for Standardized Transparent Representation in order to Extend Legal Accessibility) that produced LKIF (Legal Knowledge Interchange Format) comprising the LKIF core ontology and the LKIF rules language. The LKIF core ontology was created by reusing concepts from LRI-Core and gathering and reviewing the top legal terms from consortium partners. LKIF is likely the most reusable of the core ontologies because of its legal coverage.

Domain specific legal ontologies are also an active area of development. These focus on a particular area of law such as consumer complaints in the CContology, European VAT fraud in FFPOIROT (Financial Fraud Prevention Oriented Information Resources using Ontology Technology), ship classification in the CLIME (Computerized Legal Information Management and Explanation) ontology, and intellectual property rights and copyright in IPROnto. While these legal ontologies have application in the specific domain of law chosen, relevance beyond this is impracticable because of the subject-matter it is modelling, for example, an ontology on contracts cannot be readily applied to procedural case law. Domain specific legal ontologies are not without value, some have been applied rather than remaining at prototype stage. FFPOIROT developed the Topical Ontology of Fraud, and the Topical Ontology of VAT based on European Law and preventive practices to deal with financial fraud. Trials were conducted with CONSOB (Commissione Nazionale per la Società e la Borsa – The Italian Securities Market Commission) that generated good results. However, very little was published due to the confidentiality of the real cases used [17]. An ontology on Dutch Immigration law was developed by *Be Informed* specifically for the Dutch Immigration and Naturalisation Service. It proved to be highly effective but is proprietary and therefore inaccessible.

The development of legal ontologies has many approaches but one noticeable trait is the lack of involvement of legal experts. The majority of legal ontologies are developed using text-extraction later reviewed by legal experts [7], if at all. The limited involvement of legal experts can compromise the correctness, application and acceptance of the ontology within the legal arena.

There is also a need to look at the work undertaken on semantic standards particularly for legislative drafting. This allows for legal documents to be displayed in both human and machine readable forms. Metalex resulted from the E-POWER (European Program for an Ontology based Working Environment for Regulations and Legislation) project. It provided a generic and easily extensible framework for the coding of the structure and contents of legal documents. It was redesigned taking account of Norme in Rete [18] and Akoma Ntoso[3].

---

[3] http://www.akomantoso.org/. Akoma Ntoso was developed as part of a UNDESA project to set standards for e-Parliament services in a Pan-African context. See Palmirani & Vitali, 2011. It is also being adopted in Switzerland, the State of California, The European Parliament amongst others.

Akoma Ntoso 'is a technology-neutral XML machine-readable descriptions of parliamentary, legislative and judiciary documents...that enable addition of descriptive structure (markup) to the content of parliamentary and legislative documents' [19]. It allows management of legislative change for legal documents. Akoma Ntoso has been adopted in several jurisdictions worldwide as the XML standard for parliamentary and legislative documents.

RuleML[4] is a standard for rules knowledge representation across all industries. LegalRuleML extends RuleML in order to capture in an expressive XML language, legal norms, rules and legal knowledge to allow it to be used for legal reasoning and for semantic information of legal documents to be shared [20].

There is recognition that while all the research to date is contributing to a rich landscape of semantic solutions for the legal domain, new approaches are needed to represent regulations through the development of a regulatory ontology.

3   Challenges in Consuming Regulations

Understanding regulations has proven to be a complex task to both, non-trained human agents, and to machines. This section describes a set of challenges or difficulties in understanding a regulatory text. It categorizes them in five types based on the nature of the difficulty. It provides examples extracted from the US Code of Federal Regulations, Title 31 Chapter X - Financial Crimes Enforcement Network, Department of the Treasury, which deals with Anti Money Laundering.

3.1   References to Follow and Flesh Out

Typically, in a regulatory text the sentences aren't self-contained, they refer to content in other sections or even in other documents. This content is needed to ensure correct understanding. For example in 31 CFR 1022.210(d)(1)(iii), shown below, one needs to read/consume the content of §1022.380(a)(2) in order to understand when a *person* is considered *a money services business*.

A person that is a money services business solely because it is an agent for another money services business as set forth in §1022.380(a)(2), and the money services business for which it serves as agent [...]

Following these references can prove to be tedious, especially when one is faced with a chain of references. In the previous example, §1022.210(d)(1)(iii) redirects to §1022.380(a)(2) to understand when a *person* is considered *a money services business*. In turn, §1022.380(a)(2), as shown below, redirects to §1010.100(ff) to complete the definition of an *agent for money services business*.

A person that is a money services business solely because that person serves as an agent of another money services business, see § 1010.100(ff) of this chapter, is not required to [...]

---

[4]   http://ruleml.org/

In this example, §1010.100(ff) is a ten-paragraph section which the reader/consumer of the regulation should process to identify when *a person is a money services business solely because that person serves as an agent of another money services business* to ensure that her understanding of the sentence she started with, in 31 CFR 1022.210(d)(1)(iii), is accurate.

3.2   Definitions to Identify, Delimit and Disambiguate

Usually legal documents contain sections dedicated to define/redefine the terms and the concepts used in these documents. Naturally, regulatory documents follow this rationale. For example in the US Code of Federal Regulations, Title 31 Chapter X contains §1010.100 which is a list of *General definitions*. However, other definitions could be embedded in the body of the regulatory text. These definitions are usually made explicit by using connectors such as "means", "as set forth in", "includes", etc. But sometimes they aren't explicitly stated as such and locating them becomes a harder task.

Definitions of regulatory terms tend to be highly context-related, thus rendering the reuse of existing vocabularies, without adapting them and validating them, practically impossible. For example, a reader/consumer of 31 CFR Chapter X might be well familiar with a definition of Financial Institutions not containing telegraph companies. Conversely, in the context of prepaid access for money services businesses, telegraph companies are considered as financial institutions as stated in §1010.100(t).

When trying to delimit the coverage of a concept in a regulatory document, definitions taken from the original regulatory text are key. However, these definitions often contain terms whose definition, scope and coverage aren't necessarily clear. The reader/consumer of the regulation is then faced with a recursive search-and-understand process. For example, §1010.100(mm) defines the entity *Person* as a list of other entities considered as *Person*s for the purpose of 31 CFR Chapter X such as the entity *Indian Tribe*. If the reader/consumer of the regulation isn't clear on what is considered an *Indian Tribe* in the "spirit of this regulation", it is up to her to refer to the Indian Gaming Regulatory Act and place the definition in context.

3.3   Complex Sentences to Make Sense Of

The complexity of legalese is no secret [21] and regulations do not escape this complexity. For example, §1022.320(a)(4) on the reporting of suspicious transactions as shown below, starts with an obligation (*to identify*) followed by two imbricated assumptions (*provided that* and *so long as*) and finishes with a related possibility (*of liability depending on the nature of some relationship*).

> *(4) The **obligation to identify** and properly and timely to report a suspicious transaction rests with each money services business involved in the transaction, **provided that** no more than one report is required to be filed by the money services businesses involved in a particular transaction (**so long as** the report filed contains all relevant facts). **Whether, in addition to any liability on** its own for failure to report, a money services business [...] **may be liable for** the failure of another money services business involved in the transaction to report that transaction **depends upon the nature** of the contractual or other **relationship** between the businesses [...]*

3.4   Ambiguities to Clarify

The potential ambiguity of natural language sentences is widely recognized. A regulatory text written in natural language is certainly no exception. For example, in §1022.380(a)(2) shown below, it is not clear what *location* refers to. It could be the business of the agent, the agent's home address or the location where the registration form has to be filed.

Each foreign-located person doing business, whether or not on a regular basis or as an organized or licensed business concern, in the United States as a money services business shall designate the name and address of a person who resides in the United States and is authorized, and has agreed, to be an agent to accept service of legal process with respect to compliance with this chapter, and shall __identify the address of the location within the United States for records__ pertaining to paragraph (b)(1)(iii) of this section.

Moreover, in regulations some sentences deliberately introduce ambiguity around the meaning or the scope of certain concepts. For example, the usage of sentences such as: *unless the context otherwise requires, matter of "Facts and Circumstances", or any other similar items*, etc. introduces a deliberate opening for possibilities not captured in the text.

3.5   Exceptions to Take into Account

Whether in a concept definition or in a list of requirements, most regulations contain exceptions. Take for example, §1022.380 on registration of money services businesses. This section starts by listing the exceptions before listing the requirements. Furthermore, the difficulty in understanding listed exceptions increases when these exceptions are hidden in the body of a referenced text, as illustrated by the sentence hereafter from §1022.380.

Except as provided in paragraph (a)(2) of this section, relating to agents, each money services business [...]

To address this challenge type, a reader/consumer of the regulation needs to rely on her subject matter expertise to put these exceptions in context and ensure a correct understanding of them.

To overcome challenges when facing a regulatory text, such as the ones previously described, it is clear that Subject Matter Experts (SMEs) play an important role in consolidating and making sense of the text. We believe that this step is a key requirement preceding formal Knowledge Representation. To the best of our knowledge, state of the art approaches (as described in section 2) proposing legal ontologies are silent on the SMEs role. The remainder of this paper suggests an alternative way to creating regulatory ontologies that is characterized by the introduction of an intermediate step while going from regulation to formal ontologies. This step involves the consolidation, the disambiguation and the interpretation of regulatory text by subject matter experts using Structured Natural Language (SNL) which is SME-friendly and which has precise semantics (grounded in formal logic).

4   Interpreting Regulations with SBVR

The suggested approach relies on subject matter expertise in disambiguating and interpreting the regulatory text for the purpose of formal knowledge representation. This section describes the structured natural language used to bridge the gap between Subject Matter Experts (SMEs) and Semantic Technologies Experts (STEs) and a methodology for collaborative development of regulatory vocabulary and regulatory guidance.

4.1   Semantics of Business Vocabulary and Business Rules

Semantics of Business Vocabulary and business Rules (SBVR) [22] is an Object Management Group (OMG) specification for Business Natural Language that is grounded in ISO Common Logic. SBVR structures natural text around elements from the SBVR metamodel. The frequently used elements are:

- *Noun Concepts*, which are things in the domain of interest. For example, regulator, regulation, financial institution, etc. *Individual Noun Concepts* are a particular type of *Noun Concepts* representing actual entities or individuals. For example, Securities and Exchange Commission, RegulationW, Wells Fargo Bank, etc.
- *Verb Concepts*, which capture the relationships between Noun Concepts. For example, the *Verb Concept* "money services business submits suspicious activity report" captures the submission relationship between a money services business and a suspicious activity report.

It is also common for SBVR users to look in the text for *Keywords*, which are linguistic symbols listed in the OMG-specification. For example, the natural language representation of logical quantifiers, logical operators and modal operators are identified as *keywords* in SBVR Structured English.

Typically an SBVR document has two parts: a Vocabulary and a Rulebook. An SBVR Vocabulary is a Terminological Dictionary where entries are *Noun Concepts* and *Verb Concepts*. It also contains *definitional rules* - which constrain, in the form of alethic modalities (it is necessary that), the relationships represented by verb concepts - and related *advices of possibility*. An SBVR rulebook is a set of guidance statements containing *behavioral rules* in the form of deontic modalities (it is obligatory that) and *advices of permission/ prohibition*. An SBVR vocabulary & rulebook should be complete and consistent [23]. This is determined by three basic principles: (1) noun concepts should be explicitly defined from the text, from other authoritative sources or recognized as implicitly-understood by the SMEs; (2) only defined/recognized noun concepts may play roles in verb concepts; (3) definitional rules and behavioral rules may only be built using defined verb concepts.

SBVR does not have a normative syntax but the OMG specification describes SBVR Structured English (SBVR SE) which is a simplified version of natural English. SBVR SE relies on text styles to visually identify elements from the SBVR metamodel. In the following we adopt a similar style to express examples in SBVR. Noun concepts are underlined with a single line. Individual noun concepts are doubled underlined. **Keywords** are in a bold font face. The *verb part* of a verb concept is in italic-bold font face.

To illustrate the usage of SBVR in the context of financial regulations, take for example the definitions of currency from 31 CFR Chapter X § 1010.100(m):

The coin and paper money of the United States or of any other country that is designated as legal tender and that circulates and is customarily used and accepted as a medium of exchange in the country of issuance. Currency includes U.S. silver certificates, U.S. notes and Federal Reserve notes. Currency also includes official foreign bank notes that are customarily used and accepted as a medium of exchange in a foreign country.

The related SBVR entry is

Definition: <u>coin</u> **and** <u>paper money</u> *of* a <u>country</u> **that** *is designated as legal tender in* **the** <u>country</u> **and that** *circulates and is customarily used and accepted as a medium of exchange in* **the** <u>country</u>

Concept Type: general noun concept
General Concept: legal tender
Source: 31 CFR Chapter X § 1010.100(m)
Example: the <u>coin</u> and <u>paper money</u> *of* the <u>United States</u>
Example: U.S. silver certificates, U.S. notes and Federal Reserve notes

4.2 Disambiguation and Interpretation Approach

The objective of this approach is to rely on subject matter expertise to overcome the challenges described in section 3. SMEs produce, in SBVR, a regulatory vocabulary capturing definitions of the concepts underlying the studied regulation. The vocabulary also contains descriptions of the relationships between these concepts and constraints over these relationships. The SMEs also produce, in SBVR, regulatory guidance capturing a list of obligations and a list of prohibitions expressing the regulatory imperatives. These lists are constructed using the aforementioned vocabulary.

Figure 1 recalls the circle of understanding to illustrate the iterative disambiguation and interpretation process to which it adds the stylizing in SBVR SE activity.

Fig. 1. Iterative Interpretation of Regulations with SBVR

The stylizing activity consists of indicating which element in the SBVR metamodel a term (or set of terms) corresponds to. This is done by applying the appropriate SBVR SE font styles. The disambiguation activity consists of consolidating and understanding the text. It can require any combination of the following activities:

- Consolidate references, which implicates following reference chains and integrating required parts to produce self-contained sentences.
- Define terms/ concepts from the text itself or find appropriate definitions, which implicates delimiting concepts coverage, clarifying "confusing" terms and identifying parent concepts.
- Define unclear terms in the definitions themselves, which implicates repeating the previous activity for terms and concepts in the produced definitions (imbricated levels of disambiguation).
- Identify relationships between things represented by the terms, which requires capturing the roles played by previously defined concepts. Each relationship is captured in a verb concept wording.
- Identify constraints on these relationships, which are represented in SBVR SE by necessity-formulations.
- Identify modalities and the action(s) on which these modalities lie, which implicates navigating the list of previously defined verb concepts and identifying the ones that are modified by regulatory imperatives (obligations, prohibitions).

The clarification activity consists of relying on Subject Matter Expertise to formulate guidance when the regulatory intent is not clear (after each of the disambiguation activities).

4.3   Experimental Work

Multiple experiments to test and evaluate the relevance of this approach to compliance practitioners and ontology engineers are carried out as part of the research program of the Governance, Risk and Compliance Technology Centre (GRCTC) in University College Cork, Ireland. The following describes a completed experiment on the US Bank Secrecy Act (US BSA) and its implementing regulation Chapter X of Title 31 of the Code of Federal Regulations (31 CFR Chapter X). The scope of this experiment was limited to sections of 31 CFR Chapter X that are modified by the following Federal Register final rule: 76 FR 45403 Bank Secrecy Act Regulations - Definitions and Other Regulations Relating to Prepaid Access.

The experimental setting and supporting software environment were described and discussed in [23]. Four legal SMEs participated in the disambiguation/interpretation process. They were tasked with producing a vocabulary and a rulebook built on this vocabulary as described in the previous section. The following is a selection of excerpts from the produced SBVR interpretation of 76 FR 45403 explaining how SBVR brings regulatory knowledge closer to formal representation while being SME-friendly.

- On reference chains and producing self-contained sentences:

The regulation defines transaction accounts as "[...] transaction accounts includes accounts described in 12 U.S.C. 461(b)(1)(C) [...]". After consolidation and interpretation in SBVR, the definition of transaction accounts becomes a list as follows:

deposit accounts *on which* the depositor *or the* account holder *can make withdrawals by* transferable instrument, payment orders of withdrawal, telephone transfers [...]

- On definitions and levels of disambiguation:

A seller of prepaid access has to abide by a list of obligations. For example, **It is obligatory that a** seller of prepaid access *sells* prepaid access *offered under* a prepaid program *provided that* the prepaid access *can be used before* verification of customer identification [...]. It is clear that the noun concept verification of customer identification needs to be precisely defined. To this purpose, a related SBVR vocabulary entry is created:

Verification of customer identification
Definition: is **the** collection *of* information *about* **the** customer *including* name, date of birth, address, **and** identification number.
Source: § 1022.210(d)(1)(iv).

- On identifying, describing and constraining relationships:

A person can structure a transaction. This is captured in the following verb concept entry: a person *structures* a transaction **if** that person, [...] *for* the purpose *of evading* reporting requirements. Like the previous example, capturing this verb concept definition isn't sufficient, one needs, for example, to flesh out the definition of reporting requirements to ensure complete understanding of transaction structuring cases. Typical examples of constrained relationships consist of qualifying the noun concepts playing roles in a verb concept, for example: agreement *designates* **only one** person *to register* money services business.

- On capturing regulatory requirements:

The regulation imposes on providers of prepaid access to maintain access to a history of transactional records for five years. The verb concept provider of prepaid access *maintains access to* transactional records is modified as follows:
It is obligatory that each provider of prepaid access *maintains access to* transactional records *for* **a** period *of* five years.

5   Discussion

This work aims at bringing regulatory knowledge closer to formal representation in a subject-matter-expert-friendly way. The role of subject matter experts is central to the presented approach. Their active participation guarantees a correct and accurate representation of domain knowledge.

The on-going experimental work, described in section 4.3, highlighted the advantages of applying SBVR using the described approach and confirmed some expected shortcomings. For instance, the SBVR specification doesn't provide a technique to directly represent an exception to a rule. However, a subject matter expert drafting guidance in SBVR SE could represent an exception to an obligation as a permission related to the rulebook entry describing the aforementioned obligation. The possibility of transforming such from SBVR SE to a formal representation depends highly on the logical expressiveness of the selected machine-readable representation language.

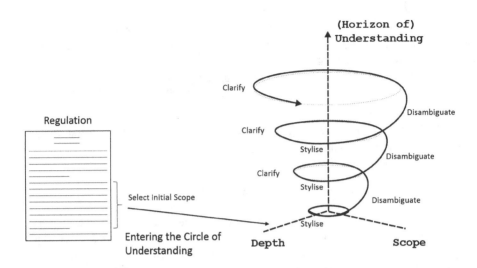

Fig. 2. Increasing Understanding of Regulations Using the Described Approach

On the formal semantics underpinning SBVR, Gordon *et al.* [26] identified two major areas where future versions of the SBVR specification could evolve. First, the under-specification of the semantics of SBVR deontic modalities, which hinders an accurate modeling of legal norms. And second, the inherited shortcomings of classical first order logic such as the lack of support to defeasibility, which precludes the formal representation of conflicts (conflictual statements). These limitations related to formal semantics have an impact on automated reasoning.

However, in the experiments described here, automated legal reasoning is not the primary intended application. We intend to highlight the need for a step preceding a "complete" formal representation of regulatory knowledge. This step consists of subject matter experts capturing regulatory intent in clearer and more accessible representations than the challenging ones provided by regulators, as described in section 3. For instance, these SBVR-based representations would provide financial services compliance officers, who don't necessarily have complete legal training, with support to make more informed decisions which are traceable back to the regulations. As illustrated in Figure 2, the usage of SBVR Structured English in the iterative manner detailed in section 4.2 and within a technical environment such as the one presented in [23], guarantees a deeper understanding of the regulations and a broader comprehension of the context while preserving clear provenance of underlying concepts.

6   Conclusion and Future Work

This paper built on the need for regulatory ontologies in the financial industry to describe an approach to represent knowledge from financial services regulatory documents in structured natural language as a step towards representing a subset of this knowledge using Semantic Web technologies. It identified a list of challenges that require human subject matter expertise in understanding regulatory text. To overcome these challenges, the paper suggests relying on subject matter experts to interpret and represent regulations using Semantics of Business Vocabulary and business Rules. The described approach was supported by a series of examples in SBVR from a completed experiment on a piece of regulation from the US Bank Secrecy Act.

This approach overcomes uncertainty and imprecision in regulations by combining Subject Matter Expertise and SBVR precision in representing domain knowledge. It is targeted at removing complexity and ambiguity from regulations and resulting policies and rules. With the underlying formal logic of SBVR guaranteeing a certain level of accuracy in knowledge representation, immediate understanding is expected to increase and communications are meant to improve. Clear provenance of the vocabulary entries and the guidance rules renders possible tracing back to the original text regulatory concepts and constraints described in resulting ontologies making the whole knowledge model auditable.

The developed vocabulary answers the need for a common and shared language as described in section 1 whereas guidance rules can be used in policies and procedures to build controls. The ultimate potential of this approach is achieved when SBVR vocabularies and rules are transformed into fully machine understandable models using for example the semantic web representation languages or more expressive/ more adequate representation languages.

Next steps will focus on transforming the vocabulary part of an SBVR SE document to formal ontologies and specifically OWL ontologies to enable several applications such as knowledge management, regulatory change management, etc. To the best of our knowledge, and to date, there is a lack of methods/tools supporting automated transformation of SBVR vocabularies to OWL. Current work is focusing on developing such methods and stressing on maximizing their automation. Due to the natural language characteristics of SBVR, full automation is not expected but a high degree of automated support is sought. Promising results were described by Kendall and Linehan in [24]. Future work will focus on leveraging Natural Language Processing techniques to assist the subject matter experts in the interpretation process as discussed in [25].

Acknowledgements. This work was supported by Enterprise Ireland (EI) and the Irish Development Authority (IDA) under the Government of Ireland Technology Centre Programme. The authors would like to acknowledge the role played the Subject Matter Experts, John Lombard, Patrick O'Sullivan and Peter O'Sullivan and thank them for their contributions. The authors would also like to thank John Hall and Donald Chapin for their help and their valuable advice.

References

1. Davies, H., Green, D.: Global Financial Regulation: The Essential Guide (Now with a Revised Introduction), Polity, UK (2013)
2. Abbott, K.W., Snidal, D.: Hard and Soft Law in International Governance. International Organization 54(3), 421–456 (2000)
3. Baldwin, R., Cave, M., Lodge, M.: Understanding regulation: theory, strategy, and practice. Oxford University Press (2011)

4. Brummer, C.: How International Financial Law Works (and How It Doesn't). Geo. LJ 99, 257 (2010)
5. Levine, R.: The governance of financial regulation: reform lessons from the recent crisis. International Review of Finance 12(1), 39–56 (2012)
6. Greene, E.F., Potiha, I.: Examining the extraterritorial reach of Dodd–Frank's Volcker rule and margin rules for uncleared swaps—a call for regulatory coordination and cooperation. Capital Markets Law Journal 7(3), 271–316 (2012)
7. Casellas, N.: Law, Governance and Technology: Legal Ontology Engineering: Methodologies, Modelling Trends, and the Ontology of Professional Judicial Knowledge, vol. 3. Springer (2011)
8. Sartor, G., Casanovas, P., Biasiotti, M.: Approaches to legal ontologies: theories, domains, methodologies, vol. 1. Springer (2011)
9. Sartor, G., et al.: Legislative XML for the Semantic Web. Springer, Heidelberg (2011)
10. Breuker, J.: Law, Ontologies and the Semantic Web: Channelling the Legal Information Flood, vol. 188. IOS Press (2009)
11. Krieger, H.-U., Declerck, T., Nedunchezhian, A.K.: MFO-The Federated Financial Ontology for the MONNET Project. In: KEOD (2012)
12. Bennett, M.: Semantics standardization for financial industry integration. In: IEEE Collaboration Technologies and Systems (CTS), May 23-27, pp. 239–445. IEEE (2011)
13. Mazando, F.H.: The Taxonomy of Global Securities: is the US Definition of a Security too Broad? Northwestern Journal of International Law & Business 33(1), 121 (2012)
14. Kendall, E.: Talk on Semantics in Finance: Addressing Looming Train Wreck in Risk Management, Regulatory Compliance and Reporting. Semantic Technologies & Business Conference, New York (2013)
15. van Kralingen, R.: A conceptual frame-based ontology for the law. In: Proceedings of the First International Workshop on Legal Ontologies (1997)
16. Valente, A., Breuker, J.: A functional ontology of law. In: Towards a Global Expert System in Law (1994)
17. Zhao, G., Leary, R.: Topical ontology of fraud. Deliverable of The FFPOIROT IP project (IST-2001-38248) Deliverable D2 (2005)
18. Biagioli, C., Francesconi, E., Spinosa, P., Taddei, M.: The Norme in Rete Project: Standards and tools for legislative drafting and legal document web publication. In: Proceedings of ICAIL Workshop on e-Government: Modelling Norms and Concepts as Key Issues, pp. 69–78 (2003)
19. Palmirani, M., Vitali, F.: Akoma-Ntoso for Legal Documents. Legislative XML for the Semantic Web, 75–100 (2011)
20. Palmirani, M., Governatori, G., Rotolo, A., Tabet, S., Boley, H., Paschke, A.: LegalRuleML: XML-based rules and norms. In: Palmirani, M. (ed.) RuleML - America 2011. LNCS, vol. 7018, pp. 298–312. Springer, Heidelberg (2011)
21. Kimble, J.: Plain English: A charter for clear writing. TM Cooley L. Rev 9 (1992)
22. Semantics of Business Vocabulary and Business Rules, Object Management Group (November 2013), http://www.omg.org/spec/SBVR/1.2
23. Abi-Lahoud, E., Hall, J., Butler, T., Chapin, D.: Interpreting Regulations with SBVR. In: Joint Proceedings of the 7th International Rule Challenge, the Special Track on Human Language Technology and the 3rd RuleML Doctoral Consortium, hosted at the 8th International Symposium on Rules (RuleML2013), Seattle, USA, July 11-13 (2013)
24. Kendall, E., Linehan, M.: Mapping SBVR to OWL2. IBM Research Paper (2013), http://domino.research.ibm.com/library/cyberdig.nsf/papers/A 9777F4EDB2552AE85257B34004C4EB3

25. Lévy, F., et al.: An Environment for the Joint Management of Written Policies and Business Rules. ICTAI (2), 142–149 (2010)
26. Gordon, T.F., Governatori, G., Rotolo, A.: Rules and norms: Requirements for rule interchange languages in the legal domain. In: Governatori, G., Hall, J., Paschke, A. (eds.) RuleML 2009. LNCS, vol. 5858, pp. 282–296. Springer, Heidelberg (2009)
27. Boella, G., Humphreys, L., Martin, M., Rossi, P., van der Torre, L.: Eunomos, a Legal Document and Knowledge Management System to Build Legal Services. In: Palmirani, M., Pagallo, U., Casanovas, P., Sartor, G. (eds.) AICOL-III 2011. LNCS, vol. 7639, pp. 131–146. Springer, Heidelberg (2012)

Conceptual Modeling of Judicial Procedures in the e-Codex Project

Enrico Francesconi[1], Ginevra Peruginelli[1], Ernst Steigenga[2],
and Daniela Tiscornia[1]

[1] ITTIG-CNR - via de' Barucci 20, Florence, Italy
{francesconi,peruginelli,tiscornia}@ittig.cnr.it
[2] Ministry of Security and Justice, The Hague, The Netherlands
e.j.steigenga@minvenj.nl

Abstract. Simplification of judicial procedures management and the possibility to file and exchange them between European Member States are essential pre-conditions to increase cross-border relations in a pan-European e-Justice area. In this paper an overview of the e-Delivery platform architecture, developed by the e-CODEX project, as well as the semantic solution conceived to transmit business documents within a scenario characterized by different languages and different legal systems, are described. A proposal for implementating such solution with semantic web technologies is described.

Keywords: e-Justice, Semantic interoperability, Knowledge representation, e-Delivery, Domain Modeling, Document Modeling.

1 Introduction

Simplification and rationalization of judicial procedures management by information and communication technologies represent one of the main goal of the current policies of the EU institutions: the aim is to reduce operating costs and procedural deadlines in the administration of Justice, to facilitate the access to cross-border judicial procedures for citizens, to create a European system of e-Justice as a cornerstone to develop a European area of freedom and security. As support for the construction of the European judicial area, the e-Justice Action Plan [1] promoted the development of the European e-Justice Portal, as well as projects aimed to create direct services for the citizens in order to facilitate access to the information in the field of justice, dematerialization of proceedings, as well as communication between judicial authorities.

In this context the e-CODEX[1] project is a Large Scale Pilot in the domain of e-Justice, aiming to implement building blocks for a system supporting cross borders judicial procedures between European Member States and to provide citizens, enterprises and legal professionals with an easier access to transnational justice. In this respect it is not intended to replace national solutions but

---

[1] e-Justice Communication via Online Data EXchange (http://www.e-codex.eu/).

P. Casanovas et al. (Eds.): AICOL IV/V 2013, LNAI 8929, pp. 202–216, 2014.

to provide standards and tools for information exchange and interoperability in the software tools, respecting the existing diversity. Transport of data and documents is a key target of the e-CODEX platform. In a transnational settings it means transport of information from one country to another, also including communication between the e-Justice Portal and national systems.

In this paper the main features of the e-CODEX system, based on semantic technologies and Web services, are summed up. In particular the relation with other similar pilots (Section 2) and the e-Delivery platform architecture (Section 3) are presented. Moreover the approach, based on document standards and semantic models, able to provide a semantic interoperability layer for message exchange are described (Sections 4, 5, 6). In particular (Section 7) such knowledge modeling approach, deployed on a specific example, is presented. Finally some conclusions and future developments are discussed (Section 8).

2   Related Projects

The e-Justice pilot represented by e-CODEX is not intended to operate in isolation but is able to benefit strongly from the experiences and results of the other Large Scale Pilots (LSPs) and also other pan-European e-Government projects. Especially with regard to the other LSPs, the e-Justice pilot aims to build on existing products and standards already created in the other projects, in particular PEPPOL, STORK and SPOCS.

PEPPOL[2] aims at enabling seamless cross-border e-Procurement, connecting communities through standard-based solutions. To this aim it enables access to the Business Document Exchange Network (BUSDOX), its standards-based IT infrastructure for metadata transport service based on OASIS BDX. It provides services for e-Procurement with standardised electronic document formats, with the aim to facilitate the pre-award and post-award procurement process. STORK[3] and SPOCS[4] are meant to allow citizens to establish new e-relations across borders. STORK, in particular, is targeted to establish a European eID Interoperability Platform; SPOCS, on the other hand, aims to support small and medium enterprises delivering services in all Member States through the provision of seamless electronic procedures by building cross-border solutions based on each country's existing systems. Both projects use the same e-Delivery solution exploiting standards in the area of Registered E-Mail (REM) using ETSI specifications (ETSI-REM) but also the generalized implementation of transportation standards based on the Web Services Stack and SOAP (OASIS ebMS).

The solutions provided by such LSPs represent the infrastructure which the e-CODEX platform is based on; in this respect, and for explicit mandate of the EU Commission, the e-CODEX platform is going to represent the convergence solution for the other LSPs.

---

[2] Pan-European Public Procurement Online (http://www.peppol.eu).

[3] Secure idenTity acrOss boRders linKed (https://www.eid-stork.eu).

[4] Simple Procedures Online for Cross-border Services (http://www.eu-spocs.eu)

3   The Architecture of the e-CODEX e-Delivery Solution

The e-CODEX platform for e-Delivery will provide facilities for cross border communication via gateways, behind which national domains should stay unchanged. It aims to implement functionalities of reliable messaging delivery between national gateways, including persistence, timestamps to track the chain between sender and receiver, evidences of delivery and acceptance, large message handling, security and encryption of messages. In Fig. 1. an e-Codex scenario is sketched related to a claim filed from a country by the victim of an offense of this country against an offender of a different country.

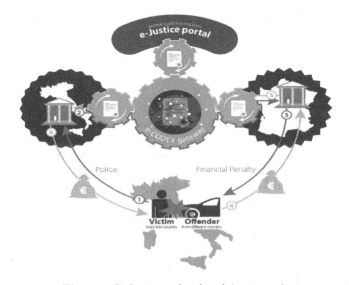

Fig. 1. e-Codex cross-border claim scenario

To guarantee such a reliable messaging between the actual endpoints located within the national domains, a so-called "circle of trust", based on legal agreements, is established and technically implemented by a "Trust-ok token". Moreover, to provide reliability and non-repudiation between endpoints, the e-Delivery convergence scenario also foresees standardized evidences based on ETSI REM specifications [2]. Gateways will be endowed with routing capabilities able to resolve gateway physical addresses and national competent courts from a central/decentral DB including national filing system IDs for integration into existing national infrastructure.

The details of the connection between national systems via gateways are sketched in Figs. 2. and 3. Fig. 2. shows a basic architecture of the e-Delivery solution, set up by national gateways which are bilaterally connected to each other, consequently there is no central hub in the middle. National gateways interconnect to the national systems respective applications by adapters (here called 'connectors') which handle the e-Codex message (eCM) format with respect to national oriented communication and formats (Fig. 3.).

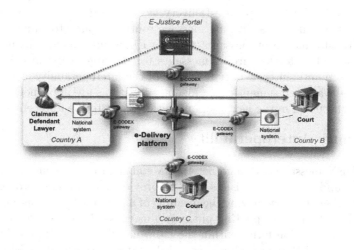

Fig. 2. e-CODEX system architecture

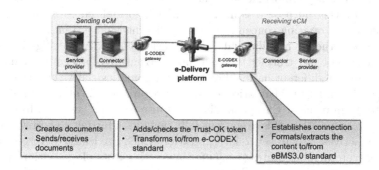

Fig. 3. Communication between national systems via connectors and gateways

The interoperability framework is, on the other hand, represented by an interoperability layer including profiles of secure and reliable transport standards, as OASIS ebMS 3.0 format for message exchange, ETSI-REM evidence format, Web services engines based on Apache Axis2[5] architecture, as well as a semantic layer necessary for negotiating concepts between different Member States and legal systems (see Section 4).

The open source product Holodeck[6] is used as basic infrastructure to implement business documents exchange using ebMS 3.0 standard. This will serve as the basis for the e-CODEX gateway. The reason for choosing this product is that it is freely usable (open source), easily extensible and natively implements an ebMS 3.0 stack.

The development of 'connectors' between national gateways and national information systems is up to each Member State. Connectors act as an interface

---

[5] http://axis.apache.org
[6] http://holodeck-b2b.sourceforge.net

between national and European e-Delivery systems, keeping national systems unchanged, nevertheless facilitating message routing. Connector functions concern the transformation of messages to/from EU format, as well as metadata and address lookup for forwarding messages to the target gateways. Similarly, format and semantic intermediary functions of the interoperability layer are developed. The way such semantic intermediary functions are implemented in the project are discussed in the next sections with respect to the foreseen use cases.

4   Semantic Interoperability

For e-CODEX message exchange between Member States, having different legal systems and traditions, it is essential to provide a semantic interoperability layer for sharing and harmonizing the meaning of national jurisdiction-dependent concepts. For the project piloting phase two use cases have been foreseen: the exchange of application forms within the EU Small Claims (SC) and European Payment Order (EPO) procedures, as ruled by the corresponding EU regulations ([3] [4] [5]). Country-dependent legal systems, as well as the diversity of languages, make legal information exchange between Member States a challenging task. For this purpose a conceptual model, formalized in an ontology, is necessary for negotiating concepts between different legal systems.

To approach the EU multilingual legal scenario complexity and align legal concepts, one cannot just transfer the conceptual structure of a legal system to another, because of different national legal contexts and legislative cultures within EU [6] [7]. A similar problem arises even with regards to the obligation of Member States to implement European Directives into national laws. Far from being a straight transposition, this process usually includes a further step in which European Directives are subject to interpretation which can lead to diverging legislation between Member States (see [6] for interesting examples). With respect to other domains where conceptual negotiations mainly pertain to linguistic aspects (as for example the e-Health domain), in the e-Justice one meanings negotiation addresses concept nuances in different legal systems and traditions. On the other hand shared interpretation of legal concepts is a pre-condition of EU regulations, which directly apply at national levels.

The literature offers different methods to approach the multilingual complexity of the European law, for example controlled vocabularies implemented in a terminology database (such as IATE, used by all the main EU Institutions), thesauri (as EUROVOC), semantic lexicons or lightweight ontologies as WordNet ([8], [9], [10]). The alignment of multilingual terminologies can be effectively obtained by using a pivot language. More expressive descriptions of concepts associated with lexical units can be represented in domain ontologies (or statute specific ontologies), representing concepts used in a specific statute (as IPROnto [11]). More general organizations of domain concepts are addressed in literature as core ontologies (as LRI-Core [12], LKIF [13] and CLO [14] for the legal domain), while foundational concepts categories, applicable to all domains, are usually addressed in top or foundational ontologies (as SUMO [15] and DOLCE [16]). Such ontologies represent conceptual systems aimed at base-concepts sharing and promoting consensus in building more specific ontologies for specific domains or activities. The integration of different lexical resources (heterogeneous because of belonging to different law systems, or expressed in different languages, or pertaining to different domains) can be carried out in different fashions: 1) generate single resources (merging); 2) compare and define correspondences and differences (mapping); 3) combine different levels of knowledge, basically interfacing lexical resources and ontologies.

The use of a pivot conceptual structure is generally recommended in order to provide a reference for negotiating concepts meaning between Member States, thus providing a layer of legal concepts harmonization in view of the creation of a pan-European judicial area. In this respect the methodological approach chosen in the e-CODEX project is to combine different levels of knowledge, where national legal concepts are reconciled or mapped towards a more general conceptual model.

5   Modeling Semantic Interoperability

e-CODEX uses a 3-levels model towards semantic interoperability: conceptual, logical and physical. The *Conceptual model* is the model for communication and harmonization. It guides and supports business and IT to create the foundation for information exchange, through reuse of experience and application of already known and used concepts. The *Logical model* is the set of data types and code lists ensuring that data definitions are derived methodologically to enhance reusability at the physical level (for e-CODEX the CCTS[7] standards are used). The *Physical model* is the syntax and data formats ensuring mutual understanding between systems of information exchanging partners (XML/XSD and PDF are example of syntax and data formats at physical layer).

5.1   Domain and Document Modeling

The three layers of abstraction introduced so far (conceptual, logical and physical) allow us to identify both the conceptual and technical (data types and syntax) building blocks for describing document types and domain concepts to be exchanged: they represent a methodological framework which is followed by e-CODEX. The main requirement of the project is that, while legal concepts at EU level have different nuances in different legal systems and traditions, e-CODEX documents, pertaining to specific legal procedures, have a structure regulated by the related directives, valid for all the Member States jurisdictions. Within such framework, proper domain and document modeling have been conceived to address the e-Justice cross-border data/documents exchange as exemplified by the foreseen use cases.

---

[7] UN/CEFACT Core Components Technical Specification. Version 3.0. Second Public Review. 16-April-2007.

The analysis of the e-CODEX use cases regulations, referred in Section 4, and of the related application forms, identified the following steps and formats for business document exchange, as implemented through the EU e-Justice portal:

- To generate and sign a PDF version of a Web filled form;
- To generate a machine readable version (typically in XML) from the same Web filled form;
- To deliver both signed PDF and XML versions of the form.

In this scenario the descriptions of both domain concepts, addressed in the use cases forms, and form instances are essential requirements for modeling the e-CODEX form generation and delivery. In particular we can distinguish between *Domain Model*, as the model of the scenario to be addressed, and *Document Model*, as the model of a document instance (in our case a form) pertaining to that scenario. Each of them can be furtherly distinguished as follows.

In a bottom-up modeling approach, the Document Model can be viewed according to two layers of abstraction, whose definitions follow those firstly given in literature in [17–19]:

- The *Document Physical Model* is the collection of the document objects viewed on the basis of their physical, domain independent, function. In e-CODEX it represents the view of a document form in terms of physical components (ex: input fields, check boxes, labels, text boxes, etc.). A specific PDF form or an HTML form are instances of the Document Physical Model.
- The *Document Logical Model* is the collection of the document objects, viewed on the basis of the human-perceptible meaning of their content. In e-CODEX it represents the view of a document form in terms of logical components: ex. Claimant, Claimant name, Claimant address, Court name, etc, as well as their values and relations. A specific XML or an RDF set of triples are instances of the Document Logical Model.

According to the same bottom-up modeling approach, the Domain Model can be viewed according to two layers of abstraction:

- The *Domain Logical Model* is the set of building blocks (data types, code lists, etc.) to describe the documents of a particular domain of interest.
- The *Domain Conceptual Model* is a semantic description of the scenario (entities and relations) of a specific domain. In e-CODEX it allows us to provide meaning to the document physical objects: it gives semantic interpretation to the document elements (physical objects) in terms of logical objects, and it can be represented by element hierarchies (XMLSchema) or ontologies (RDFS/OWL).

To sum up, we can distinguish the following modeling layers and hierarchies:

1. Domain Model
  1.a) Domain Conceptual Model;

1.b) Domain Logical Model;
2. Document Model
  2.a) Document Logical Model;
  2.b) Document Physical Model.

Fig. 4. in particular shows the relationships between Domain and Document Models.

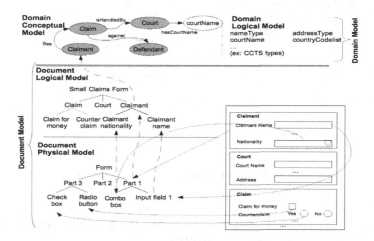

Fig. 4. Relations between Domain and Document Models

In this view, the two sub-layers of the Document Model are different levels of abstraction (physical and logical) for modeling a document instance. On the other hand, the two sub-layers of the Domain Model are the description of the scenario in terms of concepts and relations between them (Domain Conceptual Model) as well as data types (Domain Logical Model) according to which you give logical meaning to the document physical components. In other words, they are the semantic instruments to view document physical objects in terms of document logical objects.

6   Technical Implementation of the Modeling Layers

From a technical point of view two strategies for implementing the knowledge modeling proposed in Fig. 4. are being carried out, according to different degrees of complexity, so that they can be viewed in a short or long term.

6.1   Short Term Strategy

In a short term strategy, needed in e-CODEX piloting phase, the modeling layers are implemented using semantic tools with a limited degree of expressivity. According to this strategy, while the Document Physical Model is the view of an

HTML or PDF form in terms of physical objects, the Document Logical Model is the view of such objects as logical components, described by an XML file compliant to an XMLSchema representing the Domain Model including elements and relations (Domain Conceptual Model), as well as datatype (Domain Logical Model) (Fig. 5.).

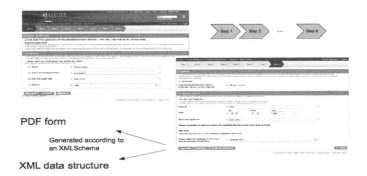

Fig. 5. Short term strategy form generation

In Tab. 1 such knowledge modeling and its technical implementation for the e-CODEX short term strategy are reported.

Table 1. e-CODEX "short term strategy" knowledge modeling

Knowledge Modeling	Technical Implementation
Domain Model	
a) Domain Conceptual Model	XMLSchema
b) Domain Logical Model	Data types, code lists (ex. CCTS or specific e-CODEX datatypes)
Document Model	
a) Document Logical Model	XML file
b) Document Physical Model	HTML or PDF forms

For implementing such modeling strategy, a 'core-team' of data modelers has been established: it is responsible for creating, editing and extending the concept of a shared semantic library. This limited amount of staff members creates the concepts based on the articulated information requirements from the use cases. A created concept is presented to a 'user council' in order to approve a concept for use. The 'user council' is formed by all stakeholders of the semantic library. Finally a 'schema creation group' has been formed, responsible to create and maintain an XML Schema based on the available semantic library.

6.2   Long Term Strategy

In a long term, e-CODEX knowledge modeling is supposed to develop a solution with a high degree of expressivity in order to describe the complexity of the scenario to be addressed and to cope with sustainability requirements. For these reasons a more complex knowledge modeling solution can be foreseen.

According to this long term solution, the Document Physical Model is the view of an HTML or PDF form in terms of physical objects, the Document Logical Model is the logical view of such objects that can be described in RDF able to represent statements over entities, including qualified relations (Fig. 6.).

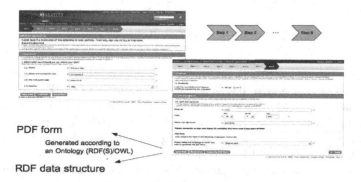

Fig. 6. Long term strategy form generation

The meaning of entities and relations can be given by an ontology (Domain Model) of classes and relations (Domain Conceptual Model) as well as datatype and codelists (Domain Logical Model). In Tab. 2 such knowledge modeling and its technical implementation for the e-CODEX long term strategy are reported.

Table 2. e-CODEX "long term strategy" knowledge modeling

Knowledge Modeling	Technical Implementation
Domain Model	
a) Domain Conceptual Model	RDFS/OWL model (ontology)
b) Domain Logical Model	Data types, code lists (ex. CCTS or specific e-CODEX datatypes)
Document Model	
a) Document Logical Model	RDF file
b) Document Physical Model	HTML or PDF forms

Differently from the short term strategy (Domain Model expressed by an XMLSchema), in the long term strategy the Domain Model is expressed using RDFS/OWL technologies, so to provide a more detailed representation of the meaning of the concepts involved and a more expressive description of the relations between them. An excerpt of concepts and qualified relations between the actors involved in the e-CODEX EPO domain is reported in Fig. 7. It represents an excerpt of the general scenario of a claim including its basic players: Claimant, Defendant and Court, as well as their mutual relationships. In the e-CODEX knowledge modeling language, it represents an excerpt of an e-CODEX Domain Model: it is composed by the Domain Conceptual Model (concepts and relationships) and the Domain Logical Model (data types, code lists, etc. associated to concepts and relationships).

Fig. 7. e-CODEX EPO Domain Model excerpt

An important goal of the Domain Model is to overcome the project finding that "all legislation seems to define its own semantics". e-CODEX noticed that currently each time a legal procedure is taken up for electronic proceeding basic legal concepts have to be analyzed and modeled to match exactly the definition in the legislation at hand. Notwithstanding the necessity for nuances in legal matters, the aforementioned legal concepts are of such generic nature that harmonization seems possible and desirable. Therefore the e-CODEX working group on semantics proposes to develop Core Legal Concepts, as a ground to develop a Domain Model, following the methodology used by the European Commission DIGIT's ISA Program[8]. The idea is to harmonize data definitions to the benefit of electronic proceedings through the introduction of Core Legal Concepts. Also, such Core Legal Concepts would enable faster electronic deployment of cross border legal procedures.

ISA has in particular provided specific recommendations for concepts identification, both in terms of format as well as design rules and management, in order to guarantee persistence and long term maintenance. As recommended by the ISA initiative[9], Core Vocabularies are to be published in multiple formats, including RDF to be useful for linked data applications. This entails that vocabulary terms have to be identified by dereferentiable http URIs.

Following such URIs pattern suggestions for vocabularies, the terms of a Core Legal Concepts vocabulary can be identified by the following hash URI namespace: `http://[URIroot]/def/CoreLegalConcepts#`, where `[URIroot]` is the domain name of the provider. For example the URI for the concept Claim, represented in the Core Legal Concept vocabulary, will be: `http://[URI root]/def/CoreLegalConcepts#Claim`; such URI will point to the latest version of related vocabulary. In order to distinguish between different versions of the same vocabulary, as well as different meaning of the same terms in different vocabulary versions, it is recommended that the version date of the vocabulary is added to the vocabulary namespace, according to the following pattern

`http://[URIroot]/def/{year}/{month}/{day}/CoreLegalConcepts#`

---

[8] DIGIT: Directorate-General for Informatics; ISA: Interoperability Solutions for European Public Administration.

[9] PwC EU Services EESV, "D3.1 – Process and Methodology for Core Vocabularies", ISA – Interoperability Solutions for European Public Administrations.

7   e-CODEX Knowledge Modeling Deployed on Example

In this section a deployment of the e-CODEX knowledge modeling architecture, based on semantic technologies, in particular on RDF(S)/OWL, is shown. A narrative example, here below, concerning a scenario about a dispute leading litigants to start a European small claim procedure, is used as example:

> *Franz von Liebensfels from Klagenfurt rented a car on the Internet for use in Portugal. Due to the existence of damage to the vehicle he decided to go to the company's office at the airport and the employee agreed to the change. The employee discovered damage to the windscreen. Mr. Liebenfels assured him that this was already there when he had collected the vehicle. The consumer subsequently saw that his credit card had been charged with the sum of 400 Euro. He decides to file a claim against Rental Car at the court of Lisbon using the European Small Claim Procedure.*

The narrative of Franz von Liebensfels and his car rental, can be generalized and summarized into a more abstract narrative as follows:

> *A claimant from a Member State files a claim against a defendant in another Member State. The claimant filed the claim at a court in the other Member State demanding reimbursement of the money taking form his credit card by the defendant.*

The two narratives at different levels of abstraction are the extensional (real case) description and intentional (generalization) model, respectively, of a small claim procedure. In the language of the e-CODEX knowledge modeling they can be, respectively, represented in terms of:

- Document Model, namely the document description of the specific case including real players and their relations, as well as the document physical template that implements the logical description of the real case;
- Domain Model, namely the description of the general scenario of a small claim procedure, including actor categories and relations.

In the e-CODEX knowledge modeling approach, the extensional description of the real case is represented by an e-CODEX Document Logical Model generated by a document template (Document Physical Model) which, in our narrative case, is a form pertaining to the Small Claim procedure, properly filled in by the Claimant. The connection between extensional and intensional representations of a Small Claim scenario stemming from our example is shown in Fig. 8., where individuals and related concepts are represented at different levels of abstraction.

Here below an RDFS/OWL description of the Court-Claimant-Defendant scenario and the RDF/XML serialization of the narrative instance of it, addressed in this paper where pre-defined URI naming conventions for concepts and documents are used, are here below respectively reported.

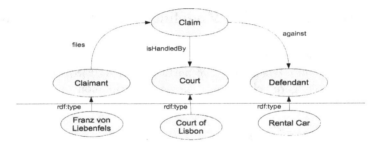

Fig. 8. Relation between extensional (Document Logical Model) and intensional (Domain Model) representations (lower and upper part, respectively) in a Small Claims scenario

Small Claims Domain Model Excerpt in RDFS-OWL/XML

```
<?xml version="1.0"?>
<rdf:RDF
    xmlns:rdf="http://www.w3.org/1999/02/22-rdf-syntax-ns#"
    xmlns:rdfs="http://www.w3.org/2000/01/rdf-schema#"
    xmlns:ESC="http://[URI root]/def/EuropeanSmallClaims#">
    <owl:Class rdf:about="ESC:Claim"/>
    <owl:Class rdf:about="ESC:Court"/>
    <owl:Class rdf:about="ESC:Claimant">
        <rdfs:subClassOf rdf:resource="ESC:Person"/>
    </owl:Class>
    <owl:Class rdf:about="ESC:Defendant">
        <rdfs:subClassOf rdf:resource="ESC:Person"/>
    </owl:Class>
    ...
    <owl:ObjectProperty rdf:about="files">
        <rdfs:comment>  [Definition of 'files' property]  </rdfs:comment>
        <rdfs:domain rdf:resource="#Claimant"/>
        <rdfs:range  rdf:resource="#Claim"/>
    </owl:ObjectProperty>
    <owl:ObjectProperty rdf:about="against">
        <rdfs:comment>  [Definition of 'against' property]   </rdfs:comment>
        <rdfs:domain rdf:resource="#Claim"/>
        <rdfs:range  rdf:resource="#Defendant"/>
    </owl:ObjectProperty>
    <owl:ObjectProperty rdf:about="isHandledBy">
        <rdfs:comment>  [Definition of 'isHandledBy' property]   </rdfs:comment>
        <rdfs:domain rdf:resource="#Claim"/>
        <rdfs:range  rdf:resource="#Court"/>
    </owl:ObjectProperty>
    ...
</rdf:RDF>
```

Small Claims Document Logical Model Excerpt in RDF/XML

```
<?xml version="1.0"?>
<rdf:RDF
    xmlns:rdf="http://www.w3.org/1999/02/22-rdf-syntax-ns#"
    xmlns:ESC="http://[URI root]/def/EuropeanSmallClaims#">
    <rdf:Description rdf:about="[FormInstanceURI]#id1">
        <rdf:type rdf:resource="ESC#Court"/>
        <ESC:hasCourtName>Court of Lisbon</ESC:hasCourtName>
        <ESC:hasCourtAddress>Rua Polo Sul 43, Lisboa</ESC:hasCourtAddress>
        <ESC:hasCourtCountry>Portugal</ESC:hasCourtCountry>
    </rdf:Description>
```

```
<rdf:Description rdf:about="[FormInstanceURI]#id2">
    <rdf:type rdf:resource="ESC#Claimant"/>
    <ESC:hasClaimantName>Franz von Liebenfels</ESC:hasClaimantName>
    <ESC:hasClaimantAddress>Museumstrasse 12,Klagenfurt</ESC:hasClaimantAddress>
    <ESC:hasClaimantCountry>Osterreich</ESC:hasClaimantCountry>
<rdf:Description rdf:about="[FormInstanceURI]#id3">
    <rdf:type rdf:resource="ESC#Defendant"/>
    <ESC:hasDefendantName>Rental Car</ESC:hasDefendantName>
    <ESC:hasDefendantAddress>Avenida Sol 1345,Lisboa</ESC:hasDefendantAddress>
    <ESC:hasDefendantCountry>Portugal</ESC:hasDefendantCountry>
</rdf:Description>
...
</rdf:RDF>
```

8 Conclusions

The e-CODEX project aims to represent an effective implementation of the current e-Justice policies of the European Commission towards e-Justice, as well as a basic framework for other pan-European e-Government projects. Legal contents representation and content transport infrastructure are the key activities currently under implementation in a scenario characterized by language and legal systems diversity. Both activities aims to create an interoperability framework based on standards and semantic tools to start and carry out judicial procedures on-line. In particular a legal knowledge modeling approach to promote semantic interoperability for e-Justice in the multilingual and multi-cultural complexity of the EU legal scenarios is proposed and implemented by RDF(S)/OWL technologies. In the next phases of the project particular attention will be payed to the implementation of a secure and reliable data exchanged system, based on evidences and circle of trust, as well as an e-Payment system for a complete on-line finalization of the judicial proceedings.

References

1. European Union Institutions and Bodies. Multi-annual European e-Justice action plan 2009-2013 (2009)
2. ETSI. Electronic signatures and infrastructures (esi); registered electronic mail (rem); part 2: Data requirements, formats and signatures for rem. Technical Report ETSI TS 102 640-2, ETSI, v.2.1.1 (2010)
3. The European Parliament and the Council of the European Union. Regulation of the european parliament and of the council of 11 july 2007 establishing a european small claims procedure (July 2009)
4. The European Parliament and the Council of the European Union. Regulation of the European Parliament and of the Council of 12 december 2006 creating a European order for payment procedure (December 2006)
5. Contini, F., Lanzara, G.F. (eds.): The Circulation of Agency in E-Justice. Interoperability and Infrastructures for European Transborder Judicial Proceedings. Law, Governance and Technology Series, vol. 13. Springer (2014)
6. Ajani, G., Lesmo, L., Boella, G., Mazzei, A., Rossi, P.: Terminological and ontological analysis of european directives: multilingualism in law. In: Proc. of Int. Conference on Artificial Intelligence and Law, pp. 43–48. ACM (2007)

7. van Laer, C.J.P.: The applicability of comparative concepts. Electronic Journal of Comparative Law, 2(2) (1998)
8. Fellbaum, C. (ed.): WordNet: An Electronic Lexical Database. MIT Press, Cambridge 1998)
9. Vossen, P. (ed.): EuroWordNet: A Multilingual Database with Lexical Semantic Networks. Kluwer Academic Publishers, Dordrecht (1998)
10. Sagri, M.-T., Tiscornia, D.: Ontology-based models of legal knowledge. In: Wang, S., et al. (eds.) ER Workshops 2004. LNCS, vol. 3289, pp. 577–588. Springer, Heidelberg (2004)
11. Delgado, J., Gallego, I., Llorente, S., Garcia, R.: Ipronto: An ontology for digital rights management. In: Proc. of the Int. Conference on Legal Knowledge and Information Systems, pp. 111–120. IOS Press (2003)
12. Breuker, J., Hoekstra, R.: Core concepts of law: taking common-sense seriously. In: Proc. of Formal Ontologies in Information Systems (2004)
13. Hoekstra, R., Breuker, J., di Bello, M., Boer, A.: Lkif core: Principled ontology development for the legal domain. In: Breuker, J., Casanovas, P., Klein, M., Francesconi, E. (eds.) Legal Ontologies and the Semantic Web. IOS Press (2009)
14. Gangemi, A., Sagri, M.-T., Tiscornia, D.: A constructive framework for legal ontologies. In: Benjamins, V.R., Casanovas, P., Breuker, J., Gangemi, A. (eds.) Law and the Semantic Web. LNCS (LNAI), vol. 3369, pp. 97–124. Springer, Heidelberg (2005)
15. Niles, J., Pease, A.: Towards a standard upper ontology. In: Proc. of the 2nd Int. Conference on Formal Ontology in Information Systems, pp. 2–9 (2001)
16. Gangemi, A., Guarino, N., Masolo, C., Oltramari, A., Schneider, L.: Sweetening ontologies with DOLCE. In: Gómez-Pérez, A., Benjamins, V.R. (eds.) EKAW 2002. LNCS (LNAI), vol. 2473, pp. 166–181. Springer, Heidelberg (2002)
17. Esposito, F., Malerba, D., Semeraro, G.: Multistrategy learning for document recognition. Applied Artificial Intelligence, an Int. Journal 8(1), 33–83 (1994)
18. Tsujimoto, S., Asada, H.: Major components of a complete text reading system. Proc. of the IEEE 80, 1133–1149 (1992)
19. Tang, Y.Y., De Yan, C., Suen, C.Y.: Document processing for automatic knowledge acquisition. IEEE Transation on Knowledge and Data Engineering 6(1), 3–20 (1994)

Organized Crime Structure Modelling for European Law Enforcement Agencies Interoperability through Ontologies

Jorge González-Conejero, Rebeca Varela Figueroa,
Juan Muñoz-Gomez, and Emma Teodoro

Institute of Law and Technology (IDT),
Universitat Autònoma de Barcelona (UAB),
Bellaterra 08193, Spain
jorge.gonzalez.conejero@uab.es
http://idt.uab.es

Abstract. Nowadays, organized crime networks share intelligence and knowledge as a fundamental asset for their members, thus making criminal organizations more global in nature and activities. Internet has consequently become the natural environment for these organizations. This evolution has put a bigger pressure in Law Enforcement Agencies (LEAs) demanding more efforts and resources in the fight against transnational organized crime. LEAs can therefore profit from international cooperation in fighting these organizations. However, differences among legal frameworks, languages and police and judicial culture may create interoperability issues. The CAPER project addressed the prevention of transnational organized crime by trying to provide the needed interoperability among the different European LEAs. In this work, we introduce a supranational Organized Crime Structure (OCS) modelled through an ontology in order to improve European LEAs Interoperability (ELIO). Results suggest that ELIO is able to provide the required interoperability features, overcoming the issues that arise in this scenario.

Keywords: Law Enforcement Agencies cooperation, transnational organized crime, knowledge acquisition, ontologies, interoperability.

1 Introduction

Nowadays, global criminals are sophisticated managers of technology [1], consequently, this high level of knowledge shown by these networks requires more efforts to be put in place by governments, Law Enforcement Agencies (LEAs) and citizens. Central networked intelligence and coordinated knowledge are fundamental assets shared within organized crime organizations. Moreover, online child pornography [2], prostitution [3] and all sorts of extortion and aggressive behaviour have been fuelled by the explosion of the Web 2.0. The Internet is not only the tool, but the condition and natural environment of organized crime. In this scenario, the work introduced in [4] suggests that utilizing information

P. Casanovas et al. (Eds.): AICOL IV/V 2013, LNAI 8929, pp. 217–231, 2014.

from multiple jurisdictions provides higher quality information about criminal networks. Furthermore, Europol[1] latest analysis [5] states that there are an estimated 3,600 organized crime groups currently active in the EU. These organizations show a tendency to be more international in nature and activity which creates an even greater need for international cooperation in fighting crime. As a result, LEAs can greatly benefit from sharing information, however, an interoperability issue arises due to EU countries having different legal frameworks as well as cultural and language differences.

The CAPER[2] project addresses the prevention of organized crime through sharing, exploitation and analysis of open and private information sources. Its main targets are: information acquisition, processing, exploitation and standardisation; integration with large scale systems, secure knowledge sharing and collaboration; and legal issues. Specifically, knowledge share and collaboration lie in an interoperability issue as we stated before.

This work is twofold. Firstly, we propose a Organized Crime Structure (OCS) based on Europol Annual Reviews and the International LEAs cooperation literature. This structure is devised to provide a common supranational structure in order to perform interoperability for European LEAs. Secondly, we also introduce an ontology, named as European LEAs Interoperability Ontology (ELIO), which models the OCS, the relationships among its concepts, the attributes and all the knowledge directly gathered from LEAs. The main idea is to ease the sharing of information related to organized crime among LEAs.

This paper is organized as follows: Section 2 briefs the problems on the definition of a conceptual structure of organized crime in Europe; Section 3 addresses International LEAs cooperation literature, annual reviews published by Europol and our proposal of organized crime structure focus on bringing interoperability among European LEAs; Section 4 sums up the related work with ontologies, legal ontologies and interoperability; Section 5 introduces ELIO and its iterative knowledge acquisition process, structure and evaluation; and finally, Section 6 points out some conclusions.

2   Problems on the Definition of a Conceptual Structure of Organized Crime in Europe

The attempt to define a conceptual structure of the field of cross-border organized crime presents mainly two problems that have to do with the very nature of organized crime. First, the lack of a consolidated definition of what organized crime is and, second, the dynamic, ever-changing nature of the phenomenon itself. When trying to define a conceptual structure of organized crime for the concrete purpose of facilitating LEAs interoperability, it is unavoidable to face a

---

[1] EUROPOL is the European Union's LEA. Home page:
http://www.europol.europa.eu/

[2] *"Collaborative   information   Acquisition   Processing   Exploitation   and   Reporting   (CAPER)   for   the   prevention   of   organized   crime"*.   Home   page:
http://www.fp7-caper.eu

third problem: the diverse conceptualization of the different crimes both from a semantic perspective and in the different legal system of the European countries.

The first problem, i.e., the lack of a non-contested clear concept of organized crime has been extensively discussed and some consensus has been reached over the years [6,7]. The first one focuses on the idea of "crime" and tries to build the concept around different categories of criminal manifestations, emphasizing the element of criminal activity [6]. This notion of organized crime takes the different crimes identified in each legal system as a starting point and analyses the existence of certain elements such as continuity, sophistication or seriousness to apply the label "organized" to a certain event. Hagan [7] suggests that this last type of occurrences be identified as "organized crime", as opposed to the cases in which there actually is an organization, which he names "Organized Crime".

The second issue arisen in the process of defining a conceptual structure of the field of organized crime is the changing nature of both the activities and the associative forms of this kind of criminals. Nowadays, every LEA, either national or international, points at cybercrime as one of the main threats our societies are exposed to [8,5,9], even though up to ten years ago this form of criminality was not even on the map of these very same agencies. Clearly, any possible conceptual structure of organized crime constructed around typologies of felonies would differ according to the moment in which it is designed. The geographical scope of the structure, moreover, would also have a relevant impact on the design itself. Although organized crime is intrinsically linked to the international dimension as it mostly occurs in cross-border circumstances, the relevance or the impact of a certain type of activity is not the same in different areas of the world, or in different countries in the same area.

Apart from these two general problems, the design of a conceptual structure in the field of organized crime for the purpose of improving the interoperability between LEAs in a transnational environment presents the additional issue of the different legal and semantic constructs that each criminal type bears in each country. This dimension implies the emergence of a semantic problem when interoperability is sought between LEAs from countries with different languages [10]. A fitting example, again in a situation of cooperation between Spanish and a British LEA, is that of money laundering. In Spain this crime is referred to as "blanqueo de capitales", which translates literally as "whitening of assets". The difference between the use of laundering or whitening is not relevant for the content of the illicit activity but how about the money-assets binomial? Does the fact that the crime in UK only refers to "money" mean that other financial products cannot be investigated? These are the problems that a simple translation cannot solve when defining a structure of organized crime.

3 Methodology for the Definition of a Conceptual Structure of Organized Crime in Europe

Once we have explained the three main problems encountered in the task of mapping organized crime in Europe for interoperability between LEAs purposes, we will present both the methodology used in an attempt to overcome these issues and the results obtained from the process. Taking into account the impossibility to identify an unambiguous concept of what organized crime is, the preferred option was to reflect upon the adequacy of the different perspectives for the concrete purpose pursued in this research, i.e., improving the interoperability capabilities of LEAs. In this regard the definition of organized crime as a series of crimes that are committed in a certain manner "is a practical way of understanding and tackling it."[3] It is true that an idea of organized crime that focuses on the criminal activities and especially on a classification of its manifestations lacks completeness, as it leaves out one of the distinctive element of this phenomenon: the organizations. Nevertheless, this methodological choice can be explained because of the horizontal cross-border nature of the characteristics that the academic literature attributes to the "organized" element of the concept. For instance, if we take the works of Hagan, Finckenauer [11] or Albanese [12], we can see, as Hagan himself summarized [7], that there are four distinctive traits of criminal organizations: the continuing organized hierarchy, the profit from illegal activities, the use of violence or threats, and the fact that they represent corruption and immunity. As mentioned before, these identifying elements are horizontal and do not depend on the specific legal system of a specific country. Although it is true that the regulation of "organized crime" varies within European Union countries, these traits are abstractions of the basic idea of criminal organization that any European LEA uses. The conclusion, therefore, is that a structure of the field of organized crime in the European Union for the purposes of improving interoperability would not benefit from an "organized"-based perspective. On the contrary, a concept that focuses on the criminal activity can have an impact on the fight against transnational forms of criminality and help improve the cooperation experiences of police officers, since this is how LEAs organize their work, their internal structures and their databases.

After it was established that the "crime"-based concept would be used to build the structure for the interoperability ontology, a second problem needed to be addressed, i.e., the changing nature of organized crime. This issue has been haunting every work on organized crime because, by its own nature, crime-whether organized or not, international or localis not a stable concept. It is not possible to define, therefore, an immutable structure to represent either the crime typologies or even the defining traits of the organizations. The only possible solution, consequently, is to design an open structure that enables the updating and adjustment of the contents, thus keeping the framework developed.

The main issue regarding the improvement of interoperability possibilities between LEAs is related to the third problem mentioned above, the different semantic and legal configuration of each crime in the different Member States of the

---

[3] This assertion can be found in the website of the newly founded "National Crime Agency", established by the UK Government and that replaced the famous Serious Organized Crime Agency (SOCA). http://www.nationalcrimeagency.gov.uk/ crime-threats Last accessed: 18/11/2013.

European Union. The first element, semantic differences, concerns not only the name of the crimes but also its internal configuration in the "working language" of each LEA. It is very unlikely; therefore, that one could be able to address it in an abstract way. The solution is then to extract that knowledge from the LEAs themselves in order to be able to include in the process the specific terms they use when referring to certain crimes. This part of the work performed in this research will be explained in Section 5.1 of this paper, under the title of knowledge acquisition. Even if these semantic issues –if not properly solved– can have a negative impact on the interoperability between LEAs in their fight against international organized crime, the different conceptualization of crimes has a bigger potential to pose serious problems to the collaboration between police authorities from different countries, problems that most certainly arise when criminals have to be brought in front of a judicial authority and the procedure followed by the LEA will be judged under national law requirements. In a situation of not-harmonized criminal law systems, such as that of EU countries, the structure used in a platform that is built with the purpose of improving the acquirement and sharing information process between European LEAs cannot be based on a particular national structure and has to take into account all the specificities of the different national structures. If the aim is to design a conceptual structure of the field of organized crime in the European framework, these national structures need to be embedded in a supranational structure. This is why in this case, and considering that the subjects of our investigation are the LEAs, the solution is to refer to EUROPOL and its definition of organized crime [4].

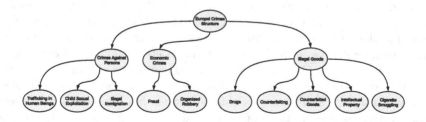

Fig. 1. Europol Organized Crime Structure (OCS)

There were several possible options to extract this structure from the work of EUROPOL. The first one was to look at the internal sections/divisions of the Database of the European agency. In order to better organize its task-forces and the intelligence gathered and obtained from national LEAs, EUROPOL organizes its database in theme-guided sub-units that were traditionally known as EUROPOL Analytical Workfiles (AWF) and have recently been renamed as Focal Points. The main problem with this approach is the fact that the decision about which broad categories become Focal Points is made at a European

---

[4] Terrorism is out of the scope of the CAPER project and therefore has been left out on the structure designed although is part of the EUROPOL activities and has been included in their reports since 2007.

level, that of EUROPOL itself, and therefore do not necessarily coincide with the national criminal law outlines. Focal Points may be very useful for structuring the intelligence gathered by national police agencies and communicated to EUROPOL; nevertheless another product seems better suited to define a general framework of the phenomenon of organized crime: the EUROPOL Reviews. These reviews can be defined as "a compilation of national annual reports on the domestic crime situation". The EUROPOL Reviews should not be mistaken for the so-called SOCTA/OCTA Reviews, the Serious and Organized Crime Threat Assessment, also published by EUROPOL on a yearly basis. The latter had also been considered as a source in the designing work but were deemed not suitable as they present the most dangerous threats Europe faces for the near future according to EUROPOL's analysis, thus leaving aside organized crime activities that continue to take place on the territory of the European Union but do not qualify as significant enough in a specific year. The EUROPOL Reviews were thought to be the most relevant documents to be studied as the main source when defining the structure. These reviews have been published for the last decade, since 2004, and have varied in their name and structure. Until 2008, they were issued as "Annual Reports" and not EUROPOL Reviews.

General trends and the crime typologies were taken into account to build the taxonomy needed to implement the interoperability ontology. As for the temporal scope of the analysis, and bearing in mind that organized crime is, as explained before, a rapidly changing phenomenon, the last three reports were studied in order to identify the "individual criminal activities" to be included[5]. The previous reports where used as a complementary source in order to define the broader categories included in the structure. We can dissect the process through an example. The criminal activity of "Trafficking in Human Beings" is reported in the 2010, 2011, and 2012 reviews as one of the relevant manifestations of organized crime in the European Union context. Taking a closer look to these last three reports, it can be noted that this crime is always reported together with two other criminal types: "child sexual exploitation" and "facilitated illegal immigration". The reason for this association lies in the subject of these criminal activities since all three of them "abuse individual's human rights", as expressed in the words of EUROPOL itself in the analysed reports. Through the reasoning explained above, the three criminal activities, all of them present in these reports, have been included together under the category "Crimes against persons". Furthermore, a complementary justification for this choice of words in the definition of the wider level can be found in the annual reports from 2005 to 2008, in which the crimes mentioned above were encompassed under the heading "Crimes against persons".

---

[5] This refers to the reports of the years 2010, 2011 and 2012 as at the time of writing this paper the 2013 report was not yet available.

4   Ontologies, Legal Knowledge and Interoperability

Interoperability is the ability of two or more systems or components to exchange information and to use the information that has been exchanged. The Semantic Web and ontologies provide the abstraction layer needed to carry out a "negotiation" or "dialog" between the participant systems to put in common concepts, vocabulary, terms, etc. Therefore, all the participants will know the meaning (not necessarily the content) of the exchanged information. For instance, an ontology-based framework devised to exchange meaningful representation of product data for collaborative environments is introduced in [13]; information exchange for different network management devices through ontologies is discussed in [14]; [15] proposes an approach devised to handle Electronic Health Records; and [16] brings together Clinical Research and Clinical Care fields through Semantic Web and ontologies.

On the other hand, legal professionals are used to consume an important part of their time searching, retrieving and managing legal information. Therefore, the organization and formalization of legal knowledge for computer processing produce many desirable features such as enhance of information search, retrieval and knowledge management. However, legal systems are complex, integrated search between the legislation of several European countries; e-government and e-administration; electronic institutions, privacy or digital rights management systems, are just some examples. Ontologies have been used successfully in these fields: in [17] the legal knowledge modelling and acquisition, the knowledge applications and the integrated applications are discussed; and [18] reviews and discusses different purpose legal ontologies.

Therefore, from the European LEAs interoperability point of view, ontologies provide these needed capabilities for exchanging meaningful information through the OCS introduced in Section 3. The main features that make ontologies suitable for LEAs interoperability are: i) ability to share common information; ii) enabling reuse of knowledge; iii) resilience to changes in the acquired knowledge; and iv) reasoning to determine interoperability.

5   European LEAs Interoperability Ontology

In this Section, the development of ELIO is discussed. The first step consists of gathering both domain and development requirements that define the ELIO build-up process. Table 1 consists of two different parts that collect these requirements. The top of the table sets the domain requirements: i) *competency questions* which set the domain, range and scope of the ontology; ii) *sources of knowledge*, which is a main point since the knowledge acquisition process is usually a bottleneck; iii) *conceptualization* of the ontology; and iv) *development approach*. On the other hand, the bottom of the table states the development stage requirements: i) *methodology* (based on METHONTOLOGY [19], On-To-Knowledge (OTK) [20], HCOME [21], UPON [22] and [23]); ii) *ontology editor*; iii) *reasoner*; and iv) *representation language* are the topics addressed in this part.

Table 1. ELIO Ontology requirements document. *Top*: Domain requirements. *Bottom*: Development requirements.

ELIO Domain Requirements	
Competency questions	Which techniques exist to commit these crimes? Which is the relation between techniques and crimes? Which are the EC* for these crimes in each country? Which are the related EC* from other countries? Which EC* belongs to a specific country? *EC stands for Essential Conditions
Sources of knowledge	Europol's Reviews International LEAs cooperation literature Expert elicitation
Purpose	Provide LEAs interoperability focusing on both EU legal frameworks and languages
Conceptualization	According to the issue of the conceptualization into [24], it is a specific ontology which represents knowledge related to a particular domain. **Domain Ontologies** provide vocabularies about concepts in a domain and their relationships, or about the theories governing the domain.
Development approach	The methodology approach begins with abstract concepts; how those concepts map to physical data is addressed later. Then, one begins with the data necessary for a specific analytic use-case, and models the concepts necessary for performing such analysis on the physical data. **(Middle-out Strategy)**
ELIO Development Requirements	
Methodology approach	The ontology development methodology is based on three main steps: 1) **preparatory stage**; 2) **development**; 3) **evaluation**
Ontology editor	**Protégé v4.3**
Reasoner	**Pellet** [25]
Representation Language	**OWL 2** [26]

There are three main points that define the process of creating an interoperability ontology in this scenario: knowledge acquisition, ontology structure and evaluation. Therefore, Section 5.1 sums up the knowledge acquisition process; Section 5.2 shows the structure that defines ELIO; and Section 5.3 addresses the evaluation literature and tests the level of interoperability.

5.1   Knowledge Acquisition

In the specific framework of each research project which deal with Law and Semantic Web issues, we use to apply a socio-legal approach [27]. It combines qualitative and quantitative methodologies depending on the sort of problems that have to be solved, the concrete objectives to be achieved, and the type of ontology that should be built up to modelling expert knowledge. The aim of this kind of approach is to provide the technology needed to solve specific end users needs.

The traditional means of knowledge acquisition is the traditional talking and question answering method which takes knowledge engineers as the intermediary of domain experts and computer systems. This method has drawbacks such as time and resources consume and prone to errors, but it is still one of the most basic knowledge acquisition methods. Fully automatic methods can not obtain totally correct and sufficient knowledge, even if acquiring knowledge, its reasonable and reliability have yet to be verified by experts. There are many works in literature addressing the knowledge acquisition issue. For instance, in [28], different families of techniques specifically devised to elicitation and analyse of knowledge acquired from experts are discussed. In our scenario, taking into account the socio-legal approach and the techniques reviewed in [28], we propose an iterative knowledge acquisition process based on five stages: elicitation, collection, analysis, modelling and validation.

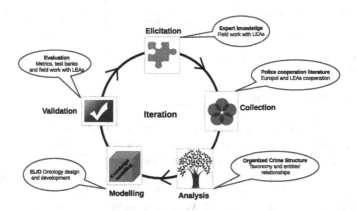

Fig. 2. Iterative knowledge acquisition process in this work lies in an iterative five process paradigm: elicitation, collection, analysis, modelling and validation.

Figure 2 depicts the five stages of the iterative knowledge acquisition process. The first one consists of expert knowledge acquisition devised to: i) validate the OCS definition; and ii) provide the essential conditions for each LEA. The second stage is based on the international LEAs cooperation literature reviewed in Section 3. The third stage carries out the analysis of the first two stages in order to define the concepts and their relationships for the ELIO definition. The fourth stage develops the model of the ontology. Finally, the last stage is focused on the validation of the interoperability provided through different bank tests and LEAs experts.

5.2 Structure

Although researchers have written much about the potential benefits of using ontologies, the design process must take into account some constraints. The design and maintenance stages related to ontologies are high resource consumption processes [29,30]. Moreover, ontologies must incorporate five main features such as clarity, coherence, extendibility, minimal encoding bias and minimal ontology commitment [31]. As a result, useful ontologies must be small enough to have reasonable design and maintenance costs and big enough to provide substantial added value for using them. In this work, we introduce a light ontology, ELIO, in order to minimize the design and maintenance resource consumption.

Table 2. Taxonomy and object property definitions for ELIO

Taxonomy	Object Property	Range
* Crimes → CrimesAgainstPersons → EconomicalCrimes → IllegalGoods	hasTechnique hasEssentialCondition	Techniques EssentialConditions
* Techniques	–	–
* EssentialConditions	hasCrime hasCountry	Crimes Countries
* Countries → EuropeanCountry → NonEuropeanCountry	–	–

The taxonomy and object properties present in ELIO are shown in Table 2. It has four main concepts represented as classes into the ontology structure: "Crimes", "Techniques", "Essential Conditions" and "Countries". Moreover, four object properties that connect elements among these classes are also defined: "hasTechnique", "hasEssentialCondition", "hasCrime" and "hasCountry". This knowledge representation enables LEAs interoperability through the OCS.

The designed architecture addresses two main issues: the development of ontologies and the maintenance process. ELIO design method falls on a two-layer paradigm depicted in Figure 3. From bottom to top, the first layer models the knowledge elicited from each source considered in Table 1 "Sources of knowledge". The upper layer joins knowledge from previous layer and suitable elements to obtain reasoning capabilities. In this architecture, changes are propagated from lower to upper layers and this structure eases the inclusion of new knowledge. Both benefits provide the flexibility feature to the ELIO Ontology.

The ontology developed in this work have different individuals that represent crimes, essential conditions, countries and techniques. Therefore, the main reason

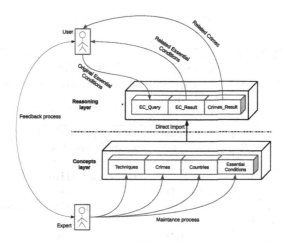

Fig. 3. ELIO architecture. Bottom layer models the knowledge acquired. The upper layer provides reasoning capabilities. In this architecture changes are propagated from lower to upper layers in order to ease the maintenance process.

for the described ontology-stack approach is to enable the detection of equivalent individuals through the modelled OCS. The interoperability is achieved in three different steps: i) modelling a special subhierarchy of essential conditions related to OCS crime instances besides the fundamental domain concepts; ii) putting these interoperability-defining subclasses in the Reasoning layer of the ontology-stack together with the logical rules; and iii) executing the reasoner against the ontology-stack. When this last step is taken, individuals in the Reasoning layer are classified into the interoperability-defining subhierarchy, according to their relation with the OCS.

5.3   Evaluation

In the literature, there are several methods specifically devised to perform an unbiased ontology evaluation. For instance, in [32] and [33] different metrics are introduced, and in [34] a comparison among different approaches is addressed. However, we leave this topic as a future work since to date we are only focused on LEAs validation. Therefore, in this Section two evaluation tests are shown as interoperability samples. The main target of ELIO is to detect the essential conditions with equivalent meaning in other countries, providing interoperability among European LEAs. The evaluation test provides the results after the execution of the reasoning algorithm over the ontology-stack introduced in the previous section to ensure that the whole essential conditions with equivalent meaning are detected.

This section shows two examples of test bank. The name of the essential conditions have been changes in order to ease the comprehension of the process, since these individuals are represented in the language of its corresponding

Table 3. Evaluation test #1

Query		
EC Individual	Object property	Crime Individual
ec-THB-CSE-G	hasCrime	childSexualExploitation
	hasCrime	traffickinginHumanBeings
	hasCountry	germany
Result		
ec-II-CSE-PR	hasCrime	illegalInmigration
	hasCrime	**childSexualExploitation**
	hasCountry	portugal
ec-THB-SP	hasCrime	**traffickinginHumanBeings**
	hasCountry	spain
ec-THB-F-FR	hasCrime	**traffickinginHumanBeings**
	hasCrime	fraud
	hasCountry	france
ec-THB-F-PR	hasCrime	**traffickinginHumanBeings**
	hasCrime	fraud
	hasCountry	portugal
ec-THB-II-PO	hasCrime	**traffickinginHumanBeings**
	hasCrime	illegalInmigration
	hasCountry	poland

country. As a result, the name pattern is "ec-C_1-C_2-...-C_n-$Country$", where C_n states that this essential condition is related to a specific crime represented for its initials and "$Country$" shows the country related to this essential condition. Table 3 shows an example when the query for an essential conditions is related to "*child sexual exploitation (CSE)*" and "*trafficking in human beings (THB)*". The results after the reasoning process are shown at the bottom of the table. All of these resultant essential conditions have at least one of the concepts present

Table 4. Evaluation test #2

Query		
EC Individual	Object property	Crime Individual
ec-CS-D-SP	hasCrime	cigarretteSmuggling
	hasCrime	drugs
	hasCountry	spain
Result		
ec-CS-D-FR	hasCrime	**cigarretteSmuggling**
	hasCrime	**drugs**
	hasCountry	france
ec-F-CS-G	hasCrime	fraud
	hasCrime	**cigarretteSmuggling**
	hasCountry	germany
ec-OR-D-SP	hasCrime	organizedRobbery
	hasCrime	**drugs**
	hasCountry	spain

in the query. Then, from a query performed for the essential conditions related to THB and CSE in Germany, ELIO answer is composed of essential conditions related to the same crimes in Portugal, Spain, France and Poland.

In addition, Table 4 shows another example of interoperability, which the query is related to *"cigarette smuggling (CS)"* and *"drugs (D)"*. Consequently, the results given by ELIO have equivalent concepts in their definition, allowing this way the interoperability among European LEAs in a similar way than the previous example.

6   Conclusions

In this work we address the issue of European LEAs interoperability in the fight against transnational organized crime. In order to do so a two steps process was put into practice. This process consists first, of the design of an Organized Crime Structure (OCS) and second of the creation of a model of the OCS into a machine-readable format: and ontology. The first part entails the extraction of the contents of Europol annual reviews on serious crime in order to define a relevant structure of the field of organized crime that improves interoperability, overcoming differences among legal systems, languages and police cultures. The second step consists of an interactive knowledge acquisition process, the design of the ontology and finally the evaluation process.

Regarding the first part of this work several methodological choices lead to the selection of Europol yearly reviews as the more suitable source for the identification of both the general abstract criminal categories and the individual criminal activities to include in the OCS in order to truthfully represent the map of the situation of organized crime in the territory of the European Union. This choice allows us to overcome the problems related to the lack of a common concept organized crime in different EU countries, the mutable nature of organized crime and the different semantic and legal configuration of each crime in each legal system

As for the second part of this work we can highlight on the first place that the knowledge acquisition process performed allowed for the extraction of the information needed to identify the essential conditions in each country. Finally we argue that the ontology based on the OCS is able to improve interoperability capabilities among EU LEAs. The evaluation process is able to determine the existence of different examples in which the equivalent essential conditions of the legal system of each country are stated within the Reasoning layer.

References

1. Goodman, M.: What Business Can Learn From Organized Crime. Harvard Business Review (November 2011), http://hbr.org/2011/11/what-business-can-learn-from-organized-crime/
2. Carr, J.: New approaches to dealing with online child pornography. In: Second Worldwide Cybersecurity Summit (WCS), pp. 1–3 (June 2011)

3. Cunningham, S., Kendall, T.: Prostitution 2.0: The changing face of sex work. Journal of Urban Economics 69, 273–287 (2011)

4. Kaza, S., Xu, J., Marshall, B., Chen, H.: Topological Analysis of Criminal Activity Networks in Multiple Jurisdictions. In: Proceedings of the 2005 National Conference on Digital Government Research. dg.o 2005, pp. 251–252. Digital Government Society of North America (2005)

5. Europol: Europol Review 2012. General Report on Europol activities (August 2013)

6. Hagan, F.E.: The Organized Crime Continuum: A Further Specification of a New Conceptual Model. Criminal Justice Review 8(2), 52–57 (1983)

7. Hagan, F.E.: "Organized crime" and "organized crime": Indeterminate problems of definition. Trends in Organized Crime 9(4), 127–137 (2006)

8. Bundeskriminalamt: National Situation Report (2011), http://www.bka.de

9. Interpol: Annual report (2012), http://www.interpol.int/News-and-media/Publications#n627

10. Benyon, J.: Law and order review, 1993 an audit of crime, policing and criminal justice issues. Centre for the Study of Public Order, University of Leicester, Leicester, England (1994)

11. Finckenauer, J.O.: Problems of definition: What is organized crime? Trends in Organized Crime 8(3), 63–83 (2005)

12. Albanese, J.S.: North American Organised Crime. Global Crime 6(1), 8–18 (2004)

13. Patil, L., Dutta, D., Sriram, R.: Ontology-Based Exchange of Product Data Semantics. IEEE Transactions on Automation Science and Engineering 2(3), 213–225 (2005)

14. Wong, A.K.Y., Ray, P., Parameswaran, N., Strassner, J.: Ontology Mapping for the Interoperability Problem in Network Management. IEEE Journal on Selected Areas in Communications 23(10), 2058–2068 (2005)

15. Berges, I., Bermudez, J., Illarramendi, A.: Toward Semantic Interoperability of Electronic Health Records. IEEE Transactions on Information Technology in Biomedicine 16(3), 424–431 (2012)

16. Laleci, G.B., Yuksel, M., Dogac, A.: Providing Semantic Interoperability between Clinical Care and Clinical Research Domains. IEEE Journal of Biomedical and Health Informatics 17(2), 356–369 (2013)

17. Breuker, J., Casanovas, P., Klein, M.C., Francesconi, E.: Law, Ontologies and the Semantic Web. IOS Press (2009)

18. Casellas, N.: Legal Ontologies. In: Legal Ontology Engineering: Methodologies, Modelling Trends, and the Ontology of Professional Judicial Knowledge. Law, Governance and Technology Series, vol. 3, pp. 109–170. Springer (2011)

19. Gómez-Pérez, A., Fernández-López, M., Corcho, O.: Ontological Engineering. With Examples from the Areas of Knowledge Management, e-Commerce and the Semantic Web. Advanced Information and Knowledge Processing. Springer (2004)

20. Sure, Y., Studer, R.: A Methodology for Ontology-Based Knowledge Management. In: Towards the Semantic Web: Ontology-driven Knowledge Management, pp. 33–46. John Wiley and Sons (2003)

21. Kotis, K., Vouros, A.: Human-centered ontology engineering: the HCOME methodology. Knowledge and Information Systems 10(1), 109–131 (2006)

22. De Nicola, A., Missikoff, M., Navigli, R.: A proposal for a unified process for ontology building: UPON. In: Andersen, K.V., Debenham, J., Wagner, R. (eds.) DEXA 2005. LNCS, vol. 3588, pp. 655–664. Springer, Heidelberg (2005)

23. Casellas, N.: Methodologies, Tools and Languages for Ontology Design. In: Legal Ontology Engineering: Methodologies, Modelling Trends, and the Ontology of Professional Judicial Knowledge. Law, Governance and Technology Series, vol. 3, pp. 57–108. Springer (2011)
24. Guarino, N.: Formal ontology in information systems, pp. 3–15. IOS Press, Amsterdam (1998)
25. Sirin, E., Parsiaa, B., Graua, B., Kalyanpura, A., Katza, Y.: Pellet: A practical owl-dl reasoner. Journal of Web Semantics: Science, Services and Agents on the World Wide Web 5(2), 51–53 (2007)
26. W3C: OWL 2 Web Ontology Language Structural Specification and Functional-Style Syntax (December 2012)
27. Casanovas, P., Casellas, N., Vallbé, J.J.: Empirically Grounded Developments of Legal Ontologies: A Socio-Legal Perspective. In: Approaches to Legal Ontologies: Theories, Domains and Methodologies, pp. 49–68. Springer (2011)
28. Hua, J.: Study on Knowledge Acquisition Techniques. In: Second International Symposium on Intelligent Information Technology Application, pp. 181–185. IEEE (2008)
29. Lambe, P.: Organising knowledge: taxonomies, knowledge and organizational effectiveness. Oxford and Chandos Publishing (2007)
30. Hepp, M.: Possible Ontologies. How Reality Constrains the Development of Relevant Ontologies. IEEE Internet Computing 11(1), 90–96 (2007)
31. Gruber, T.R.: A translation approach to portable ontology specifications. Knowledge Acquisition 5(2), 199–220 (1993)
32. Guarino, N., Welty, C.A.: An Overview of OntoClean. In: Handbook on Ontologies. Springer (2009)
33. Mostowfi, F., Fotouhi, F.: Improving Quality of Ontology: An Ontology Transformation Approach. In: Proceedings 22nd International Conference on Data Engineering Workshops, p. 61 (2006)
34. Tartir, S., Arpinar, I.B., Sheth, A.P.: Ontological Evaluation and Validation. In: Theory and Applications of Ontology: Computer Applications, pp. 115–130. Springer (2010)

Harnessing Content and Context
for Enhanced Decision Making

Paulo Novais[1], Davide Carneiro[1], Francisco Andrade[2], and José Neves[1]

[1] Department of Informatics/CCTC, University of Minho, Braga, Portugal
{pjon,dcarneiro,jneves}@di.uminho.pt
[2] Law School, University of Minho, Braga, Portugal
fandrade@direito.uminho.pt

Abstract. In a time in which a significant amount of interpersonal interactions take place online, one must enquire to which extent are these milieus suitable for supporting the complexity of our communication. This is especially important in more sensitive domains, such as the one of Online Dispute Resolution, in which inefficient communication environments may result in misunderstandings, poor decisions or the escalation of the conflict. The conflict manager, in particular, may find his skills severely diminished, namely in what concerns the accurate perception of the state of the parties. In this paper the development of a rich communication framework is detailed that conveys contextual information about their users, harnessed from the transparent analysis of their behaviour while communicating. Using it, the conflict manager may not only better perceive the conflict and how it affects each party but also take better contextualized decisions, closer to the ones taken in face-to-face settings.

Keywords: Online Dispute Resolution, Context-aware Computing, Stress, Fatigue.

1   Introduction

The surroundings or the circumstances in which a given event or occurrence takes place is known as *Context*. It allows one to correctly understand and interpret said occurrence. Taking the field of linguistics as an example, context may refer to the set of information that is relevant to fully understand a text. This information may be very varied and include the identity of things named in the text (e.g. people, places) as well as many other aspects such as birth dates, geographical locations or temporal location. In fact, different contexts may provide completely different interpretations for the same text. Although most of the times in an unconscious way, in our daily living we constantly rely on contextual information.

The importance of Context goes as far as shaping ourselves and who we are. The knowledge acquired in the ongoing process of learning that is our life after all, comes with a strong social, cultural and physical context [2]. This bound is so strong that *cognition* cannot be separated from *context*, i.e., *knowing* is inseparable from *activity, people, culture, language* or *time*.

P. Casanovas et al. (Eds.): AICOL IV/V 2013, LNAI 8929, pp. 232–246, 2014.
© Springer-Verlag Berlin Heidelberg 2014

This relationship goes as far as individuals exhibiting different cognitive and reasoning processes under different contexts: the context in which they are inserted provides the symbols and values that individuals will use [21]. Thus, no individual can be accurately and absolutely defined without a notion of context, i.e., we cannot say to another person "I behave like this" and expect to really be like that all the time. Instead, we should say "In a scenario with these conditions, I would *probably* behave like this".

Conflicts and their resolution, as many other processes we engage in daily, have a strong contextual background. Frequently, most of the meaning present in the underlying communication processes stems not from the words used but from accessory information that helps understanding the real meaning or purpose of the words. This includes aspects such as the body language, the gestures, the posture, the emotional response or the tone of voice.

The importance of context in conflict resolution has been noted in one of the earliest documents written about Online Dispute Resolution (ODR), by Etan Katsh [25]:

> "Context can influence the approach of the neutral, the choice of process, and the behavior and attitudes of disputants. In any environment, context can affect the kinds of disputes that are likely to arise and also affect who the parties are who are likely to be involved in the dispute. Context implicitly feeds us information about the extent or nature of the injury as well as how the injury or dispute is perceived by those involved. Context situates a dispute in a particular time and place, and we react and adjust accordingly as the parameters of the environment become clear to us."

Contextual information is indeed important for the involved parties (disputants and conflict managers) to perceive the conflict in its whole, with its peculiarities, subjectivities and particular views. This paper looks at this topic, from a Computer Science perspective. It addresses different the types of context that are meaningful in the conflict resolution arena and it compares traditional ways of acquiring such information with more innovative ones, put forward by the authors in the last years. Finally, the paper briefly describes a conflict resolution platform that implements these ideas.

2   Context-Aware Computing: Harnessing Content and Context

In the last years, context has acquired such an importance that it even gave birth to a new field in Computer Science: Context-aware Computing. It refers to systems that are aware of the state and surroundings of their users, and that are able to adapt their behaviour according to changes in their context [36]. The knowledge about the environment or the user may be relatively simple (e.g. the network to which the device is connected, the devices in the proximity) or may be more complex and even built from assumptions about the user's current situation.

In its early days, context was much focused on the issues related to user location, encouraged mostly by the rapid emergence of mobile devices and applications who could make particular use of this information [20]. The hype revolved around applications that would provide services personalized according to their user's location. Nevertheless, in the last years the notion of context has widened significantly, and now refers not only to where the users are but also to *who* the users are, *what* the users are doing, *when* they are doing it and *with who* they are doing it. All this information allows the system to infer *why* the users are doing it and this is essential in the task of providing personalized, meaningful and useful services.

These novel and increasingly complex contextual models provide the support for applications that are able to adapt interfaces, improve information retrieval techniques, target services more efficiently or use implicit user-interaction techniques [9].

Given the contextual richness of today's environments, information about users' context can be acquired from many different sources: from the objects the user is interacting with (e.g. title or topic of an e-book being read), inferred from the activity being performed (e.g. nature/objective of the activity), provided explicitly (e.g. social tagging, comments, bookmarks) or can be acquired from hard or soft sensors, just to name a few examples.

The potentialities of using information describing the user's context are also wide. Indeed, Many applications have already been developed that build on such notions to provide high-value and innovative services. Most of them are based on learning the patterns of the users on the environment, their preferences and their habits. This is fundamental in order for personalized services to be provided [4].

A good example of this is Magitti: an activity-aware leisure guide running on a smartphone [7]. It is essentially a recommendation service that has a model of the preferences of the user. This model is continuously built from their current situation or their past behaviours. This allows Magitti to infer the current activity of the user in order to filter and rank information items that may be of interest.

Under a different perspective, the Hearsay service developed as part of the GLOSS project [29] allows users to pick up small notes left for them in the environment. It makes sure users will find the message only in a correct context (e.g. right person in the right place at the right time). The same approach is applied to other applications, providing a structured link between environment and behaviour to improve utility and usability [19].

Other projects have also focused on acquiring contextual information from the observation of the user. [11] claim that user interactions with everyday productivity applications (e.g. word processors, Web browsers) provide rich contextual information that can be leveraged to support just-in-time access to task-relevant information. Besides discussing the requirements for such systems and developing a general architecture, the authors present Watson, a system which gathers contextual information in the form of the text of the document the user is manipulating in order to proactively retrieve documents from distributed information repositories.

There is also a marked interest on the use of context-aware computing in the medical field, particularly in alleviating the tasks of medical practitioners or on supporting patients. [5] presents the design of a context-aware pill container and a context-aware hospital bed, both of which react and adapt according to what is happening in their context. The system is able to, among other tasks, point out the location of the correct medicine to take or verify if it is the correct patient receiving it.

On a different field, [34] define an innovative communication framework that incorporates Augmented Reality techniques. Users can dynamically attach newly created digital information such as voice, notes or photographs to the physical environment, through wearable computers as well as normal computers. Attached data is stored with contextual tags such as location IDs and object IDs that are obtained by wearable sensors, so the same or other wearable users can notice them when they come to the same environment. The approach implemented has a role that is similar to the one that Post-it notes play in community messaging.

A particular interest lies also on the development of context-aware systems that can be carried by the user, generally in small devices such as smartphones or video cameras. This is empowered by the functionalities and potential of current mobile devices, rich in sensors, computational power and communication capabilities. [31] address the challenge of organizing our ever-growing collections of digital photos, consequence of the enormous rise in popularity of digital cameras. To achieve it, the authors developed the MediAssist project, which uses date/time and GPS location for the organization of personal collections. The project retrieves photos of known objects (e.g. buildings, monuments) using both location information and content-based retrieval tools from the AceToolbox, allowing to improve information search and retrieval when compared to more traditional approaches.

3   Contextualizing Conflicts

As detailed in the previous section, Context-aware Computing has been used for many different purposes in the last years. Nonetheless, no initiatives can be found at the moment that use such approaches in the domain of conflict resolution.

After acknowledging this fact our research team, with a strong background on Ambient Intelligence, decided to tackle this challenge. In the last years we have been taking steps towards the development of a context-aware conflict resolution platform that can not only provide personalized and proactive support but also be sensitive to the state of the users. This section describes some of the approaches implemented that allow software agents and human parties to contextualize and fully understand a conflict.

In fully understanding a conflict, one of the most challenging tasks is to realistically perceive the boundaries of the case, i.e., which outcome would be realistic and relatively consensual and which outcome would be nonsense. In order for parties to build a realistic view on the conflict, they must be aware of a few notions such as the boundary possibilities (the best/worst possible cases) as well as the most likely outcome, according to the characteristics of the case, the norms or past experiences. In this sense, a case-based approach was developed [3] that computes the values of concepts such as the BATNA (Best Alternative to a Negotiated Agreement), WATNA (Worst Alternative to a Negotiated Agreement), MLATNA (Most Likely Alternative to a Negotiated Agreement) or the ZOPA (Zone of Possible Agreement) [30,22,39]. With these values parties can realistically frame their conflict and understand what their possibilities are.

Nonetheless, one of the best ways for someone to understand a present event that is taking place or going to take place in the future, is to analyse a similar event that took place in the past, under a similar context. In Computer Science, the field of Case-based Reasoning deals with approaches that are based on this notion. In this field, a case represents a past experience, teaching a past lesson, properly contextualized [26]. Depending on the domain of application, a case should contain a description of the state of the world when the event occurred (i.e. the problem description), the derived solution for the problem and/or the achieved outcome (i.e. the description of the state of the world after the case occurred) [18]. The individual can analyse the past event, with all its characteristics and within its contextual framework to fully understand it. Then, he will be in a better position to understand a similar event that may occur in the future.

The analysis of past cases is also of interest in the domain of conflict resolution. Disputant parties involved in a conflict resolution process may leverage on past cases, under a similar context, to better perceive the present. This allows the parties to calmly understand the important aspects of the past cases, being better prepared to fully understand their own and gaining a more realistic and objective view on the case. For example, if a given party is too greedy but notices that the outcome he desires has never happened in the past, he may conclude that such outcome is unlikely and even unrealistic. He may then reconsider his objectives and go for a more middle-ground solution. In order to achieve such objectives, case-retrieval approaches have been implemented by our research team that provide parties with an intuitive way to perceive their conflict and their chances in the conflict through the analysis of past similar cases [12].

Another very important aspect when contextualizing a conflict is the personal conflict handling style of each individual: it determines the expected behaviour of an individual before a conflict. Each individual has their own style, which may however vary according to variables such as the identity of the other parties, the nature of the conflict, or even the level of escalation of the conflict. Knowing the conflict handling style of the parties is especially important for the mediator, who can better prepare the strategy for the resolution of the conflict. Gathering and providing evidence about the party's conflict handling style may also be used by the mediator as a way to put pressure and to lead the parties into changing an undesired behaviour: a mediator may show to a party that he is being an obstacle to the successful resolution of the conflict by being too greedy or competitive. We have addressed this issue by devising an approach to infer the conflict handling style without the use of the traditional questionnaires. We rather look at the proposals exchanged by the parties during the resolution of the conflict and, framed in the boundaries of the case, build a notion of how greedy, cooperative or benevolent the parties are [3]. All this is done in real-time and in a non-invasive way, providing the mediator with very important contextual information about the case and the parties.

In seeking to contextualize a conflict an its setting, the level of escalation should not be forgotten. It describes how confrontational, violent, painful or "less comfortable" a conflict is. Desirably, parties involved in a conflict are in a cooperative and compliant state. However, in general this is not the case. If not adequately managed, the conflict may escalate to increasingly worse states, in which individuals passively or actively resist to proposals or commands, with the aggravating of the escalation potentially resulting in violence. Therefore, it is mandatory that the conflict manager is able to perceive signs of escalation of the conflict. Typically he can do so through cues such as the tone of voice, signs of inflammatory speech, gestures, body postures or the use of particular words. It is the responsibility of the mediator not only perceives these signs but to act in order to prevent the conflict to escalate further. Nonetheless, this may result particularly difficult in an online setting, as the mediator lacks the non-verbal cues that show how a person is feeling. Thus, alternatives should be devised that can inform the mediator of the level of escalation of the conflict when he is not meeting the parties face-to-face.

One of the possible approaches, prototyped by our research team in the last two years, does so by measuring the level of stress of the participants. Stress is a universal phenomenon that affects virtually our whole life. Low levels of stress make us soft, depressive and with lack of motivation, while continued high levels of stress may result in exhaustion and breakdowns [14]. The resolution of a conflict may be a particularly stressful process for many reasons, including the potential emotional charge (mostly when the parties had a prior relationship that may affect the process), the significant amount of gains or losses involved, the fear of the novelty or the unknown, mostly when parties engage in such a process for the first time. In the short-term, high levels of stress lead to clouded-mind, poor decision-making, irritability, lack of judgement or violence. Failing to control these effects may jeopardize the whole conflict resolution process and even the relationship among the parties involved.

It is thus important that the mediator is able to perceive signs of stress evidenced by the parties, and act accordingly in order to mitigate its effects by performing the necessary changes in the conflict resolution process. In this sense, an approach was developed by our research team that quantifies the level of stress of an individual from the analysis of their interaction patterns with common technological devices such as smartphones or computers. It is a non-invasive and non-intrusive approach that does not rely on questionnaires but on the sheer observation of the individual's behaviours [14]. Further details on this approach are given further ahead.

A similar approach is used to quantify the level of fatigue of the individuals [32]. Fatigue is a particular feeling of tiredness in which individuals experience lack of energy, lethargy or languidness. Two main forms of fatigue can be identified: physical fatigue, which defines a temporary physical inability of a muscle to function normally, and mental fatigue, which is the temporary inability to maintain optimal cognitive performance. Both types of fatigue have their negative consequences. However, in the context of a conflict resolution process, mental fatigue may be more disturbing.

Mental fatigue may be caused by sleep deprivation, long periods of work, mental stress or over-stimulation, among others. A mentally fatigued individual will have a reduced cognitive capacity, namely in terms of memory, attention and decision making. A fatigued individual participating in a conflict resolution process may take poor decisions, be unable to keep up with the process or even become irritable and uncooperative. It is therefore important that the mediator is able to detect early signs of fatigue through the behaviours and attitudes of the parties and act accordingly, namely by making pauses or by rescheduling the continuation of the process. Once again, the mediator is responsible for detecting such signs and warn the parties or take actions that prevent them from taking bad decisions, such as making a pause or resuming on another time. On the other hand, parties themselves could profit from some kind of notification, provided directly by the platform, letting them know that they may not be in their best shape to take binding or relevant decisions.

To sum up, in order to properly contextualize a conflict and its resolution, we look at its important boundary values, at past similar cases, at the conflict handling style of the parties and at the levels of stress and fatigue of these same parties. All of this is incorporated into a single conflict resolution platform, designated UMCourt. Moreover, all this information is compiled in a non-invasive and non-intrusive way, in line with the Context-aware Computing philosophy. While some of these aspects have already been addressed in detail in the past, the quantification of stress and fatigue, by being the most recent, innovative and interesting, are detailed further ahead in this paper.

4   Traditional Approaches on Context Acquisition

Stress, fatigue or emotional state are extremely important in describing the inner state of an individual, which ultimately affects all their conscious and unconscious actions and decisions. These personal, subjective and conscious experiences have known effects at the level of psycho-physiological expressions, biological reactions, mental states, mood, temperament, disposition, personality, motivation and, ultimately, health and well-being. Hence the significance of their study, which is very complex, involving fields such as psychology, philosophy, neurology, physiology, or medicine.

Traditionally, two main approaches may be followed to study such phenomena. The field of psychology relies more on the use of questionnaires or surveys, whereas the field of medicine relies on different kinds of sensors. Each of these approaches has advantages and disadvantages of their own.

Surveys are generally seen as a cost-effective way of gathering large amounts of information. They do not require much effort from the researcher and often have standardized answers that make it simple to compile data [1]. They are eminently practical and may be carried out by the researcher or by any other person without significantly affecting its validity and reliability. However, surveys also suffer from a number of problems that go beyond the traditional ones related to question constructing and wording [33]. Namely, surveys are particularly inadequate to understand some complex issues such as emotions, behaviours or feelings. They are based on the individual's perception of rather subjective perceptions such as *good, poor, high* or *low*. People can also hide information, lie voluntarily, or unconsciously depreciate/overvalue certain signs [28]. It is nearly impossible for the researcher to detect such behaviours. Finally, when developing the questionnaire or survey, researchers make their own decisions and assumptions as to what is or is not important. Even if the individual finds some aspect of importance, they may not express it if it is not mentioned in the questionnaire or if it is not mentioned appropriately.

The medical field developed a highly accurate approach on the problem, based on a wide range of different sensors that measure changes on physiological or neurological features of the human body, affected by fatigue, stress or emotions. Currently, one of the most accurate indicators in use is cortisol [38], measured in the saliva, hair or blood. It is particularly useful to measure the level of stress of human beings, since this hormone is released in response to this symptom.

Other approaches on the problem may also be followed using other sensors or combinations of sensors. The Galvanic Skin Response measures the electrical conductance of the skin, which varies with its moisture level. This is of interest since the sweat glands are controlled by the sympathetic nervous system, so skin conductance is used as an indication of psychological or physiological arousal, which may happen due to stress or fatigue. The temperature of the skin, the hearth rate, or the respiratory rate are also key indicators for the study of stress or emotions [6,23,24]. In particular, hearth rate variability, the physiological phenomenon of variation in the time interval between heartbeats [10], has been used increasingly to study stress as it is highly related to it [8].

The significant emergence of biofeedback tools in the last years is also noteworthy. They provide combined feedback about many of the body's functions, using instruments that analyse brainwaves, muscle tone, skin conductance, heart rate or pain perception [37]. The study of brainwaves is particularly interesting as it provides clues about the inner state of an individual, in aspects such as fatigue, stress level, arousal or emotional state. Additionally, biofeedback tools can be used to improve certain aspects or habits of the daily living, as they allow perceiving changes in the body and mind affected by such habits [27].

In general, such sensor based-approaches can be deemed highly accurate and are used not only to assess the state of individuals, but also as a base to perform medical treatments and interventions. Their use is thus unquestionable and unparalleled in the medical arena.

Nevertheless, in the context of this paper we must look at both approaches, questionnaires and physiological sensors, from the point of view of someone who intends to build a context-aware conflict resolution platform. Thus, one must ask to which extent are questionnaires or physiological sensors suitable approaches to assess the user's state in a Virtual Environment. We argue that they do not constitute suitable approaches. Let us look at both approaches and their potential disadvantages.

When individuals use questionnaires to describe themselves and their behaviour, they sometimes do not fit in any of the four answers of the multiple choice question. They may choose not to answer, or select the option they may think to be closest to what they would do. Moreover, they have doubts quantifying some of the other answers. While some of the concepts used, such as *never* or *always*, are easy to understand, others such as *often* or *occasionally* are unclear. Additionally, the individuals that undertake this process end up behaving differently when the process is under its way. Indeed, they may assume that they are going to act in a given way while they are filling in the questionnaire, but, under the pressure of the proceedings, individuals may behave in a different way.

When the same individuals use physiological sensors, they are not comfortable or are even refusing to use sensors to which they are connected constantly, seriously limiting their movements. Moreover, they may not be entirely sure about which information will be collected and what it will be used for. Hence, they are reluctant and the sheer use of such devices seems to stress them and to deflect their attention from the conflict resolution itself.

It results clear that none of these approaches looks reasonable to quantify the state of individuals in a conflict resolution process conducted online. Hence, the following sections depict a new paradigm in which the behaviour of the individuals is analysed in order to infer information about their inner state. Indeed, phenomena such as stress, fatigue or emotions affect not only our physiology but also our behaviours. If we have a way of identifying and measuring behaviours, and if we have a way to relate given behaviours to given states, we may be able to infer the inner state of an individual through the observation of their conduct.

5   Acquiring Contextual Features from Behavioural Analysis

The study of stress or fatigue, including their causes and symptoms, has been a topic of disciplines such as Medicine or Psychology. Traditionally, data about users is acquired either through self-reporting mechanisms (generally questionnaires), or through the use of physiological sensors. As seen above, both have disadvantages of their own.

In that sense, this section puts forward a new approach based on behavioural analysis. The key idea is to observe, in a non-invasive way, the behaviour of the individuals, and map certain known behaviours to specific states or changes in these states. This approach can thus be put beneath the umbrella of Behavioral Biometrics [40], and results in a multi-modal approach on the problem of behavioural analysis, where individuals' symptoms are taken as input. Such approaches yield accuracy rates that exceed their unimodal counterparts [17].

Specifically, we target behaviours that can be observed commonly in a typical scenario of use of an ODR platform without the need to use additional or invasive sensors. In that sense, we consider aspects such as the movement patterns of the user in the environment and the patterns of interaction with devices such as computers (through the mouse and the keyboard), smartphones or touch screens. From these devices, a wide range of features can be extracted that characterize behaviours that, as our previous studies conclude, are significantly affected by stress [14,32].

The following features are extracted from the mentioned devices:

- Touch pattern - this information is acquired from touch screens with support for touch intensity. It represents the way the pressure changes over time, during a touch;
- Touch accuracy - the relationship between touches in active controls versus touches in passive areas (e.g. areas without controls, empty areas), where touches are pointless. This information is acquired from touch screens;
- Touch intensity - the amount of pressure exerted by the finger on the touch screen. It is analysed in terms of the maximum, minimum and average intensity of each touch;
- Touch duration - the time span between the beginning and the end of the touch event. This data is acquired from devices with touch screens;
- Amount of movement – its evaluation is provided by the INT3-horus framework. The image-processing stack uses the principles established by [15], and uses image difference techniques to evaluate the amount of movement between two consecutive frames [16];
- Acceleration - the acceleration is measured from accelerometers integrated or fitted into the mobile devices, the keyboard or the mouse;
- Score - this feature quantifies how well the individual performs on the several tasks he was assigned;
- Stressed touches - this feature describes which touches are classified as stressed, according to the shape of the intensity curve;
- Key Down Time - the timespan between the pressing down and the release of a key, i.e., for how long was a given key pressed;
- Time Between Keys - how long did the individual take to press another key after the previous one was released;
- Mouse Velocity - the distance travelled by the mouse (in pixels) over the time (in milliseconds);
- Mouse Acceleration - the increase in velocity of the mouse (in pixels/milliseconds) over the time (in milliseconds);
- Time Between Clicks - the time span between each two consecutive clicks;
- Double Click Duration - the duration of a double click event, whenever this time span is inferior to 200 milliseconds. Wider time spans are not considered as double clicks;
- Average Excess of Distance – the excess of distance, in average, that the pointer travels between each two consecutive clicks of the mouse, when compared to the straight line between the same points which represents the shortest (more efficient) path;
- Average Distance of the Mouse to the Straight Line - it measures the average distance of the mouse, between each two consecutive clicks, to the straight line defined by the two consecutive clicks;

- Distance of the Mouse to the Straight Line - this feature is similar to the previous one. However, it returns the total of the distance travelled by the mouse rather than a computed average of the mouse's trajectory;
- Signed Sum of Angles – here the aim is to determine how much the movement of the mouse turns more to the right or to the left;
- Absolute Sum of Angles - this feature is quite similar to the previous one. However, it seeks to find how much the mouse turned in absolute terms, i.e., without considering the direction;
- Distance Between Clicks – it stands for the distance travelled by the mouse (in pixels) between each two consecutive clicks.

These features ensure two things. First of all, their number and different sources increase the availability of sources of data at all times, i.e., the user may stop interacting with the computer but start interacting with the smartphone, and we can acquire meaningful data to characterize their context nonetheless. Secondly, these features provide insights into different modalities affected by stress, namely the physical (through the movement pattern for example), the behavioural (through changes in the typing rhythm) or the cognitive (through measures of score). Its multi-modal nature provides this approach with higher accuracy than uni-modal ones [17].

In the last two years we have been using this approach to study and understand how stress and fatigue influence our interaction and our behaviour in such environments. Essentially, under small periods of acute stress people become more efficient and perform their tasks quickly and more efficiently. This efficiency tends to drop significantly after some time, depending on the intensity of the stressors. Fatigue, on the other hand, starts to become noticeable through a decrease in the performance and a generalized slowness in the interaction with the devices (e.g. slower mouse velocity, larger keydown time) [14,32], [35]. These findings are not surprising nor exceptionally revealing. Indeed, the importance of this study lies not on such conclusions but on the datasets and models trained that depict how people behave when under certain conditions, that can be used to build an environment that is sensitive to its users' state, in real time.

This classification, in real time, can be crucial for the mediator to accurately understand the state of the parties and take appropriate actions. If the mediator notices that one of the parties is showing significant signs of fatigue, they may advise that party to make a pause, go for a walk or they may even decide to resume the process on the following day. Moreover, a fast and significant change in the level of stress may be indicative of a sudden escalation of the conflict. In this scenario, the mediator may decide to calm down the parties by making a pause, changing the subject or even interrupt the direct contact between the parties. To make this kind of decisions, the mediator may take into consideration raw data and an explanation of its meaning. Figure 1, for example, depicts a steady increase in the keydown time (from 80ms to 100ms) in the period of a few hours, a clear and known sign of increasing fatigue.

Nonetheless, the potentially large amount of information to consider in each instant and the complex inter-relation between the several sources of information

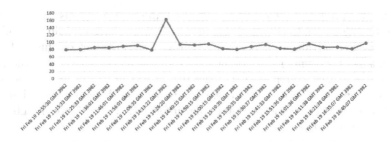

Fig. 1. Evolution of the keydown time for a user during a few hours of the day: it increases steadily as fatigue settles in

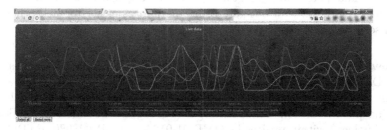

Fig. 2. Evolution of the level of stress of a user through time (dashed red line), the quality of the information (dashed orange line) and the different features available and their contribution to the level of stress

may make it difficult for the mediator to draw the correct conclusions in due time. Moreover, it may overload the mediator with additional tasks and shift their focus of attention from the conflict. In that sense, more refined interfaces are available that provide the high-level knowledge compiled in each instant (Figure 2). This makes it easier for the mediator to perceive the state of the parties. Nonetheless, the mediator still has the possibility to access the more detailed data or even the raw data depicted in Figure 2 ,whenever they feel it is necessary for a better perception of the state of the parties.

6   Conclusion

Communication tools are turning our daily interactions colder and dehumanized. Our communication process, generally very rich in accessory information, is being significantly simplified to rely on words alone. This leaves aside very important contextual information such as the tone of our voice, the rhythm of speech, our gestures, our body language or our facial expressions. Indeed, many of the times, this contextual information contains more meaning than words themselves. This can be particularly worrying in a domain such as the one of conflict resolution, in which very important decisions are taken based on the communication between the parties.

Mediators and disputant parties must take important decisions "in the dark", without having access to the whole gamut of information that would be available in a face-to-face setting. In that sense, we described a conflict resolution framework that has been under development in the last years. Apart from the more traditional services developed in the past (e.g. information retrieval, generation of solutions), we have now been focusing on improving the communication layer of the framework, by including contextual information that is meaningful in a conflict resolution process.

Information and Communication technologies, that have been deemed to bring people closer together, seem to do the exact opposite in some aspects. It is our conviction that approaches such as the one described in this paper, focused on sensing and perceiving the state of the parties, will result in online tools for conflict resolution that actually bring the parties closer to each other, in a human sense rather than in purely practical one.

Acknowledgements. This work is part-funded by ERDF - European Regional Development Fund through the COMPETE Programme (operational programme for competitiveness) and by National Funds through the FCT – Fundação para a Ciência e a Tecnologia (Portuguese Foundation for Science and Technology) within project FCOMP-01-0124-FEDER-028980 (PTDC/EEI-SII/1386/2012) and project PEst-OE/EEI/UI0752/2014.

References

1. Ackroyd, S., Hughes, J.A.: Data Collection in Context. Longman (1981)
2. Anderson, J.R., Reder, L.M., Simon, H.A.: Situated learning and education. Educational Researcher 25(4), 5–11 (1996)
3. Andrade, F., Novais, P., Carneiro, D., Zeleznikow, J., Neves, J.: Using bATNAs and wATNAs in online dispute resolution. In: Nakakoji, K., Murakami, Y., McCready, E. (eds.) JSAI-isAI 2009. LNCS, vol. 6284, pp. 5–18. Springer, Heidelberg (2010)
4. Aztiria, A., Izaguirre, A., Augusto, J.C.: Learning patterns in ambient intelligence environments: a survey. Artificial Intelligence Review 34(1), 35–51 (2010)
5. Bardram, J.E.: Applications of context-aware computing in hospital work: examples and design principles. In: SAC 2004 Proceedings of the 2004 ACM Symposium on Applied Computing, pp. 1574–1579. ACM, New York (2004)
6. Barreto, A., Zhai, J., Adjouadi, M.: Non-intrusive physiological monitoring for automated stress detection in human-computer interaction. In: Lew, M., Sebe, N., Huang, T.S., Bakker, E.M. (eds.) HCI 2007. LNCS, vol. 4796, pp. 29–38. Springer, Heidelberg (2007)
7. Bellotti, V.: The Magitti activity-aware leisure guide. Social Brain Forum, Tokyo Institute of Technology, Tokyo, Japan (2008)
8. Bernardi, L., Wdowczyk-Szulc, J., Valenti, C., Castoldi, S., Passino, C., Spadacini, G., Sleight, P.: Effects of controlled breathing, mental activity and mental stress with or without verbalization on heart rate variability. Journal of the American College of Cardiology 35(6), 1462–1469 (2000)
9. Bolchini, C., Curino, C.A., Quintarelli, E., Schreiber, F.A., Tanca, L.: A data-oriented survey of context models. ACM 36(4), 19–26 (2007), doi:10.1145/1361348.1361353, ISSN 0163-5808

10. Brüser, C., Stadlthanner, K., de Waele, S., Leonhardt, S.: Adaptive Beat-to-Beat Heart Rate Estimation in Ballistocardiograms. IEEE Transactions on Information Technology in Biomedicine 15(5), 778–786 (2011), doi:10.1109/TITB.2011.2128337
11. Budzik, J., Hammond, K.J.: IUI 2000 Proceedings of the 5th international conference on Intelligent user interfaces, pp. 44–51. ACM, New York (2000)
12. Carneiro, D., Novais, P., Andrade, F., Zeleznikow, J., Neves, J.: Using case-based reasoning to support alternative dispute resolution. In: de Leon F. de Carvalho, A.P., Rodríguez-González, S., De Paz Santana, J.F., Rodríguez, J.M.C. (eds.) Distributed Computing and Artificial Intelligence. AISC, vol. 79, pp. 123–130. Springer, Heidelberg (2010)
13. Carneiro, D., Gomes, M., Novais, P., Andrade, F., Neves, J.: Automatic Classification of Personal Conflict Styles in Conflict Resolution, vol. 235, pp. 43–52. IOS Press - Frontiers in Artificial Intelligence and Applications (2011) ISBN 978-1-60750-980-6
14. Carneiro, D., Carlos, C.J., Novais, P., Fernández-Caballero, A., Neves, J.: Multimodal Behavioural Analysis for Non-invasive Stress Detection. Expert Systems with Applications 39(18), 13376–13389 (2012)
15. Castillo, J.C., Rivas-Casado, A., Fernández-Caballero, A., López, M.T., Martínez-Tomás, R.: A multisensory monitoring and interpretation framework based on the model-view-controller paradigm. In: Proceedings of the 4th International Workshop on the Interplay between Natural and Artificial Computation, vol. 1, pp. 441–450 (2011)
16. Fernández-Caballero, A., Castillo, J.C., Martínez-Cantos, J., Martínez-Tomás, R.: Optical flow or image subtraction in human detection from infrared camera on mobile robot. Robotics and Autonomous Systems 58(12), 1273–1281 (2010)
17. D'Mello, S., Kory, J.: Consistent but modest: a meta-analysis on unimodal and multimodal affect detection accuracies from 30 studies. In: Proceedings of the 14th ACM International Conference on Multimodal Interaction, ICMI 2012, pp. 31–38. ACM, New York (2012)
18. David, B.S.: Principles for case representation in a case-based aiding system for lesson planning. In: Proceedings of the Workshop on Case-Based Reasoning, Washington (1991)
19. Dearle, A., Kirby, G.N.C., Morrison, R., McCarthy, A., Mullen, K., Yang, Y., Connor, R.C.H., Welen, P., Wilson, A.: Architectural Support for Global Smart Spaces. In: Chen, M.-S., Chrysanthis, P.K., Sloman, M., Zaslavsky, A. (eds.) MDM 2003. LNCS, vol. 2574, pp. 153–164. Springer, Heidelberg (2003)
20. Dey, A.K.: Understanding and Using Context. Personal Ubiquitous Computing 5(1), 4–7 (2001), doi:10.1007/s007790170019
21. Eysenck, M.W., Keane, M.T.: Psychology Press, 1st edn. (2005) ISBN 978-1841693590
22. Fisher, R., Ury, W.: Getting To Yes: Negotiating Agreement Without Giving In. Houghton Mifflin, Boston (1981)
23. Healey, J.A., Picard, R.W.: Detecting stress during real-world driving tasks using physiological sensors. IEEE Transactions on Intelligent Transportation Systems 6(2) (2005)
24. Jovanov, E., O'Donnell Lords, A., Raskovic, D., Cox, P.G., Adhami, R., Andrasik, F.: Engineering in Medicine and Biology Magazine. IEEE 22(3), 49–55 (2003)
25. Katsh, E.: The Online Ombuds Office: Adapting Dispute Resolution to Cyberspace, http://www.umass.edu/dispute/ncair/katsh.htm (last accessed in January, 2013)

26. Kolodner, J.L.: Case-based Reasoning. Morgan Kaufmann Publishers (1993)
27. Lubar, J.F.: Discourse on the development of EEG diagnostics and biofeedback for attention-deficit/hyperactivity disorders. Biofeedback and Self-regulation 16(3), 201–225 (1991)
28. Milne, J.: Questionnaires: advantages and disadvantages. Evaluation cookbook (1999)
29. Munro, A., Welen, P., Wilson, A.: Interaction Archetypes. GLOSS Consortium Report D4 (2001)
30. Notini, J.: Effective Alternatives Analysi in Mediation: "BATNA/WATNA" Analysis Demystified (2005), http://www.mediate.com/articles/notini1.cfm (last accessed November 2013)
31. O'Hare, N., Gurrin, C., Jones, G.J.F., Smeaton, A.F.: Combination of content analysis and context features for digital photograph retrieval. In: 2nd European Workshop on the Integration of Knowledge, Semantics and Digital Media Technology (EWIMT 2005), pp. 323–328 (2005)
32. Pimenta, A., Carneiro, D., Novais, P., Neves, J.: Monitoring mental fatigue through the analysis of keyboard and mouse interaction patterns. In: Pan, J.-S., Polycarpou, M.M., Woźniak, M., de Carvalho, A.C.P.L.F., Quintián, H., Corchado, E. (eds.) HAIS 2013. LNCS, vol. 8073, pp. 222–231. Springer, Heidelberg (2013)
33. Popper, K.: The Logic of Scientific Discovery (2004) (reprinted) by Routledge, Taylor & Francis (1959)
34. Rekimoto, J., Ayatsuka, Y., Hayashi, K.: Wearable Computers. In: Second International Symposium on Digest of Papers (1998), doi:10.1109/ISWC.1998.729531
35. Rodrigues, M., Gonçalves, S., Carneiro, D., Novais, P., Fdez-Riverola, F.: Keystrokes and clicks: Measuring stress on E-learning students. In: Casillas, J., Martínez-López, F.J., Vicari, R., De la Prieta, F. (eds.) Management Intelligent Systems. AISC, vol. 220, pp. 119–126. Springer, Heidelberg (2013)
36. Satyanarayanan, M.: IEEE Pervasive Computing: From the Editor in Chief - Challenges in Implementing a Context-Aware System. IEEE Distributed Systems Online 3(9) (2002)
37. Schwartz, M.S., Andrasik, F.E.: Biofeedback: A practitioner's guide. Guilford Press (2003)
38. Staufenbiel, S.M., Penninx., B.W.J.H., Spijker, A.T., Elzinga, B.M., van Rossum, E.F.C.: Hair cortisol, stress exposure, and mental health in humans: A systematic review. Psychoneuroendocrinology 38(8), 1220–1235 (2013) ISSN 0306-4530
39. Steenbergen, W.: Rationalizing Dispute Resolution: From best alternative to the most likely one. In: Proceedings 3rd ODR workshop, Brussels (2005)
40. Yampolskiy, R.V., Govindaraju, V.: Behavioural biometrics: a survey and classification. International Journal of Biometrics 1, 81–113 (2008)

Consumedia. Functionalities, Emotion Detection and Automation of Services in a ODR Platform

Josep Suquet[1], Pompeu Casanovas[1,2], Xavier Binefa[3], Oriol Martínez[3], Adrià Ruiz[3], and Jordi Ceballos[4]

[1] Institute of Law and Technology, Universitat Autònoma de Barcelona, Barcelona, Spain
[2] Centre for Applied Social Research, Royal Melbourne Institute of Technology, Melbourne, Australia
{josep.suquet,pompeu.casanovas}@uab.cat
[3] Department of Information and Communication Technologies, Universitat Pompeu Fabra, Barcelona, Spain
{xavier.binefa,oriol.martinez,adria.ruiz}@upf.edu
[4] Informática y Comunicaciones Avanzadas, Barcelona, Spain
jordi.ceballos@grupoica.com

Abstract. This paper presents a legal and technological approach to online mediation. It shows the technologies that are usually employed in this field and presents the prototype of Consumedia, an online mediation platform, as well as its functionalities and technological architecture. Moreover, it uncovers the technology implemented as regards the recognition of emotions in the mediation room. Furthermore, it considers that an online mediation platform may automatically provide the parties with all the required documentation of the process. Thus, it unveils the documents that an online mediation platform should automatically provide to the disputants.

Keywords: Online mediation, ODR, emotion detection, ontologies, automation, documents, phases, B2B, B2C.

1   Introduction: The Regulatory Framework

Alternative Dispute Resolution (ADR) refers to out-of-court mechanisms of redress and usually includes negotiation, mediation, recommendation and arbitration. Mediation is a relational justice mechanism, because the parties who suffer a conflict are empowered so that they may solve it by themselves through a collaborative conduct [1]. Accordingly, mediation is a structured process whereby a third impartial party assists the contending parties without imposing or proposing a solution to the conflict, but puts his efforts in bringing the parties together in order to solve the conflict by themselves.

Mediation may be undertaken as a conflict resolution mechanism in a wide array of domains, such as business-to-consumers disputes (B2C), business-to-business disputes (B2B), family law, labor law, etc. Yet, mediation mechanisms should be fostered for low-value disputes because this may be the only available option for redress. Indeed,

P. Casanovas et al. (Eds.): AICOL IV/V 2013, LNAI 8929, pp. 247–260, 2014.

European surveys show that the access to courts is usually too complex and costly for low-value disputes and consequently very few European consumers bring a conflict to a court [2]. To allow reaching solutions as regards this type of controversies, mediation procedures in this field should be kept as simple as possible. For instance, consumer conflicts are usually of a similar nature involving many consumers against few traders arising in certain domains such as air transport (conflicts with air companies), telecom/telephony and utilities [3].

It has been said that this field is particularly suitable for the introduction of Information and Communication Technologies (ICT) and fully supports automated mediation procedures because the similarity of conflicts may enable mediation providers to standardize a great deal the treatment of information in the consumer redress arena [3].

Online Dispute Resolution (ODR) is usually understood as any ADR mechanism in which ICT play a significant role [4]. Today, the areas of both ADR and ODR are very active. On the one hand, the ODR academic community is developing both a theoretical and practical approach towards ODR [5]. On the other hand, the legislative authorities are fostering the use of ITC in the dispute resolution domain. The EU Directive 2013/11 (Directive on consumer ADR) will provide a common regime in the EU for consensual, advisory and determinative ADR mechanisms. This Directive has supposed a shift in the EU approach. While the EC Directive on mediation in civil and commercial matters applies to mediation processes only, the Directive on consumer ADR is devoted to applying to adjudication and recommendation services as well as mediation services. Moreover, the EU Regulation 2013/524 on consumer ODR is intended to create an online platform. This could develop into an interactive website offering a single point of access to consumers and traders who seek to resolve their dispute out-of-court for cross-border e-commerce transactions.

Act 5/2012 of 5 July, of mediation in Civil and Commercial Maters is the legislative instrument that transposes the EC Directive on Mediation in Spain. This entitles the parties to conduct partly or totally mediations online. Indeed, party autonomy is recognized as it sets out that parties may agree that all or some of the acts of mediation are carried out electronically (art. 24.1). It establishes that a mediation process consisting of a claim not exceeding 600 € shall be developed preferably by electronic means, unless its use is not possible for either party (art. 24.2).

Furthermore, Act 5/2012 mandates the Government to develop an online simplified mediation procedure for disputes involving purely claims for payment (Final Disposition 7). This procedure is mandated to last no more than a month, although this period may be extended by agreement of the parties. Accordingly, the Spanish legislation will enshrine two different online mediation procedures: a general procedure and a simplified procedure for low-value complaints. Recently, the Royal Decree 980/2013, has set up the simplified online mediation procedure.

After these introductory remarks, the following section briefly describes some of the technologies that may be used in online mediations today. Then, section 3 presents the functionalities and the architecture of Consumedia, a pilot online mediation platform. Section 4 unveils the implementation of an automatic facial expression recognition mechanism that may help overcome the loss of information in comparison to in-person mediations. Then, section 5 considers further aspects that can be automated in a platform and focuses on the documentation that an online mediation platform could automatically provide to the parties. Since this is an aspect that may be regulated by the Law, section 5.2 analyses the Spanish legislation regarding civil and commercial mediation. Finally, section 6 provides some conclusions.

2   Technologies for Online Mediation

Many technologies can be used in the context of online mediation procedures or other methods of ODR. Some technologies may, though, be specifically indicated to support the mediation process, and may be addressed to support the process itself or to support the parties of the process [4]. In the first case, technology is dedicated to the administration of the process in order to guarantee a smooth running of it. For example, it can enable the automatic flow monitoring of the process, controlling the sequence of the process, timing and participation of the parties, or it may register the cases, performing a digital transcript of them. In addition, technology can structure information like the claim request, offer, counteroffer or final agreement, particularly through electronic forms.

The so-called Agreement Technologies (AT) is a new area of research that aims to discuss the theory and practice of computer systems in which agents negotiate and take agreements on behalf of human users [6]. Furthermore, Negotiation and Decision Support Services (NDSS) can assist the parties in formulating their positions and quantifying different aspects of the dispute. Here, Family Winner and Asset Divider are two most remarkable, interest-based, computer programs developed in the field of family law in Australia. They attribute the conflicting items at stake to each of the parties according to the evaluation that each party has performed [7, 8].

Technology can also help create automated negotiation, blind-bidding services, which may be complementary or substitutive to online mediation processes. In these cases, the parties agree that their pecuniary dispute can be solved by the use of a software device. The parties who have an economic dispute agree on an economic margin (e.g., 1000 €) and make a few rounds of confidential negotiations with different confidential bids. If the parties fail to agree on an economic margin, the software does not communicate to the other party the various economic amounts. If the parties' bid fall within the margin, the software solves the controversy by diving the margin between the parties. Thus, if party A offers 900 and party B requests 1,000 and the parties have previously established a margin of 100, the system resolves the dispute by dividing the net and subtracting it with equal parts: A will pay 950 and B will receive 950. In these cases, the technology presents a determinative role [9].

Mediation technologies can assist both the mediator and the parties to the dispute and can also be classified as communication technologies and case management technologies. Communication technologies can assist them with basic communication tools such as emails or SMS, whereas other technologies can enhance the creative process with 2.0 tools such as wikis or virtual maps of the conflict. In addition, there are other cases in which technology enables the mediation process to be conducted online [4]. The most sophisticated cases are online mediation platforms. These may include synchronous tools such as video conferencing or audio conferencing, instant messaging or chats or asynchronous tools such as email, discussion forums, online forms, or electronic boards. The following paragraphs precisely show a prototype of one of these mediation platforms, in which the authors have been devoting their efforts in the lasts years, and the functionalities implemented.

3   Consumedia: A User Case

Consumedia is a research project funded by the Spanish Ministry of Science and Innovation, which aims at implementing an online mediation platform for both B2C and B2B disputes. The Consumedia project is a follow-up of other research projects such as Ontomedia [10,11] and, specifically, the White Book of Mediation in Catalonia (WBMC). The WBMC was an in-depth research project funded by the Government of Catalonia which aimed at analyzing mediation as a means of redress in a wide-array of areas such as B2C, B2B, family, labor or healthcare, and to assess the degree of implementation of mediation in Catalonia. The WBMC has proved to be a very fertile environment for knowledge acquisition. In particular, one of the chapters was devoted to technologies and mediation ([4] and a prototype of a online mediation platform was created [12]. Accordingly, the WBMC considered that mediations could also take benefit from the implementation of ICT, and it recommended "the promotion and implementation of online mediation (ODR) in the various fields of mediation, significantly in areas such as in Consumer Law (…)" [13].

The Consumedia platform provides both asynchronous and synchronous communication tools, with text-based and audio-visual facilities. The online mediation platform implements electronic messaging and a public and a private chat. It also provides an electronic board where parties can write any eventual arguments or drawings. Moreover, it contains a mediation room with a videoconference facility. This shows both the claimant and the respondent, which are horizontally placed on the top of the screen; the mediator is displayed underneath. In addition, one of the technological tools derives from the use of automatic facial expression detection, as it is further explained in section 4. The platform also provides back-office facilities such as a repository of pending and current mediation cases, a historical database of mediation processes already conducted, and the description of the different cases. Figure 1 shows the different use cases, and which user profiles can access each of them.

Use case	Administrator	Mediator	Claimant/Respondent
To appoint a mediator to a new case	X		
To create, modify or delete a mediation case	X	X	
To accept the appointment of a new case to a mediator		X	
To reject the appointment of a new case to a mediator		X	
To change the current phase of mediation		X	
To edit private notes in a mediation case		X	X
To manage shared documents in a mediation case		X	X
To manage mediation rooms with videoconference facilities		X	
To make a videoconference call		X	X
To visualize a videoconference already made		X	
To browse emotion automatic recognition in a videoconference already made		X	
To check text from automatic audio transcription of a videoconference already made		X	
To create a new claim			X

Fig. 1. Use case and user profiles

The technical architecture of the application is shown in figure 2. The layers of the software architecture are based on ontologies and semantics, multimedia analysis and the mediation platform.

- Ontologies and semantics: Set of ontologies for online mediation that allow semantic search according to mediation databases.
- Multimedia analysis: Set of multimedia processes that offer the following advanced features.
    o Emotion recognition from automatic analysis of real-time video.
    o Speech analysis, which can detect the language used by each partner, and automatically transform the corresponding audio transcription into text.
    o Visual browsing based in a timeline, which permits to visualize a videoconference already made, the content of the chat and the recognition of emotions.
- Mediation platform: It consists of a web application where users can access and which consists of the following components:
    o PostgreSQL database where all the information managed by the application is stored. It is a free and open source database, which does not require the payment of use licenses in a business environment.
    o Storage of audios and videos on a file system.
    o Portal developed under the LifeRay Java platform, which allows the creation of different security profiles, which are the administrator, mediator and claimant/respondent.
    o Videoconferencing system developed with Red5 Media Server, and customized to the needs of the project.

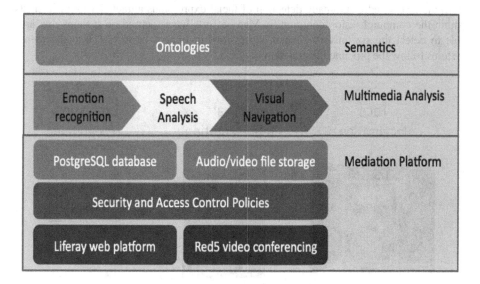

Fig. 2. Architecture of Consumedia

4   Emotion Detection in Consumedia

Communication processes over online platforms (such as chat, e-mail, video-conference or even virtual worlds) have been extensively studied, specifically for online education platforms. One of their main drawbacks is the amount of information that it is lost in comparison with in-person communication processes. For instance, in the context of online learning platforms, it has been shown to decrease student connectivity, engagement and retention [14]. However, other studies have shown the intensifying power of emotions when the attention keeps focused on specific topics in which humans feel involved [15]. Given that face is one of the main human non-verbal communication channels, the analysis of the set of facial gestures performed by a subject in a given situation can be used to improve the viewer perception of the communication process. This can partially mitigate the loss of information in computer-mediated communications.

In online mediation, where the communication process takes place in a videoconference room, emotion analysis through facial expressions detection and classification can be used to highlight or summarize the affective events that occur in the communication process, providing tools to the mediator to improve its management.

4.1   Automatic Facial Expression Recognition

In his earlier psychological studies [16], Paul Ekman showed that there are six universal facial expressions that share the same meaning across all human cultures. The facial expressions defined by Ekman were: disgust, fear, happiness, surprise, sadness and angry. As we have explained, the automatic detection of these prototypical facial expressions can provide very useful information to index, summarize or understand an online mediation process.

In the last decade, automatic detection of facial expressions attracted a lot of interest within the computer vision community. Many efforts have focused on systems that are able to detect the six universal facial expressions in images or videos. Usually, these systems follow the pipeline illustrated in Figure 3.

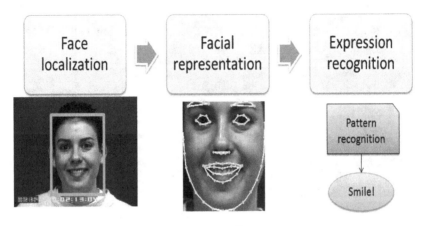

Fig. 3. Pipeline overview of an automatic facial expression recognition system

Given an image, we need to locate first where the face is. For this purpose, we can use methods such as the Viola & Jones algorithm [17] implemented in most of modern digital cameras. Once spotted, a numerical representation of the face can be extracted. This representation encodes the information. For instance, we can automatically obtain the location of different facial landmark points (eyes, mouth borders, nose, eyebrows...) and use their coordinates in the image as a facial representation. Finally, pattern recognition models are applied over these representations in order to recognize which facial expression occurs in the image of the performing subject.

For the Consumedia platform, we have used the method proposed in [18] to automatically detect the six universal expressions described above. This method follows a similar pipeline to detect facial expressions. Given a frame from a subject recorded during a mediation process, a set of 66 facial landmark points are extracted using the method proposed in [19]. Then, regression techniques are used in order to map the coordinates of all points into a number from 0 to 1 denoting the intensity of each facial expression. These regression models are trained with the CK+ Facial Expression Database [20]. For more technical details, the reader is referred to the original paper. Figure 4 shows a typical result obtained applying this method.

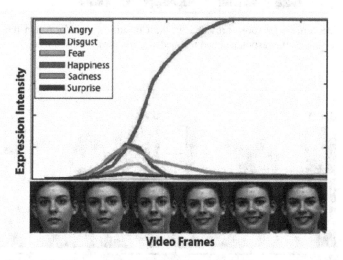

Fig. 4. Result after applying the method in [18] to a video containing a smiling subject. Each colored line represents the estimated intensity of each given facial expression along a sequence.

4.2   Use of Facial Expression Information in the Consumedia Platform

The Consumedia platform offers a mediation room with a videoconference facility (see Figure 5) that besides the mediation task also records and analyses the audio-visual streams of the claimant and the respondent. The analysis involves the facial expression recognition system, and an automatic voice transcription system as well.

The information obtained through this kind of analysis can be applied to summarize the mediation process. For instance, on the Consumedia platform, it is used to indexing and also navigating the multimedia content gathered from the videoconference facility. Navigation trough the recorded contents can be enhanced using a colored bar, in which each color represents a specific emotional state.

Fig. 5. Videoconference room provided on the Consumedia platform. On the left, windows where the claimant, the respondent and the mediator (on top) are visualized.

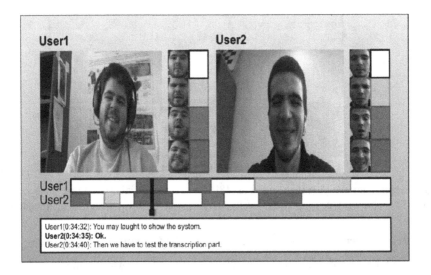

Fig. 6. Prototype for offline navigation along the videoconference content using facial expression information analysis. The black line over the bar indicates an instance of time and it is used to navigate through the recorded content.

5   A Further Step towards the Automation of Services

5.1   Ontologies in Online Mediation Procedures

Mediation is a technique that may be developed according to a diversity of processes and procedures that may vary from the specific field in which it is conducted.

The Law might regulate some of the aspects of the process, detail the phases (e.g., pre-mediation, mediation and post-mediation phases) or, instead, the flowchart of the process may be left it for the autonomy of the mediation provider. The WBMC recommends the implementation of ICT in mediation processes, but it warns about the importance of implementing protocols and the standardization of processes before any tool is built [13].

The online mediation proceedings entail heterogeneous actions from the parties (claimant and requested party) as well as from the mediator (third party) and the mediation provider. These actions may include a request for mediation, a request for conducting a mediation process online, an answer from the requested party to enter into a mediation process, or a request for a suspension of a mediation process. The mediator or the mediation provider may also require the parties to inform about several aspects, such as to correct eventual errors or omissions in the mediation request.

In this landscape, ontologies may play a central role [21,22]. They may be implemented in an online mediation platform for the automation of several aspects of the proceedings, such as the following:

i) The assignment of a given ODR/ADR provider and a mediator from the text of the claims made by the consumer claimant. From the consumer complaint expressed in natural language, an ontology might appoint a specific mediator and a provider to a new case. In this case, different "tag" words may serve to identify a provider or a mediator. In particular, one of the most useful cases can be attributing a conflict to either a B2C or a B2B entity. The functionality here, would be based on keywords to assign the case to one or the other domain (e.g, words such as "consumption", "journey", "phone", "gas", "light", "electricity", "airplane ticket", could be words usually associated with a consumer relationship). However, this may only be a guide to assign a case to an ODR provider, which must subsequently ensure that the assignment was actually correct.

ii) To provide the applicable law to the relationship involved. Similarly to the former case, from the claim made by the claimant, an ontology might help detail the applicable legislation. There may be some words such as "car, vehicle, car repair shop," that would be assigned to some legal instruments (such as the Spanish Royal Decree 1457/1986, of 10 January, on the industrial and service delivery in the vehicle repair shops, or Law 40/2002, of 14 November, regulating the parking of vehicles). Moreover, using the words "trip cancellation, flight, passenger boarding, denial, delay, accident, airline" the search engine could provide information on applicable legal instruments such as Regulation 261/2004 of 11 February 2004, that establishes common rules on compensation and assistance to air passengers or Regulation 2027/97, on air carrier liability in respect of the carriage of passengers and their baggage by air, to name only a few.

iii) An ontology may also be implemented so that it can automatically provide to the parties with the documentation required in each phase of the mediation process. This requires defining all the phases and deadlines of the mediation process, and correspondingly to implement a protocol of mediation. The following paragraphs focus on this aspect. They study the legislation of mediation in civil and commercial aspects and ascertain the different phases that may be subject to those domains. The final goal is to show what documents should, an automated platform consider in a mediation process.

5.2   Automated Mediation Process: Actions and Documents

The Spanish Act 5/2012, on civil and commercial mediation establishes some phases that the parties and the mediator must observe when conducting a mediation process. The following phases should also be maintained in online mediation proceedings. These phases encompass an initial and informative session (Article 17) as well as a constitutive session, the last one being incorporated into a formal act that has to be manually or digitally signed by the parties and the mediator (Article 19.2). Moreover, the parties may undertake several mediation sessions such as private mediation sessions (also named *caucus*) or joint mediation sessions. Furthermore, according to this legislative instrument, the procedure ends with a final act that will determine its conclusion and, where appropriate, reflect the agreements reached in a clear and understandable way; or, alternatively, its termination for any other reason (Article 22.3). The Law also sets out that the parties and the mediator must manually or digitally sign the final act and that one copy shall be given to the parties.

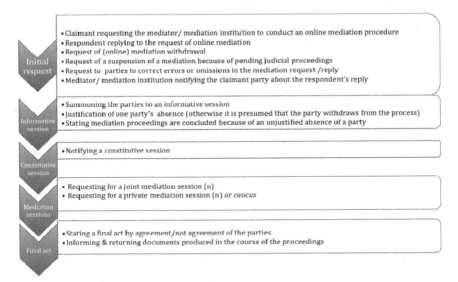

Fig. 7. Actions by phases in a general online mediation process

In order to be able to develop an ontology that could automatically provide the parties with all necessary information documents that should be exchanged between the parties and the mediator, a thorough analysis of Act 5/2012 and the regulation that develops this instrument (RD 980/2013) has been undertaken. It should be noted that the latter legal instrument sets out a simplified online mediation procedure but a general online mediation procedure is still not regulated. According to this study, the online mediation platform should provide the parties with a wide array of documentation included in the different actions that the parties and the mediator may perform in the mediation proceedings. This documentation should be available in a clear, readable and in a standardized way.

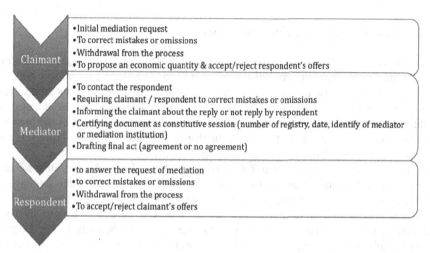

Fig. 8. List of actions to be undertaken by standard forms in a simplified online mediation procedure (according to each party)

The following figure shows the actions that should be allowed by the diverse documents uploaded into the platform.

Furthermore, one of the novelties of Act 5/2012 and, subsequently, RD 980/2013 is the possibility to create simplified online mediation procedures for low-value, pecuniary disputes that must not last more than a month. In such cases, the documentation should be simplified. Even if the phases of the mediation process (Informative session, constitutive, mediation sessions and terminative sessions) should be followed, the RD 980/2013 is not clear about how these sessions should be conducted.   In any case, this scheme creates a software assisted negotiation mechanism, rather than an online mediation process, because it allows the parties to make different offers and counteroffers, assisted by the technology, until they agree or not on a solution. The economic quantity that the claimant claims and the quantity that the respondent is willing to accept are to be offered through simple online forms.

The simplified online mediation procedure begins with an initial mediation request. Then, the mediator contacts the respondent and asks him or her whether he or she accepts entering into the process. The mediator may also require both parties to correct omissions or mistakes in both the request form and the reply form. After the respondent replies accepting to enter into the simplified process, the mediator will provide the parties with a document certifying the constitutive session, generating a file number and determining the beginning of the proceedings (art. 37 RD 980/2013). Moreover, the mediator should provide a document certifying proceedings are properly conducted, with the number of registry, date and time of presentation, identity of mediation and of the mediation institution (art. 34 RD 980/2013). Finally, the mediator or the mediation institution should also provide a document containing the agreement or the lack of agreement, and which has the value of a final act. According to this, the platform should provide several standard documents to the parties, shown in figure 8.

6   Conclusions and Further Work

Mediation is a structured process whereby a third impartial party assists the contending parties without imposing or proposing a solution to the conflict. The mediator puts his efforts in bringing the parties together so that they can solve the conflict by themselves. This process may be facilitated through the use of ICT, which can be devoted to facilitating the communication between the parties and the management of the process. This has been named the "fourth" party in the dispute, because ODR assistance sets and independent framework for the dispute be solved [23, 24].

A. Lodder [25] and Lodder and Zeleznikow [26] have shown the need to take into account a "fith" additional party — the technology provider. This means that singling out who is providing the platform, and who is taking the responsibility for the overall mediation processes which have to be run onto it, matters. In the case of Consumedia, the fifth party is not only the consortium (I+D) who is constructing the tools, but the Council of the Chambers of Commerce of Catalonia as well. In principle, the platform will be used to provide an ODR B2B mediation service, depending on the Council. Therefore, the Council's mediators and representatives have been involved in the knowledge acquisition process, the definition of functionalities, and the walkthrough testing since the very beginning. This intended prudence, seeking fairness and the "sense of reality" in the construction of computing tools, is encompassed as well by the present trends on AI and ODR [27].

Online mediation platforms can enable the parties to conduct a mediation process semi-automatically. For one thing, Consumedia stays in a classical way to implement such processes, for that it offers an electronic messaging system, along with a public and a private chat. It also provides an electronic board where parties can write any eventual arguments or drawings, with a mediation room and a videoconference facility.

One of the most innovative aspects derives from the use in the mediation room of an Automatic Facial Expression Recognition system. Here, the system records and analyses the audio-visual streams that belong to the claimant and the respondent, both in terms of facial expression recognition as well as voice transcription. For privacy and data protection compliance, the metadata are recorded and can be just visualized and used only by the mediator, and only for professional purposes, to recall the case and improve her performance.

Yet, in the search for a fully automated mediation process this paper considers a further step. Here, one of the functionalities of an online mediation platform can be to automatically provide the parties with all the necessary procedural documents that the Law might require. Foreseeing all the necessary information activities between the parties of the process (claimant and respondent), as well as the mediator and the mediation institution, is not an easy task. In order to do that, the work of legal and computer scholars lean on the analysis of the applicable legislation to standardize and restructure the legal knowledge [28]. It could ultimately determine the diverse documents that should be uploaded into the platform.

This study analyses Act 5/2012, regulating civil and commercial mediation in Spain and its Royal Decree 980/2013, regulating a simplified online procedure for claims for payment. The latter instrument also provides for an assisted negotiation mechanism. The analyses of these diverse instruments serve to identify, structure and prepare the actions and the documents that the Consumedia platform should upload when implementing fully automated online mediation services, in the next future. This will be checked too at the benchmark (with selected cases) in the upcoming testing phase of the project.

Acknowledgements. Consumedia IPT-2011-1015-430000, CROWDSOURCING: DER 2012- 39492 -C02 -01.

References

1. Casanovas, P., Poblet, M.: Concepts and Fields of Relational Justice. In: Casanovas, P., Sartor, G., Casellas, N., Rubino, R. (eds.) Computable Models of the Law. LNCS (LNAI), vol. 4884, pp. 323–339. Springer, Heidelberg (2008)
2. European Commission: Consumer attitudes towards cross-border trade and consumer protection. Analytical report (Flash Eurobarometer) (European Commission EB Series 358 (2013)
3. Barral Viñals, I., Suquet Capdevila, J.: La mediación en el ámbito de consumo. In: Casanovas, P., Magre, J., Lauroba, M.E. (eds.) Libro Blanco de la Mediación en Cataluña, pp. 301–370. Departament de Justícia, Generalitat de Catalunya, Huygens (2011), http://www.llibreblancmediacio.com/
4. Poblet, M., Noriega, P., Suquet, J., Gabarró, S., Redorta, J.: Tecnologías para la mediación en línea, estado del arte, usos y propuestas. In: Casanovas, P., Magre, J., Lauroba, M.E. (eds.) Libro Blanco de la Mediación en Catalunya, pp. 943–985. Departament de Justícia, Generalitat de Catalunya, Huygens (2011), http://www.llibreblancmediacio.com
5. Abdel Wahab, M.S., Katsh, E., Rainey, D.: Online Dispute Resolution: Theory and Practice. A Treatise on Technology and Dispute Resolution. Eleven International Publishing, The Hague (2012)
6. Ossowski, S. (ed.): Agreement Technologies. Law, Governance and Technology Series, pp. vii–viii. Springer, Dordrecht (2013)
7. Bellucci, E., Zeleznikow, J.: Developing negotiation decision support systems that support mediators: A case study of the Family Winner system. Artificial Intelligence and Law 13(2), 233–271 (2005)
8. Bellucci, E.: Software developed for use in Family Mediation: Asset Divider. In: Poblet, M., Schild, U., Zeleznikow, J. (eds.) Legal and Negotiation Decision Support Systems (LDSS 2009): a Post conference Workshop at the 12th International Conference on Artificial Intelligence and Law, Huygens, Barcelona, pp. 55–69 (2009)
9. Comité Européen de Normalisation (CEN), CEN Workshop Agreement: Standardization of Online Dispute Resolution Tools, ICS 03.120.10; Ref. No.: CWA 16026: 2009 E (2009) ftp://cenftp1.cenorm.be/PUBLIC/CWAs/STAND-ODR/CWA16026_STANDODR.pdf (Last accessed February 25, 2014)

10. Poblet, M., Casanovas, P., López Cobo, J.L.: The Ontomedia Project: Relational Law, ODR, Multimedia. In: Bourcier, D., Casanovas, P., Dulong de Rosnay, M., Maracke, C. (eds.) Intelligent Multimedia. Managing Creative Works in a Digital World, pp. 349–364. European Publishing Academic Press, Florence (2010)

11. Poblet, M., Casanovas, P., López-Cobo, J.L., Casellas, N.: ODR, Ontologies and Web 2.0. Journal of Universal Computer Science, J.UCS 17(4), 618–634 (2011)

12. Noriega, P., López de Toro, C.: Towards a Platform for Online Mediation. In: Proceedings of the Workshop on Legal and Negotiation Decision Support Systems (LDSS 2009) in conjunction with ICAIL 2009, pp. 67–75. CEUR Workshop Proceedings CEUR-WS.org/Vol-482/, Barcelona (2009)

13. Casanovas, P., Magre, J., Lauroba, M.E.: Conclusions and Recommendations. In: Casanovas, P., Magre, J., Lauroba, M.E. (eds.) Libro Blanco de la Mediación en Catalunya, Huygens, Barcelona, pp. 1161–1173 (2011), `http://www.llibreblancmediacio.com`

14. Betts, K.: Lost in translation: Importance of effective communication in online education. Online Journal of Distance Learning Administration 12(2) (2009)

15. Poblet, M., Casanovas, P.: Emotions in ODR. The International Review of Law Computers & Technology 21(2), 145–156 (2007)

16. Ekman, P., Friesen, W.V.: Constants across cultures in the face and emotion. Journal of Personality and Social Psychology (1971)

17. Viola, P., Jones, M.: Robust Real-time Object Detection. International Journal of Computer Vision (2001)

18. Ruiz, A., Binefa, X.: Modeling facial expressions dynamics with Gaussian Process Regression. In: Artificial Intelligence Research and Development, Proceedings of the 15th International Conference of the Catalan Association for Artificial Intelligence (2012)

19. Saragih, J.M., Lucey, S., Cohn, J.F.: Face alignment through subspace constrained mean-shifts. In: IEEE 12th International Conference on Computer Vision (2009)

20. Kanade, K., Cohn, J., Tian, Y.: Comprehensive database for facial expression analysis. In: Automatic Face and Gesture Recognition (2000)

21. Poblet, M., Suquet, J., Roig, A., González-Conejero, J.: Building Semantic Interoperability for European Civil Proceedings Online. In: Contini, F., Lanzara, G.F. (eds.) The Circulation of Agency in e-Justice: Interoperability and Infrastructures for European Transborder Judicial Proceedings, pp. 287–308. Springer, Dordrecht (2013)

22. Studer, R., Benjamins, V., Fensel, D.: Knowledge engineering: Principles and methods. Data Knowledge Engineering 25(1), 161–197 (1998)

23. Katsh, E., Rifkin, J.: Online Dispute Resolution: Resolving Conflicts in Cyberspace. Jossey-Bass, San Francisco (2001)

24. Katsh, E., Choi, D.: Online dispute resolution: technology as the 'fourth party'. Papers and Proceedings of the 2003 United Nations Forum on ODR, Geneva, June 30-July 1. National Center for Technology and Dispute Resolution (2003)

25. Lodder, A.: The Third Party and Beyond. An Analysis of the Different Parties, in particular The Fifth, Involved in Online Dispute Resolution. Information & Communications Technology Law 15(2), 143–155 (2006)

26. Lodder, A., Zeleznikow, J.: Enhanced Dispute Resolution Through the Use of Information Technology. Cambridge University Press, Cambridge (2010)

27. Carneiro, D., Novais, P., Andrade, F., Zeleznikow, J., Neves, J.: Online dispute resolution: an artificial intelligence perspective. Artificial Intelligence Review 41(2), 211–240 (2014)

28. Susskind, R.: Legal informatics - a personal appraisal of context and progress. European Journal of Law and Technology 1(1) (2010)

Crowdsourcing Tools for Disaster Management: A Review of Platforms and Methods

Marta Poblet[1,2], Esteban García-Cuesta[3], and Pompeu Casanovas[1,2]

[1] RMIT University, Melbourne, Australia
marta.pobletbalcell@rmit.edu.au
[2] Institute of Law and Technology, Universitat Autónoma de Barcelona, Bellaterra, Spain
pompeu.casanovas@uab.cat
[3] iSOCO, Madrid, Spain
egarcia@isoco.com

Abstract. Recent advances on information technologies and communications, coupled with the advent of the social media applications have fuelled a new landscape of emergency and disaster response systems by enabling affected citizens to generate georeferenced real time information on critical events. The identification and analysis of such events is not straightforward and the application of crowdsourcing methods or automatic tools is needed for that purpose. Whereas crowdsourcing makes emphasis on the resources of people to produce, aggregate, or filter original data, automatic tools make use of information retrieval techniques to analyze publicly available information. This paper reviews a set of online tools and platforms implemented in recent years which are currently being applied in the area of emergency management and proposes a taxonomy for its categorization.

Keywords: emergency management, disaster management, crowdsourcing, crowdsensing, micro-tasking, platforms, mobile apps.

1 Introduction

Mobile technologies and social media have transformed the landscape of emergency management and disaster response by enabling disaster affected citizens to produce real time, local information on critical events. Hurricane Sandy offers one of the most recent examples of large volumes of user-generated data: "social media use during Hurricane Sandy produced a 'haystack' of half-a-million Instagram photos and 20 million tweets" [1]. The growing interest on how to leverage social media for disaster management comes as no surprise, nor the number of platforms and tools that aim at making sense of this vast amount of crowdsourced data for emergency management and response. These initiatives come from multiple domains: governments, companies, not-for-profit organizations, volunteer and technical communities, etc. In 2012, the American Red Cross launched the Digital Operations Center, a social media-monitoring platform dedicated to humanitarian relief [2]. In Australia, the Government Crisis Coordination Centre (CCC),

P. Casanovas et al. (Eds.): AICOL IV/V 2013, LNAI 8929, pp. 261–274, 2014.

an all-hazards management facility supporting protective security, counter terrorism, pandemics, and other natural hazards, has recently started to monitor social media as a new source of data from which crisis coordinators can obtain awareness of developing situations [3]. A number of digital volunteer organizations (i.e. the Standby Task Force, Humanity Road, and Open Crisis) have integrated social media monitoring in their workflows when cooperating with large humanitarian organizations in disaster relief operations.

Two different technology approaches to disaster management can be identified from the literature review: (i) data oriented; (ii) communication oriented. Data oriented approaches rely on intensive aggregation, mining, and processing of unstructured data sourced from different social media (e.g. Twitter, Facebook, Instagram, etc.) to generate early alerts. An example of such approach is the Australian Emergency Situation Awareness (ESA) system [4,3]. ESA is a platform for emergency situation awareness which captures and analyzes messages from different sources, not to replace existing procedures and information sources but to provide additional data with many potential applications: pre-incident activity, near real time notification of incidents, or community response to emergency warning [3]. This approach has proved to be faster than other traditional meteorological warning systems [5]. In that study the authors claimed that the system provides two minutes delayed alert improving the six minutes delay of the Japan Meteorological Agency (JMA) and with a 93% of accuracy. Other similar studies have been also performed showing similar capabilities [6,7].

The second approach aims at enhancing communication between people and disaster management systems by allowing seamless interaction between them. One example of this type of collaboration is the NetQuakes[1] project promoted by the U.S. Geological Survey, which aims to get a denser and more uniform spacing of measurements by using a cheaper Wi-Fi capable seismograph and asking volunteers to send information through their private networks.

Somewhat halfway between these two approaches, there is a set of hybrid platforms and tools leveraging people's workforce in the different tasks of a disaster management lifecycle. This paper aims at offering a general overview of technological solutions that are currently applied in the area of emergency management and have in common the use of data generated and/or processed by large numbers of citizens via social media and social networks. By focusing and classifying different solutions based on their origin, methods, functionalities, and prospective end users we can outline a number of different models to address crowdsourced emergency management. In section 2 we introduce new trends combining local information with global response. Section 3 offers an overview of crowdsourcing definitions and roles and Section 4 puts those roles into the context of the disaster management cycle. Section 5 analyzes the features and functionalities of platforms and mobile applications and proposes a classification. The paper concludes by stressing the need for further research on crowdsourcing roles models matching the needs of each phase of the disaster management cycle.

---

[1] http://earthquake.usgs.gov/monitoring/netquakes/

2   Local Information and Global Response

The velocity, variety, and volume of social media information—as a particular type of big data—can be leveraged in all phases of an emergency management lifecycle. Increasingly, emergency organizations are embracing social media and mobile apps to issue alerts and provide updates for incidents (i.e. the official Facebook and Twitter accounts from fire services, rescue and civil protection organizations, etc.).With 241 million monthly active users, more than 35 languages supported, and over 500 million tweets sent per day, Twitter is perhaps the most popular outlet when it comes to disseminate disaster-related information. A growing literature on methods to mine Twitter data for disaster management confirms this emerging trend [8,9,10,11].

In contrast, this trend is not always matched by the monitoring of social media by emergency organizations, and it is frequent to read in the official profiles that accounts are "not monitored 24/7", so that the usual 000 or 999 telephone numbers should be dialed instead. Operational barriers to adopt a proactive role have been explored by recent research [12,13]. Apart from the fact that, in emergency situations, heavy usage of communication networks may cause traffic disruptions and compromise the delivery of updated information, the underlying assumption is that reliable information only travels in one direction: from authorities to citizenry [14].

The platforms reviewed in this paper challenge this notion in two different senses: (i) typically, they consider affected populations as first responders in an emergency situation, so that critical information can actually flow in two directions and facilitate peer-to-peer disaster management networks; (ii) they also empower online volunteers and organizations to offer a global response by allowing their participation in a number of tasks: social media monitoring, data filtering, tagging, geolocation of events, etc. By including the citizens ("the crowd") into the platforms, either by providing information about the disaster or as volunteers for performing specific tasks, they are able to extract global knowledge and trigger a global response based on the local information.

3   Crowdsourcing: the Power of Crowd

The term crowdsourcing was first coined by Jeff Howe in 2006 when referring to "the act of taking a job traditionally performed by a designated agent (usually an employee) and outsourcing it to an undefined, generally large group of people in the form of an open call" [15]. Since Howe's first definition, which finds its roots in the open software movement [16] different crowdsourcing categories, dimensions, and typologies have recently been discussed in the literature [17, 18, 19, 20, 21, 22, 23, 24]. Other studies consider crowdsourcing as part of the broader paradigm of collective intelligence [25] and review the similarities, overlapping and gaps between human computation, crowdsourcing, social computing and data mining [26].

The three key elements intersecting in Web-based crowdsourcing are the crowd, the outsourcing model, and advanced Internet technologies [23]. According to their definition, "crowdsourcing is a sourcing model in which organizations use predominantly advanced Internet technologies to harness the efforts of a virtual crowd to perform specific organizational tasks" [23]. Chamales also refers to "crowdsourcing technology" as an enabler to bring together a "distributed workforce of individuals" [27].

The size and composition of the crowd can also help to determinate whether the crowdsourced effort is unbounded (anyone can participate) or bounded to "a small number of trusted individuals" [28]. In this line, Prpic et al. [29] have distinguished different types of "crowd capital" generation based on the "crowd capabilities" of organizations as they engage with the dispersed knowledge of individuals (i.e. public crowd, public crowd curated, and captive crowd).

We can further distinguish the role of the crowd based on the type of data being processed and the level of participation involved. This leads to four types of crowd-sourcing roles based on: (i) type of data processed (raw, semi-structured, and struc-tured data), (ii) participants' level of involvement (passive or active) and, (iii) skills required to fulfill the assigned task (basic or specialized skills). Figure 1 below shows these roles based on how the crowd is involved in the knowledge chain.

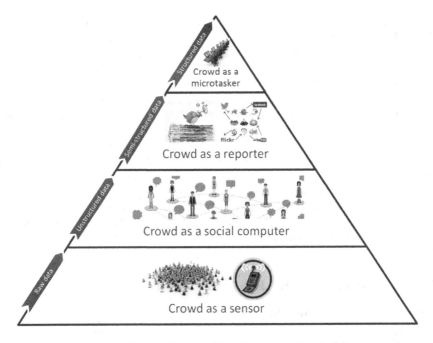

Fig. 1. Crowdsourcing roles based on users' involvement and level of data processing

The lower tiers of the pyramid represent users who generate raw or unstructured data by the mere use of mobile phones, tablets, etc. (crowd as a sensor) or their either occa-sional or regular use of social media (crowd as a social computer). In contrast, the two top tiers include users with an explicit, conscious use of a priori knowledge to achieve a specific goal (crowd as a reporter and crowd as a microtasker). Moving from lower to higher levels in the pyramid also implies a shift in the quality of the obtained data. From a knowledge generation and data processing point of view we are ranging from raw data, unstructured data, or semi-structured data, to structured data (which also become inter-preted data resulting from the execution of the process). Whereof, lower roles in the py-ramid produce raw data and higher roles high valued data which are related with the action of solving a specific problem (e.g. labeling an image). Such a categorization also implies different levels of effort by the crowds:

i) Crowd as sensors: people generate raw data just because some processes are automatically performed by sensor-enabled mobile devices (e.g. processes run in the backend by GIS receivers, accelerometers, gyroscopes, magnetometers, etc.) which can be later on used for a purpose (i.e. mobile phone coordinates for positional triangulation, traffic flow estimates, etc.). This type of data collection has been defined elsewhere as "opportunistic crowdsourcing" [30]. Opportunistic crowdsourcing requires very low data processing capabilities (if any) on the side of participants and is the most passive role in the contributing information chain.

ii) Crowd as social computers: people generate unstructured data mostly by using social media platforms for their own communication purposes (e.g. sharing contents or socializing in social media). Social media users do not process information in any specific form, but these data can later be reused to extract semantically structured information. As in i) there is no explicit participatory effort in any crowdsourced initiative or project.

iii) Crowd as reporters: people offer first-hand, real-time information on events as they are unfolding (e.g. they tweet about a hurricane making landfall and the reporting damages in a specific location). This user-generated content included valuable metadata added by users themselves (e.g. hashtags) than can be used as semi-structured, preprocessed data.

iv) Crowd as microtaskers: people generate structured, high quality, interpreted data by performing some specific tasks over raw data (e.g. labeling images, adding coordinates, tagging reports with categories, etc.). This role requires an active participation of users in the effort and it may exploit special skills or require different levels of previous training.

4   The Role of the Crowd in the Disaster Management Cycle

The UN-SPIDER glossary defines the disaster management cycle as "the complete set of phases related to disasters and their management" [31]. While disaster relief agencies and organizations may conceptualize the disaster management phases differently, most models generally include mitigation, preparedness, response, and recovery.

According to definitions by the United Nations Office for Disaster Risk Reduction, mitigation refers to "the lessening or limitation of the adverse impacts of hazards and related disasters"; preparedness includes "the knowledge and capacities developed by governments, professional response and recovery organizations, communities and individuals to anticipate, respond to, and recover from, the impacts of likely, imminent or current hazard events or conditions"; response involves "the provision of emergency services and public assistance during or immediately after a disaster in order to save lives, reduce health impacts, ensure public safety and meet the basic subsistence needs of the people affected"; recovery includes "the restoration, and improvement where appropriate, of facilities, livelihoods and living conditions of disaster-affected communities, including efforts to reduce disaster risk factors" [32].

Even if, in practice, disasters tend to unfold in a continuum and the phases of the cycle may sometimes be difficult to isolate, the four crowdsourcing roles we have delimited can be applicable to the different phases of the cycle. This association can be valuable in order to identify specific persons whenever we are at a specific phase of the disaster. Thus, the role of the "crowd as a sensor" is especially relevant in the preparedness and training phases when sensors can provide critical information of events or sub-events for different geographical locations and at large scale [33, 34, 35, 36]. People may contribute data either inadvertently or by explicit consent: while GPS location services require users' explicit permission of access on both Android and iOS systems, other location sensors such as accelerometers and gyroscopes do not [37].

The role of the crowd as a "social computer" and as a "reporter" may be critical in the other three steps of the lifecycle (response, recovery, and mitigation) where people and organizations (citizens, volunteer groups, and emergency authorities) can engage in multi-way information sharing and provide near-real time updates on the events as they occur [38]. Given the amount of information that people share during a disaster, leveraging social media information becomes most relevant to facilitate situational awareness during an emergency [3]. Yet, there are a number of issues when using social media information: trustworthiness of the sources, veracity and accuracy of information, and privacy. Some of these issues are easier to handle as the crowd actively take the role of a "reporter". In that case, as people tend to be already identified, verifying the reported information and therefore both the trustworthiness of the source and the verification process are less problematic. People who report can be part of the verification process. Efficient methods to do it by applying simple recruiter reward and punishment approach have been proposed and tested [39,40].

The role of the crowd as a "microtasker" is relevant when producing and analyzing structured data. Table 3 summarizes how the different types of crowdsourcing roles described above relate to the different phases of the emergency management cycle:

Table 1. Crowdsourcing roles and disaster management cycles

	Crowd as a sensor	Crowd as a social computer	Crowd as a reporter	Crowd as a microtasker
Preparedness	●			●
Response		●	●	●
Recovery		●	●	●
Mitigation	●	●	●	

5   Crowdsourcing Tools and Disaster Management Phases

In this section we present the different tools already available in the disaster management domain. We have classified them by establishing a set of dimensions related with the main characteristics of the tools. A preliminary list of tools was extracted from previous research on mobile technologies applied to governance [41]. Additional tools were then added through ongoing research on related sources and initiatives and turned out into the elaboration of a matrix. The four basic criteria for inclusion in the final list are: (i) the tool has been designed to be used on one or more phases of the emergency management cycle or, alternatively, it is applicable in this domain; (ii) the tool leverages at least one of the crowdsourcing roles described in Section 3 (crowds as sensors, social computers, reporters, or micro-taskers) as part of the emergency management process; (iii) the tool is currently available to end users; (iii) the tool includes information such as technical documentation, demos or uses cases.

Our review includes a total of 25 tools (16 disaster management platforms and 9 mobile apps) addressing different aspects of the disaster management cycle (DMC). The analysis does not include Mobile Data Collection Systems (MDCS) that are intended to collect specific information from targeted audiences via pre-designed surveys. In this regard, previous research on MDCS has shown that, from an initial list of 36 solutions, there are up to 24 tools currently available for use in humanitarian relief interventions [42].[2] While MDCS are relevant to our research, the platforms reviewed here have a broader scope and typically include additional functionalities (i.e. data aggregation, data filtering, data clustering, analytics, etc.). In fact, most MDCS could be integrated into DMC platforms as part of the data collection process (i.e. FrontlineSMS and Ushahidi have already worked together to push incoming SMS to the Ushahidi and Crowdmap platforms). Similarly, our analysis does not consider the 250 emergency-related applications available in Google Play already reviewed in recent research [43].

5.1   Taxonomy of Crowdsourcing Tools

We have classified the different platforms and mobile apps upon the next four major characteristics: i) the phase of the management disaster cycle where it better applies to, ii) the availability of the tool and its source code, iii) the main core functionalities, and the iv) crowdsourcing role types:

- Management crisis lifecycle step: which one of the four phases of the disaster management cycle the tool applies to (mitigation, preparedness, response, and recovery).
- Availability of the tool: how the tool is made it available and under which license (open source license, commercial license).
- Core functionalities: which are the main functionalities that the tool is offering. We have identified the following subclasses:
  - Information Retrieval (IR): the tool provides some functionalities to perform text analysis in order to obtain useful information from natural language sentences (structured or unstructured) or raw text (e.g. entity recognition).
  - Data collection: the tool enables data collection from any device connected to the platform. It also provides data management functionalities on the data collected. This dimension is closely linked to the roles of the crowd as a sensor and as a social computer.

---

[2] The NOMAD report includes a tool matrix of MDCS with different parameters: form features, synchronization, interoperability and connectivity, hardware requirements and capabilities supported, and system features and platform characteristics [42].

- Data filtering: the tool displays different filtering options over the data collected. The filtering can be done by keywords, by location, or by any other predefined filter.
- Data tagging: the tool provides tagging functionalities to facilitate the categorization of the collected data. This dimension is closely related with the crowdsourcing role of the crowd as a social computer.
- Mapping and navigation: the tool allows plotting geographic information related with the collected data in a map. It also may allow using this data for navigating in the map and retrieve data based on its geolocation. This dimension is closely related with the crowdsourcing role of the crowd as a sensor.
- Volunteer management tools: the tool comes with a dedicated module to manage the participation of digital or field volunteers (or both).

- Crowdsourcing roles: the tool provides a framework for a particular crowdsourcing role, as in:

  - Crowd as a sensor: the tool enables the collection of data from multiple devices, including mobile handsets, and each of these devices provides some local information which can be either automatically generated (run by sensors in the background) or human generated.
  - Crowd as a social computer: the tool provides some applications or human computer interfaces enabling the users to collect data from social media and engage in social conversation if needed.
  - Crowd as a reporter: the tools provides a platform where people can offer first-hand information on events as they are unfolding and allow the identification of a reporter versus an occasional user in order to preserve trustworthiness.
  - Crowd as a microtasker: the tool provides applications or human computer interfaces for the execution of specific processing tasks by users. These tasks differ from the previous ones in that they that they exploit some specific knowledge and may also require a training phase to accomplish them.

5.2   Main Findings

Online Platforms. As regards online platforms, the majority of the solutions reviewed primarily support response and recovery-based efforts. Generally, the primary focus is on single, event-based, location-specific, and dynamically-evolving scenarios that trigger an urgent response and the need for verified facts [44]. Nevertheless, most of the platforms could also be applicable in the mitigation and preparedness phases, especially those who have developed dedicated modules (i.e. Sahana contains different modules for organization registry, human resources, inventory, assets, etc. which focus on the mitigation and preparedness phases; OpenIR maps ecological risks revealed by infrared satellite data to identify vulnerable areas and support its emergency management). Crowdcrafting and ArcGIS enable developers and users to build custom applications or create and run projects that could also focus on mitigation and preparedness. Since social media information can also be leveraged at any stage of the emergency management cycle (i.e. at the preparedness and training phase, by constantly monitoring information to spot and follow emergency situations, or at the response phase, by communicating real-time between citizens or citizens and authorities) it is difficult to constrain potential uses of the platforms that include social media functionalities (i.e. Ushahidi and CrisisTracker) to just one phase.

Most of the platforms reviewed (10 out of 16) are either open source, have some open source components or can be used for free. As per core functionalities, the most common ones are data collection (12 instances) and data filtering and tagging (11 instances); up to 10 tools offer map and navigation functionalities and 6 of them include some module to manage volunteer effort.

Mobile Applications. The market for disaster management apps has remarkably expanded in the last few years [45,46]. However, even if these apps provide real-time information and updates georeferenced in storm maps, satellite images, and weather forecasts, the information flow remains one way, since it is delivered by the US National Hurricane Center or the US National Weather Service. In contrast, the apps listed in table 4 below tap into user-generated contents to supply updated information to both response organizations (i.e. UN or FEMA) and citizens.

As it is the case with online platforms, mobile applications reviewed here address the response phase of the disaster cycle (although four of them are also applicable in preparedness and one in recovery). Three of the platforms reviewed come with open source licenses and the remaining eight can be used for free.

As per core functionalities, the vast majority of the platforms allow data collection (8) and have mapping/navigation functionalities (9), while a few of them provide data filtering (4) and data tagging (4) functionalities. More specifically, Geopictures, UN Assign or FemaApp allow users to upload and share geo-tagged pictures, Pushpin and Vespucci are editing apps intended to facilitate edition and contribution of new data to OpenStreetMap. OSMTracker allows track logging and quick (voice) waypoint annotations when driving a car or on a bicycle, and OSMAnd is a map and navigation application with access to OpenStreetMap data that also offers both online and offline routing, with optical and voice guidance, for cars, bikes, and pedestrians. Jointly uses group messaging, social circles, and tasks lists to facilitate self-organization of local communities in disaster relief efforts. Fulcrum offers a suite of dedicated apps for disaster response (i.e. damage report, disaster shelter assessment, evacuee information, or post storm building damage report). Stormpins turns its users into local reporter by enabling them to share pin alerts with local TV, emergency managers and local communities. EmergencyAU, finally, also enables its users to upload pictures, videos, and comments about breaking emergencies.

Table 2. Crowdsourcing roles and disaster management cycles

6   Conclusion

Our motivation in developing a typology of crowdsourcing roles and reviewing state-of-the-art platforms and applications dealing with disaster and crisis control management was to stimulate new directions of research in the area of crowdsourced social media information applied to crisis events. While there is an emerging body of literature in this direction, comparative research on the current state of the art of tools and its functionalities is still scarce. In addition, we have found little connection between platform development and research in ontologies for disaster management, even if there are some synergies than could be explored further.

In this paper we have focused on the identification of a set of dimensions which we believe that characterize well the domain and we have classified a representative set of tools which are already available. Enriching platforms to structure their content as usable and reusable knowledge is related to contextual, ethical and legal problems that we put aside in this paper. We have shown that empowering online volunteers and organizations to offer a global response means including citizens as main players triggering such a response. Our pyramid clusters crowdsourcing roles based on users' involvement and level of data processing.

This faces new regulatory challenges in an emerging field. Privacy, data protection and security matter when we realize that accidents, earthquakes or bushfires hit people in states with a great diversity of legal and political systems. Principles, values and norms to be applied to platforms, and the processing of the information provided bottom-up by volunteers can be analyzed to the light of the relational perspective on law [48] and justice [47] aiming at fostering, empowering, and protecting citizens' participation and not only legal compliance. But liability in social media monitoring, tagging and filtering events cannot be ignored either [48]. Future research will further develop in greater detail the emergence of crowdsourcing typologies and types of regulation as they are currently being enabled by the new generation of mobile technology tools.

Acknowledgements. The research leading to this paper has been supported by two research grants from the Spanish Ministry of Economy and Competitiveness (MINECO) to the projects "CrowdCrissControl" (IPT-2012-0968-390000) and "Crowsourcing: instrumentos semánticos para el desarrollo de la participación y la mediación online (DER 2012- 39492 -C02 -01)".

References

1. Meier, P.: Verily: Crowdsourced Verification for Disaster Response. iRevolution (2013), http://irevolution.net/2013/02/19/verily-crowdsourcing-evidence/
2. Fox, Z.: Red Cross Launches Social Media Disaster Response Center. Mashable (March 8, 2012), http://mashable.com/2012/03/07/red-cross-digital-operations-center/
3. Cameron, M., Power, P., Robinson, B., Yin, J.: Emergency Situation Awareness from Twitter for Crisis Management. In: Proceedings of SWDM 2012 Workshop held jointly with WWW (2012)

4. Yin, J., Karimi, S., Robinson, B., Cameron, M.: ESA: Emergency Situation Awareness via Microbloggers. In: CIKM 2012, Maui, HI, USA, October 29-November 2 (2012), http://www.ict.csiro.au/staff/jie.yin/files/de0418-yin-CIKM12.pdf

5. Sakaki, T., Okazaki, M., Matsuo, Y.: Tweet Analysis for Real-Time Event Detection and Earthquake Reporting System Development. IEEE Transactions on Knowledge and Data Engineering 25(4), 919–931 (2013)

6. Caragea, C., McNeese, N., Jaiswal, A., Traylor, G., Kim, H., Mitra, P., Wu, D., Tapia, A., Giles, L., Jansen, B.: Classifying text messages for the haiti earthquake. In: Proceedings of the 8th International ISCRAM Conference, ISCRAM 2011 (2011)

7. Li, J., Rao, H.: Twitter as a rapid response news service: An exploration in the context of the 2008 China earthquake. The Electronic Journal of Information Systems in Developing Countries 42(4), 1–22 (2010)

8. Chowdhury, S.R., Imran, M., Asghar, M.R., Amer-Yahia, S., Castillo, C.: Tweet4act: Using incident-specific profiles for classifying crisis-related messages. In: The 10th International Conference on Information Systems for Crisis Response and Management, ISCRAM (2013), http://chato.cl/papers/roy_chowdhury_imran_rizwan_asghar_amer-yahia_castillo_2013_tweet4act_classify_crisis_messages.pdf

9. Imran, M., Elbassuoni, S., Castillo, C., Diaz, F., Meier, P.: Extracting information nuggets from disaster-related messages in social media. In: The 10th International Conference on Information Systems for Crisis Response and Management (ISCRAM 2013) (2013), http://chato.cl/papers/imran_elbassuoni_castillo_diaz_meier_2013_extracting_information_nuggets_disasters.pdf

10. Robinson, B., Power, R., Cameron, M.: A sensitive twitter earthquake detector. In: Proceedings of the 22nd International Conference on World Wide Web Companion, WWW 2013 Companion, pp. 999–1002 (2013), http://www2013.wwwconference.org/companion/p999.pdf

11. Power, R., Robinson, B., Ratcli, D.: Finding Fires with Twitter. In: Proceedings of the Australasian Language Technology Association Workshop, pp. 80–89 (2013), http://aclweb.org/anthology//U/U13/U13-1011.pdf

12. Beneito-Montagut, R., Anson, S., Shaw, D., Brewster, C.: Resilience: Two case studies on governmental social media use for emergency communication. In: Proceedings of the Information Systems for Crisis Response and Management Conference, ISCRAM 2013, May 12-15 (2013), http://www.disaster20.eu/wordpress/wp-content/uploads/2013/06/Beneito-Montagut_ISCRAM13.pdf

13. Anderson, M.: Integrating social media into traditional management command and control structures: the square peg into the round hole. In: Emergency Media & Public Affairs Conference, Melbourne, May 8 (2012), http://hpe.com.au/empa/proceedings/Martin%20Anderson.pdf

14. Poblet, M.: Spread the word: the value of local information in disaster response. The Conversation (January 17, 2013), http://theconversation.com/spread-the-word-the-value-of-local-information-in-disaster-response-11626

15. Howe, J.: The rise of crowdsourcing. Wired (June 14, 2006), http://www.wired.com/wired/archive/14.06/crowds.html

16. Howe, J.: Crowdsourcing: how the power of the crowd is driving the future of business. Crown Publishing Group, NY (2009)

17. Doan, A., Ramakrishnan, R., Halevy, A.Y.: Crowdsourcing systems on the World-Wide Web. Communications of the ACM 54(4), 86–96 (2011)
18. Schenk, E., Guittard, C.: Towards a characterization of crowdsourcing practices. Journal of Innovation Economics 7/1 (2011), http://www.cairn.info/revue-journal-of-innovation-economics-2011-1-page-93.htm
19. Geiger, D., Seedorf, S., Schulze, T., Nickerson, R.C., Schader, M.: Managing the Crowd: Towards a Taxonomy of Crowdsourcing Processes. In: AMCIS-Proceedings of the Seventeenth Americas Conference on Information Systems, http://schader.bwl.uni-mannheim.de/fileadmin/files/schader/files/publikationen/Geiger_et_al._-_2011_-_Managing_the_Crowd_Towards_a_Taxonomy_of_Crowdsourcing_Processes.pdf
20. Estellés-Arolas, E., González-Ladrón-de-Guevara, F.: Towards an integrated crowdsourcing definition. Journal of Information Science 38(2), 189–200 (2012)
21. Zhao, Y., Zhu, Q.: Evaluation on crowdsourcing research: Current status and future direction. Information Systems Frontiers, 1–18 (2012)
22. Haklay, M.: Citizen Science and Volunteered Geographic Information: Overview and Typology of Participation. In: Sui, D., Elwoold, S., Goodchild, M. (eds.) Crowdsourcing Geographic Science, pp. 105–122. Springer, Netherlands (2013)
23. Saxton, G.D., Onook, O., Kishore, R.: Rules of Crowdsourcing: Models, Issues, and Systems of Control Information Systems Management 30(1), 2–20 (2013)
24. Hetmank, L.: Components and Functions of Crowdsourcing Systems: A Systematic Literature Review. In: Wirtschaftsinformatik Proceedings 2013. Paper 4 (2013), http://www.wi2013.de/proceedings/WI2013%20-%20Track%201%20-%20Hetmank.pdf
25. Malone, T.W., Laubacher, R., Dellarocas, C.N.: Harnessing crowds: Mapping the genome of collective intelligence. MIT Sloan Research Paper 4732-09 (2009), http://18.7.29.232/handle/1721.1/66259
26. Quinn, J.A., Bederson, B.B.: Human Computation: A survey and Taxonomy of a Growing Field. In: CHI Conference, Vancouver, BC, Canada, May 7-12 (2011), http://alexquinn.org/papers/Human%20Computation,%20A%20Survey%20and%20Taxonomy%20of%20a%20Growing%20Field%20CHI%202011.pdf
27. Chamales, G.: Towards trustworthy social media and crowdsourcing. Wilson Center Commons Lab (2013), http://www.wilsoncenter.org/sites/default/files/TowardsTrustworthySocialMedia_FINAL.pdf
28. Meier, P.: Why Bounded Crowdsourcing is Important for Crisis Mapping and Beyond. iRevolution (2011), http://irevolution.net/2011/12/07/why-bounded-crowdsourcing/
29. Prpic, J., Shukla, P.: The Theory of Crowd Capital. In: Proceedings of the Hawaii International Conference on Systems Sciences #46, Maui, Hawaii, USA. IEEE Computer Society Press (January 2013), http://ssrn.com/abstract=2193115
30. Chatzimilioudis, G., Konstantinidis, A., Laoudias, C., Zeinalipour-Yazti, D.: Crowdsourcing with smartphones. IEEE Internet Computing 16(5), 36–44 (2012)
31. UN-SPIDER: Glossary, Disaster Management (2014), http://www.un-spider.org/glossary/disaster-management-cycle
32. UNISDR. Terminology (2009), http://www.unisdr.org/
33. Sheik Dawood, M., Suganya, J., Karthika Devi, R., Athisha, G.: A Review on Wireless Sensor Network Protocol for Disaster Management. International Journal of Computer Applications Technology and Research 2(2), 141–146 (2013)

34. Radianti, J., Granmo, O., Bouhmala, N., Sarshar, P., Yazidi, A., Gonzalez, J.: Crowd Models for Emergency Evacuation: A Review Targeting Human-Centered Sensing. In: 46th Hawaii International Conference on System Sciences (HICSS), pp. 156–165 (2013)
35. Kjaergaard, M.B., Wirz, M., Roggen, D., Troster, G.: Detecting pedestrian flocks by fusion of multi-modal sensors in mobile phones. In: Proceedings of the 2012 ACM Conference on Ubiquitous Computing, pp. 240–249 (2012)
36. Boulos, M.N.K., Resch, B., Crowley, D.M., Breslin, J.G., Sohn, G., Burtner, R., Pike, W.A., Jezierski, E., Slayer Chuang, K.-Y.: Crowdsourcing, citizen sensing and sensor web technologies for public and environmental health surveillance and crisis management: trends, OGC standards and application examples. International Journal of Health Geographics 10, 67, 1–29 (2011), `http://www.ij-healthgeographics.com/content/pdf/1476-072X-10-67.pdf`
37. Liu, M.: A Study of Mobile Sensing Using Smartphones. International Journal of Distributed Sensor Networks, 1–11, Art. ID 272916 (2013), `http://dx.doi.org/10.1155/2013/272916`
38. Chon, Y., Lane, N.D., Li, F.: H. Cha, Zhao, F.: Automatically characterizing places with opportunistic crowdsensing using smartphones. In: Proceedings of the 2012 ACM Conference on Ubiquitous Computing, pp. 481–490 (2012)
39. Naroditskiy, V., Rahwan, I., Cebrian, M., Jennings, N.R.: Verification in Referral-Based Crowdsourcing. PLoS One 7(10) e45924 (2012), doi:10.1371/journal.pone.0045924
40. Tang, J., Cebrian, M., Giacobe, N.A., Kim, H.W., Kim, T., Wickert, D.B.: Reflecting on the DARPA Red Balloon Challenge. Communications of the ACM 54(4), 78–85 (2011)
41. Poblet, M.: Rule of Law on the Go: New Developments of Mobile Governance. Journal of Universal Computer Science 17(3), 498–512 (2011), `http://www.jucs.org/jucs_17_3/rule_of_law_on_the/jucs_17_03_0498_0512_poblet.pdf`
42. Jung, C.: Data collection mobile systems: A review of the current state of the field (2011), `http://www.parkdatabase.org/files/documents/nomad_mdc_research.pdf`
43. Gómez, D., Bernardos, A.M., Portillo, J.I., Tarrío, P., Casar, J.R.: A Review on Mobile Applications for Citizen Emergency Management. In: Corchado, J.M., et al. (eds.) PAAMS 2013. CCIS, vol. 365, pp. 190–201. Springer, Heidelberg (2013)
44. Coppola, D.P.: Introduction to international disaster management. Elsevier, Burlington (2011)
45. Poblet, M. (ed.): Mobile Technologies for Conflict Management: Online Dispute Resolution, Governance, Participation. Springer, Berlin (2011)
46. Shih, F., Seneviratne, O., Liccardi, I., Patton, E., Meier, P., Castillo, C.: Democratizing mobile app development for disaster management. In: AIPP 2013, Joint Proceedings of the Workshop on AI Problems and Approaches for Intelligent Environments and Workshop on Semantic Cities, pp. 39–42 (2013)
47. Casanovas, P., Poblet, M.: Concepts and Fields of Relational Justice. In: Casanovas, P., Sartor, G., Casellas, N., Rubino, R. (eds.) Computable Models of the Law. LNCS (LNAI), vol. 4884, pp. 323–339. Springer, Heidelberg (2008)
48. Casanovas, P., Poblet, M.: The Future of Law: Relational Law and next Generation of Web Services. In: Fernández-Barrera, M., et al. (eds.) The Future of Law and Technology: Looking into the Future. Selected Essays, pp. 137–156. European Press Academic Publishing, Florence (2009)

A Method for Defining Human-Machine Micro-task Workflows for Gathering Legal Information

Nuno Luz[1], Nuno Silva[1], and Paulo Novais[2]

[1] GECAD (Knowledge Engineering and Decision Support Group), Polytechnic of Porto
Porto, Portugal
{nmalu,nps}@isep.ipp.pt
[2] CCTC (Computer Science and Technology Center), University of Minho Braga, Portugal
pjon@di.uminho.pt

Abstract. With the growing popularity of micro-task crowdsourcing platforms, new workflow-based micro-task crowdsourcing approaches are starting to emerge. Such workflows occur in legal, political and conflict resolution domains as well, presenting new challenges, namely in micro-task specification and human-machine interaction, which result mostly from the flow of unstructured data. Domain ontologies provide the structure and semantics required to describe the data flowing throughout the workflow in a way understandable to both humans and machines. This paper presents a method for the construction of micro-task workflows from legal domain ontologies. The method is currently being employed in the context of the UMCourt project in order to formulate information retrieval and conflict resolution workflows.

Keywords: Legal Crowdsourcing, Micro-Tasks, Workflows, Relational Law.

1   Introduction

Several experiments in different domains have shown that micro-task crowdsourcing has great potential for solving large scale problems that are often difficult for computers to solve automatically, on their own [1]. These problems usually require a degree of creativity or just common sense plus some background knowledge [2, 3]. The interpretation and recognition of images and natural language are two examples of these kinds of problems.

Crowdsourcing platforms like Mechanical Turk, CloudCrowd, ShortTask and Crowd-Flower are widely used for tasks such as (i) categorization and classification, (ii) data collection (e.g., finding a website address), (iii) moderation and tagging of images, (iv) surveys, (v) transcription from multimedia content (e.g., audio, video and images), and (vi) text translation.

More recently, a special interest in employing crowdsourcing towards solving complex tasks has emerged [4–9]. Following the trend of the current crowdsourcing platforms, which feature the execution of single micro-tasks, this interest has led to the emergence of new approaches built upon workflows of micro-tasks. The modelling of

P. Casanovas et al. (Eds.): AICOL IV/V 2013, LNAI 8929, pp. 275–289, 2014.
© Springer-Verlag Berlin Heidelberg 2014

such workflows allows the crowdsourcing of a new kind of more complex tasks (e.g., selecting and buying a video camera, recommending points of interest), which require the ordered execution of multiple types of micro-tasks.

Among these complex tasks are mediation processes often employed in relational law, which focuses on "justice produced through cooperative behavior, agreement, negotiation, or dialogue among actors in a post-conflict situation" [10]. The ordered execution of micro-tasks by individuals and groups selected from crowds not only results in cooperative solutions, but can also be used to implement conflict resolution and negotiation strategies in a wide scale. As a form of collective intelligence, the resulting data can be interpreted as a wide scale consensus or truth regarding a specific domain or topic, relevant to the law or case under scrutiny.

Micro-task workflows present new challenges at different dimensions of the crowdsourcing process, namely in micro-task specification and human-machine interaction [4, 5]. In particular, micro-task workflow approaches like CrowdForge [5], Jabberwocky [4] and Turkomatic [6] employ divide-and-conquer and map reduce strategies to build workflows. This usually involves workflows that include tasks for (i) the partitioning of the complex task (partition tasks), (ii) the execution of the partitioned tasks (map tasks), and (iii) the aggregation of results (reduce tasks).

However, in most cases, task (or micro-task) responses are unstructured and in natural language. Furthermore, micro-task interfaces are built using markup languages that contain little or no meta-data, making it difficult for machine micro-tasks to be included in the workflow.

The unstructured nature of micro-tasks in terms of domain representation makes it difficult (i) for task requesters not familiar with the crowdsourcing platform to build complex micro-task workflows and (ii) to include machine workers in the workflow execution process [11]. Furthermore, while some of the micro-tasks in the workflow are better performed by humans, others are better performed by a machine, which is seldom explicitly defined.

As stated by Obrst et al. [12], ontologies "represent the best answer to the demand for intelligent systems that operate closer to the human conceptual level". Domain ontologies are not only able to describe the domain knowledge, but also to describe workflow micro-tasks and the data flowing through them in a way understandable to both humans and machines.

Considering these, a method for the construction of human-machine micro-task workflow ontologies is proposed. Although the method is intended for the construction of crowdsourced micro-task workflows, it can be employed to build workflow ontologies for other types of applications. Possible domains of application include legal information retrieval and legal conflict resolution [13–16]. In the particular case of mediation in relational law, the essential requirements are (i) to harness structured and semantically enriched information (ii) from a crowd or group of actors. While current crowdsourcing approaches, like CrowdForge and Jabberwocky, tackle the distribution and crowdsourcing of micro-tasks, the resulting data is often found poorly structured or in natural language.

In this sense, the ultimate goal of this method is to define a set of ground rules for the assisted construction of workflow definition ontologies from domain ontologies.

A top-level workflow definition ontology is presented, upon which any workflow execution and task distribution engine can be implemented. The resulting workflow definition ontology defines the domain and rules for each task, along with the relationships between the input and output data in and between tasks.

The following sections of this paper start with a brief overview of crowdsourcing terminology and ontology-related background knowledge. Section 3 describes the proposed workflow construction method in four parts: (i) domain ontologies, (ii) the Onto2Flow ontology, (iii) micro-task specification, and (iv) workflow specification. Finally, conclusions are given along with some remarks on the future directions of this work.

2   Background Knowledge

2.1   Micro-task Workflows in Crowdsourcing

The terminology employed in the crowdsourcing domain often varies from platform to platform. In the context of this paper, a *job* (or a complex task) contains a workflow of *tasks* (or micro-tasks), along with all the data required for its execution. Micro-tasks (e.g., tag an image), as seen by the crowdsourcing community, have one or more *units* of work as input (e.g., the images to be tagged). Each of these units will be assigned to one or more *workers*, which must then submit a *response* (e.g., the tagging of the image). Multiple *assignments* of the same unit to different workers allow redundancy and quality improvements of the overall result after the aggregation of responses is performed.

Furthermore, the aggregation of responses often takes into account units for which a correct response is already known. These units are often referred to as *reference units*. Workers that give incorrect or invalid responses to reference units suffer credibility penalties, and their responses have significantly less impact in the final result.

Typically, in crowdsourcing platforms such as Mechanical Turk, human workers choose whether to perform the specific task according to the given (often monetary) *reward*. In some cases, the *requesters* of the task may require workers with certain expertise and *qualifications*, which are given after the worker successfully solves a qualification task.

Through the analysis of the evolution of crowdsourcing platforms, it is possible to conclude that an effort towards structured (sets of) tasks is being made. While early crowdsourcing platforms such as MTurk, CrowdFlower, MicroWorkers and Cloud-Crowd have added template construction features, more recent platforms and frameworks such as CrowdForge, Jabberwocky, Turkomatic and Turkit have tackled this emerging need through different workflow representations and construction strategies.

Table 1 presents a comparison of several crowdsourcing approaches. Each approach is compared according to five different dimensions. These dimensions reflect if the approach (i) relies on its own crowd or in multiple (possibly external) crowds, (ii) supports complex tasks, (iii) employs any task construction strategy, (iv) employs worker and result assessment strategies, and (v) employs result aggregation strategies when redundancy (multiple responses for the same unit) is found.

Table 1. Comparison of crowdsourcing platforms.

System	Relies on	Complex Tasks	Task Strategy	Worker Assessment	Aggregation
MTurk	Self	No	Task Templates	Qualification Tests	Manual
CrowdFlower	Several	No	Task Templates	Gold Units	Yes
ShortTask	Self	No	Task Templates	Manual	Manual
MicroWorkers	Self	No	Task Templates	Manual	N/A
CloudCrowd	Self	-	-	Credential Tests and Credibility	-
CrowdForge	MTurk	Workflows	Map Reduce	(MTurks')	Yes
Jabberwocky	Self/Several	Workflows	Map Reduce	User Profiles	Yes
Turkomatic	MTurk	Workflows	Divide and Conquer	(MTurks')	Yes (Workers)
Turkit	MTurk	Workflows	Crash and Rerun	(MTurks')	Yes (Workers)

2.2 Ontologies in Description Logics

In this paper, Description Logics (DL) knowledge bases and ontologies with $\mathcal{ALCOQ}$ expressivity are considered (see fig. 1). A DL knowledge base contains a TBox (terminological box) and an ABox (assertion box) [17], where the TBox contains all the concepts and relationships that define a specific domain, and the ABox contains the instances or individuals defined according to the elements in the TBox. It is assumed that ontology is synonym of TBox.

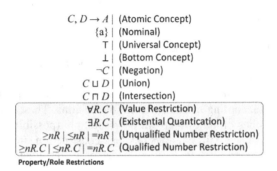

Fig. 1. TBox concept description syntax and rules with $\mathcal{ALCOQ}$ expressivity

Each concept (e.g., C, D) is defined according to other concepts (e.g., $C \sqcup D$) and property restrictions (e.g., $\exists R.D$) that define the necessary (e.g., $C \sqsubseteq \exists R.D$), and necessary and sufficient (e.g., $C \equiv \exists R.D$) conditions for an individual to be an instance of the concept.

There are two main types of properties: object properties and data-type properties. While object properties relate instances (or individuals) with other instances, data-type properties relate instances with "primitive" type values (e.g., integer, string, double, date, time).

3   The Workflow Specification Method

Micro-tasks, whether they involve physical actions or not, can be seen as a process that, in a specific context, results in the emergence of new data (responses) from the presentation of other particular pieces of data (units) to a worker. Analogously, a workflow of micro-tasks is the continuous ordered increment of new (different types of) data, in a specific context or domain.

The proposed method suggests that micro-task responses correspond to new instances of concepts (or classes) in the domain ontology, associated with input (unit) instances of domain ontology concepts. Thereafter, a micro-task can be considered to be *the instantiation of domain classes and the specification of new relationships between instances* according to the domain ontology. A workflow of micro-tasks is then considered as *the incremental instantiation of the domain ontology according to its structure and semantics.*

With the assumption that domain ontologies represent the structure and semantics of the data that must be presented and retrieved from workers during the execution of a task, workflow ontologies extend both the Onto2Flow and domain ontologies (see fig. 2).

Workflow ontologies are instantiated and executed by a workflow engine that is able to interpret the ontology according to the ground rules established by the proposed method. During the workflow execution, the input is given as an ABox described by the domain ontology. The output of the workflow will be described by the domain ontology and, in some situations, operational concepts and properties of the workflow ontology.

The ground rules established by the proposed method must be employed during the workflow construction step (1) and followed during the instantiation and execution step (2).

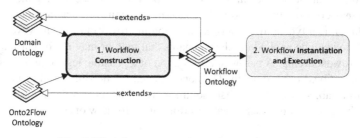

Fig. 2. Workflow construction and execution steps

3.1   Domain Ontologies

Workflow ontologies capture the tasks/operations of a certain process, and the dynamic nature of a domain. The static structure and semantics of the specific knowledge domain, on the other hand, are captured by domain ontologies in the form of concepts and their relations.

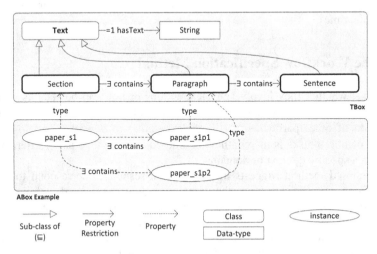

Fig. 3. The document ontology (TBox only) with a possible example ABox (or instantiation). The TBox is an adaptation from the DoCO (Document Components) ontology[1].

Unlike workflow ontologies, domain ontologies are very common and accessible. Inclusively, their structure can be analysed and employed in the construction of workflow ontologies.

Consider the document ontology and example ABox presented in fig. 3. The graph structure of the TBox defines the known properties of instances in the ABox. Following the restrictions specified in this structure, the incremental filling of the ABox is possible through the execution of several atomic operations (micro-tasks). In the specific case of the document ontology, an initial ABox with English sections may be supplied as input to the workflow, resulting in translated Portuguese sections. Since the ontology contains the semantics for the subdivision of sections, some of the micro-tasks may consider their subdivision into smaller units (e.g., paragraphs).

Translation is a typical domain of application in crowdsourcing, however, the proposed method can be applied in other domains that may or may not be currently in the scope of crowdsourcing. A partial simplification of a possible legal ontology, depicted in fig. 4, describes such a domain. The concepts and relationships in this ontology can be used to establish workflows that inquire a crowd about past legal cases (e.g., abusive discharge cases) in order to gather information for new ones.

---

[1]   DoCO: http://purl.org/spar/doco/

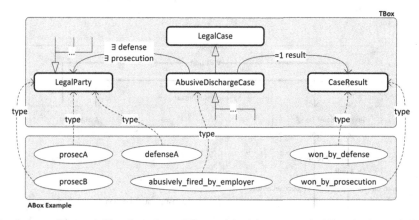

Fig. 4. A possible partial legal ontology (TBox only) with an example ABox (or instantiation)

3.2 The Onto2Flow Ontology

The Onto2Flow ontology captures the structure and semantics of workflows (see fig. 5). The main concepts are: Task, Assignment, Requester and Worker (either Machine or Person). This ontology is further extended and its concepts refined in the workflow ontology as required by the domain of application.

Assignments correspond to the execution of a task by a worker, for a single unit of work. The properties that define the domain of a task are:

- *unit* – defines the set of instances (class) that constitute the input of the task (only one property restriction allowed);
- *unitContext* – defines the input context classes of the task;
- *response* – defines the set of instances (class) that constitute the output of the task (only one property restriction allowed);
- *responseContext* – defines the output context classes of the task.

The different types of atomic operations (or micro-tasks) that can be performed are specified in the ontology through sub-classes of Task. As presented in fig. 5, the Onto2Flow ontology currently defines four atomic operations associated with the classes: CreateAndFillTask, FillTask, SelectionTask and AggregationTask.

CreateAndFillTask instances will result in new instances of the response class, for which all data-type property values will be requested to the worker.

A FillTask will request data-type property values for already existent instances of the unit class.

SelectionTask instances will result in the definition of new relationships between already existent instances in the ABox, i.e., no new response instances will be created. Instead, they will be selected by the worker from a set of possible responses.

If more than one assignment per unit is demanded, the execution of the task will result in several possible Output ABoxes for each unit. In these situations, an aggregation of the responses must be performed through an AggregationTask. AggregationTasks consider the context, unit and response classes of the previous task. Furthermore, any

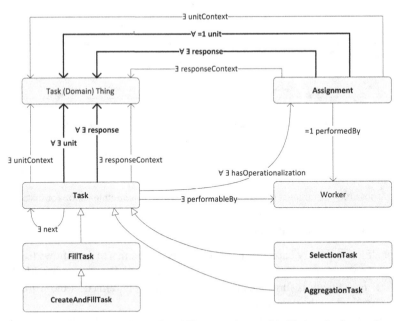

Fig. 5. Partial representation of the Onto2Flow ontology with Task sub-classes for atomic operations

number of AggregationTask sub-classes may be included in order to implement different aggregation strategies (e.g., majority voting, assessment-based).

Requesters may define the set of workers that may participate in the task through the *performableBy* property. In order to restrict or create worker roles, new Worker (Person or Machine) sub-classes may be created with restrictions applied to their properties (e.g., ∃country.{portugal}).

The *performedBy* property is established only after the worker accepts to participate in the task.

3.3 Defining Micro-tasks

A workflow ontology describes a workflow that can be instantiated multiple times. The workflow ontology must import and extend the Onto2Flow ontology. Domain concepts must either be defined in the workflow ontology, or imported from a domain ontology (as depicted in fig. 2). The following parts of this document assume that domain concepts are always imported from an external domain ontology.
In order to build the workflow ontology, the requester must extend the Task, Assignment and Worker classes from the Onto2Flow ontology, and any class from the domain ontology. Fig. 6 depicts the ontological structure of a simple micro-task.

A micro-task specification is an explicit partial TBox in the workflow ontology with, at least, the following terminological axioms:

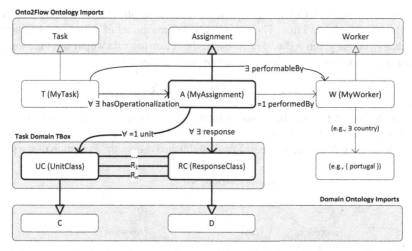

Fig. 6. Structure of a basic micro-task in a workflow. Relationships between T and UC/RC were omitted from the figure since they are similar to those between A and UC/RC.

- $T \sqsubseteq Task \sqcap \exists response.RC \sqcap \forall response.RC \sqcap \exists assignment.A$
- $A \sqsubseteq Assignment \sqcap \exists response.RC \sqcap \forall response.RC$
- $RC \sqsubseteq D$

The specification of an UC is not mandatory and is done through the following terminological axioms:

- $T \sqsubseteq \exists unit.UC \sqcap \forall unit.UC \land A \sqsubseteq= 1unit.UC \sqcap \forall unit.UC$
- $UC \sqsubseteq C$
- $UC \sqsubseteq \exists R.RC$ or $RC \sqsubseteq \exists R.UC$ (optional)

C and D are classes in the domain ontology. UC represents the subset of C instances that constitute the input of T. RC represents instances of D, which are output of T. If no additional property restrictions are defined on UC, any instance of C in the input ABox is also considered to be an instance of UC.

A establishes an n-ary relationship between UC and RC, which reflects the operational semantics of all R. R are object property restrictions (from properties and restrictions typically present in the domain ontology) that establish a direct correspondence between UC and RC (or vice-versa).

If the requester needs to select specific target workers for the task, a sub-class of Worker ($W \sqsubseteq Worker \sqcap C_1 \sqcap C_2 \sqcap ... \sqcap C_n$, where C represents a property restriction onto the W class) must be created.

The Task Domain TBox represents a partial copy of the domain ontology containing only the necessary classes and relationships: those required as input and those for which new instances and relationships will be established.

Unit Context Classes. In some situations, the requester needs to provide additional contextual information, given through related domain classes, to the worker. For these tasks, unit context classes (*UCCs*) may be specified. The set of all UC, UCCs, and their relationships form the Input TBox. The Input TBox defines the set of rules that will filter the input data from the given ABox. For instance, the following rule would filter the input of the task according to the Input TBox structure presented in fig. 7:

$$\forall x \forall y \big(D(x) \land C(y) \land S(x,y) \rightarrow UCC(x) \land UC(y) \big)$$

Any number of *UCCs* may be included in the Input TBox, with any type of relationships between them and to/from the *UC* or *RC*.

Relationships to/from the *RC* (e.g. *T*) are established during the execution of the task.

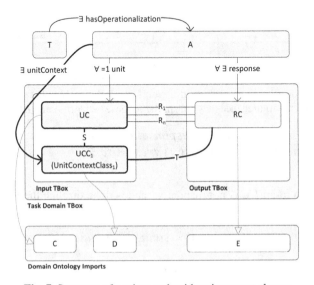

Fig. 7. Structure of a micro-task with unit context classes

Response Context Classes. Response context classes (*RCCs*) establish property restrictions onto the *RC* that must be followed by the worker (as in fig. 8). An *RCC* represents a subset of input instances (it is a sub-class of an *UCC*) that were chosen by the worker as property values for an *RC* instance. The mandatory sub-class-of relationship between the *RCC* and the *UCC* is considered a dependency.

Dependency relationships indicate that instances of the *UCC* are candidate instances of the *RCC*. In this sense, the worker will have to select which instances of *UCC* will become instances of *RCC*, related to *RC* through the property in the specified restriction, *U*.

An *RCC* is defined through an *UCC* (where *E* is a class from the domain ontology) as $RC \sqsubseteq \exists U.RCC \land RCC \sqsubseteq UCC \sqcap E \land UCC \sqsubseteq E$.

Dependencies on the Response Class. When building SelectionTask and FillTask tasks, the requester must establish a dependency between the *RC* and one of the Input TBox classes (the *UC* or an *UCC*) (see fig. 9 for an example with a SelectionTask).

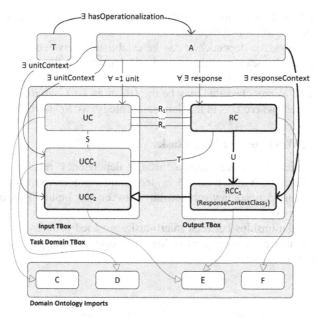

Fig. 8. Structure of a micro-task with response context classes for establishing *RC* property restrictions

For SelectionTask tasks, the dependency must be established between the *RC* and an *UCC*. Analogous to paths established through RCCs, it means that the worker will have to select the appropriate *RC* instance(s) from the set of instances given by the *UCC*. The selected instance(s) will become the response of the assignment.

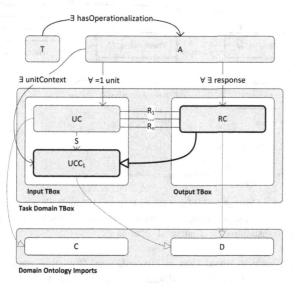

Fig. 9. Structure of a SelectionTask task with a dependency between the *RC* and an *UCC*

An *RC* dependency for SelectionTasks is defined (where D is a class from the domain ontology) as $RC \sqsubseteq UCC \sqcap D \wedge UCC \sqsubseteq D$.

For FillTask tasks, the dependency can be established between the *RC* and either the *UC* or an *UCC*. It means that the worker will have to fill the data-type properties for existent instances of the *UC* or *UCC*.

Considering *IC* as any class that may be the *UC* or an *UCC*, an *RC* dependency for FillTasks is defined (where *D* is in the domain ontology) as $RC \sqsubseteq IC \sqcap D \wedge IC \sqsubseteq D$.

3.4   Defining Workflows of Micro-tasks

Workflows of micro-tasks are defined through dependency relationships between Task Domain TBoxes. A micro-task *A* is dependent (or follows) a micro-task *B* if there is at least one dependency relationship between the Input TBox of *A* and the Task Domain TBox of *B*.

Dependency relationships between micro-tasks are used to infer the *next* relationship and to optimize the resulting workflow. The optimization process includes the parallelization of independent tasks.

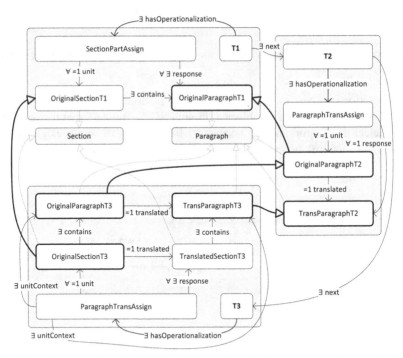

Fig. 10. Example of a CreateAndFillTask micro-task workflow built according to the translation ontology that (i) partitions sections into paragraph, (ii) translates paragraphs, and (iii) assembles paragraph translations into translated sections

Fig. 10 depicts a section translation micro-task workflow that applies a divide-and-conquer strategy. The Section and Paragraph domain classes from the translation ontology, along with their relationships, are exploited in the workflow ontology in order to define each of the CreateAndFillTask tasks *T1*, *T2* and *T3*. Dependencies exist between *T2* and *T1*, between *T3* and *T2*, and between *T3* and *T1*. The transitive closure of the inter-task dependency relation results in the workflow structure reflected by the next relationship. In this case, it results in the sequence: *T1*, *T2*, *T3*.

The information retrieval workflow presented in fig. 11 is built from the partial legal ontology in fig. 4. It depicts a situation where an expert is assessing the possibilities to take legal action against a company on behalf of a customer [18].

The first task, *T1*, is a CreateAndFillTask micro-task that asks an entity·or crowd for instances of abusive discharge cases. For each given case, the worker(s) must also fill all datatype properties of the AbusiveDischargeCase concept. The second task, *T2*, is a CreateAndFillTask micro-task where the entity or crowd must, for each case previously submitted, provide information on the defence and prosecution parties involved. Finally, on task *T3*, workers submit information on reported abuses for each case submitted in *T1*.

These types of information retrieval workflows allow legal parties to collect information on previous instances of legal procedures. The retrieved information is structured and enriched with the semantics of legal domain ontologies.

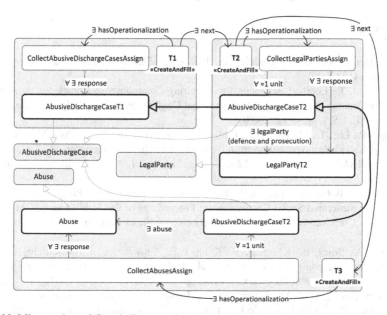

Fig. 11. Micro-task workflow built according to a legal ontology that asks for (i) abusive discharge cases, (ii) the legal parties involved in each case and (iii) reported abuses in each case

4   Conclusions and Future Work

The proposed method tackles the challenge of building micro-task workflows while promoting human-machine cooperation through high-level, declarative and semantically explicit domain ontology models. Although the process of manually building micro-task workflows requires some degree of domain expertise and knowledge of the Onto2Flow ontology, the ground rules for creating an assisted workflow construction process were defined. Since domain ontologies are interpretable by humans and machines, micro-tasks can be solved by either or both human and machine workers.

Future work includes the creation of an assisted micro-task workflow construction process, which automates the construction of workflows through the detection of ontology patterns and their aggregation into different strategies. The evolution of the Onto2Flow ontology, in order to assimilate concepts often found in workflow definition languages, is also considered. Furthermore, the proposed method is being employed in the context of the UMCourt project [13, 14] in order to further evaluate its impact in legal use cases.

Acknowledgements. This work is part-funded by FEDER Funds, by the ERDF (European Regional Development Fund) through the COMPETE Programme (operational programme for competitiveness) and by National Funds through the FCT (Portuguese Foundation for Science and Technology) within the project FCOMP-01-0124-FEDER-028980 (PTDC/EEI-SII/1386/2012). The work of Nuno Luz is supported by the doctoral grant SFRH/BD/70302/2010.

References

1. Von Ahn, L.: Human Computation. In: 46th ACM IEEE Design Automation Conference, pp. 418–419 (2009)
2. Chklovski, T.: Learner: A System for Acquiring Commonsense Knowledge by Analogy. In: Proceedings of the 2nd ACM International Conference on Knowledge Capture, Sanibel Island, FL, USA, pp. 4–12 (2003)
3. Singh, P., Lin, T., Mueller, E.T., Lim, G., Perkins, T., Zhu, W.L.: Open Mind Common Sense: Knowledge Acquisition from the General Public. In: Meersman, R., Tari, Z. (eds.) CoopIS/DOA/ODBASE 2002. LNCS, vol. 2519, pp. 1223–1237. Springer, Heidelberg (2002)
4. Ahmad, S., Battle, A., Malkani, Z., Kamvar, S.: The Jabberwocky Programming Environment for Structured Social Computing. In: Proceedings of the 24th Annual ACM Symposium on User Interface Software and Technology, Santa Barbara, CA, USA, pp. 53–64 (2011)
5. Kittur, A., Smus, B., Khamkar, S., Kraut, R.E.: Crowdforge: Crowdsourcing Complex Work. In: Proceedings of the 24th Annual ACM Symposium on User Interface Software and Technology, Santa Barbara, CA, USA, pp. 43–52 (2011)
6. Kulkarni, A.P., Can, M., Hartmann, B.: Turkomatic: Automatic Recursive Task and Workflow Design for Mechanical Turk. In: CHI 2011 Extended Abstracts on Human Factors in Computing Systems, Vancouver, BC, Canada, pp. 2053–2058 (2011)
7. Little, G., Chilton, L.B., Goldman, M., Miller, R.C.: Turkit: Human Computation Algorithms on Mechanical Turk. In: Proceedings of the 23rd Annual ACM Symposium on User Interface Software and Technology, New York, NY, USA, pp. 57–66 (2010)
8. Luz, N., Silva, N., Maio, P., Novais, P.: Ontology Alignment through Argumentation. In: 2012 AAAI Spring Symposium: Wisdom of the Crowd (2012)
9. Sarasua, C., Simperl, E., Noy, N.F.: CROWDMAP: Crowdsourcing Ontology Alignment with Microtasks. In: Cudré-Mauroux, P., et al. (eds.) ISWC 2012, Part I. LNCS, vol. 7649, pp. 525–541. Springer, Heidelberg (2012)

10. Casanovas, P.: The Future of Law: Relational Justice, Web Services and Second-generation Semantic Web. Legal Information and Communication Tech 7, 137–156 (2009)
11. Quinn, A.J., Bederson, B.B.: Human Computation: A Survey and Taxonomy of a Growing Field. In: Proceedings of the SIGCHI Conference on Human Factors in Computing Systems, pp. 1403–1412. ACM, New York (2011)
12. Obrst, L., Liu, H., Wray, R.: Ontologies for Corporate Web Applications. AI Magazine 24, 49 (2003)
13. Carneiro, D., Novais, P., Andrade, F., Zeleznikow, J., Neves, J.: Using Case-Based Reasoning and Principled Negotiation to provide decision support for dispute resolution. Knowledge and Information Systems 36, 789–826 (2013)
14. Novais, P., Carneiro, D., Gomes, M., Neves, J.: The relationship between stress and conflict handling style in an ODR environment. In: Motomura, Y., Butler, A., Bekki, D. (eds.) JSAI-isAI 2012. LNCS, vol. 7856, pp. 125–140. Springer, Heidelberg (2013)
15. Casanovas, P.: Legal crowdsourcing and relational law: What the semantic web can do for legal education. Journal of the Australasian Law Teachers Association 5, 159–176 (2012)
16. Poblet, M., Casanovas, P., Cobo, J.M.L., Casellas, N.: ODR, Ontologies, and Web 2.0. 0. Journal of Universal Computer Science 17, 618–634 (2011)
17. Baader, F., Calvanese, D., McGuinness, D.L., Nardi, D., Patel-Schneider, P.F.: The Description Logic Handbook: Theory, Implementation, and Applications, 2nd edn. Cambridge University Press (2007)
18. Gangemi, A., Presutti, V., Blomqvist, E.: The Computational Ontology Perspective: Design Patterns for Web Ontologies. In: Approaches to Legal Ontologies, pp. 201–217. Springer (2011)

Author Index